Advances in Solid State Physics
Volume 48

Advances in Solid State Physics

Advances in Solid State Physics is a book series with a history of about 50 years. It contains the invited lectures presented at the Spring Meetings of the "Arbeitskreis Festkörperphysik" of the "Deutsche Physikalische Gesellschaft", held in March of each year. The invited talks are intended to reflect the most recent achievements of researchers working in the field both in Germany and worldwide. Thus the volume of the series represents a continuous documentation of most recent developments in what can be considered as one of the most important and active fields of modern physics. Since the majority of invited talks are usually given by young researchers at the start of their career, the articles can also be considered as indicating important future developments.

The speakers of the invited lectures and of the symposia are asked to contribute to the yearly volumes with the written version of their lecture at the forthcoming Spring Meeting of the Deutsche Physikalische Gesellschaft by the Series Editor. Colored figures are available in the online version for some of the articles.

Advances in Solid State Physics is addressed to all scientists at universities and in industry who wish to obtain an overview and to keep informed on the latest developments in solid state physics. The language of publication is English.

Series Editor

Prof. Dr. Rolf Haug

Abteilung Nanostrukturen
Institut für Festkörperphysik
Universität Hannover
Appelstr. 2
30167 Hannover
Germany

haug@nano.uni-hannover.de

Rolf Haug (Ed.)

Advances in Solid State Physics

48

With 182 Figures and 13 Tables

Springer

Prof. Dr. Rolf Haug
Abteilung Nanostrukturen
Institut für Festkörperphysik
Universität Hannover
Appelstr. 2
30167 Hannover
Germany
haug@nano.uni-hannover.de

ISSN print edition: 1438-4329
ISSN electronic edition: 1617-5034

ISBN: 978-3-642-09940-3 e-ISBN: 978-3-540-85859-1

DOI 10.1007/978-3-540-85859-1

Physics and Astronomy Classification Scheme (PACS): 60.00; 70.00; 80.00

Cover concept using a background picture by Dr. Ralf Stannarius, Faculty of Physics and Earth Sciences, Institute of Experimental Physics I, University of Leipzig, Germany

Cover design: WMXDesign GmbH, Heidelberg

Printed on acid-free paper

9 8 7 6 5 4 3 2 1

springer.com

Preface

The 2008 spring meeting of the *Arbeitskreis Festkörperphysik* was held in Berlin between February 24 and 29, 2008 in conjunction with the 72. Annual Meeting of the *Deutsche Physikalische Gesellschaft*. It was the last spring meeting of the *Arbeitskreis Festkörperphysik* because in the meantime it changed its name to *Sektion kondensierte Materie* which better shows that it also includes soft materials and biological systems. Therefore, the next spring meeting in Dresden will be organized by the *Sektion kondensierte Materie* of the *Deutsche Physikalische Gesellschaft*. The number of participants of this year's meeting in Berlin exceeded 5600 and there were more than 4600 scientific contributions. With these numbers this meeting was the largest physics meeting in Europe and among the largest physics meetings in the world in 2008.

The present volume, 48 of the *Advances in Solid State Physics* contains the written version of a large number of the invited talks in Berlin and gives a nice overview of the present status of condensed matter physics. Low-dimensional systems are dominating the field and especially nanowires and quantum dots. In recent years one learned how to produce nanowires directly during a growth process. Therefore, a number of articles is related to such nanowires. In nanoparticles and quantum dots the dimensionality is further reduced and we learn more and more how to produce such systems and what effects result from the confinement in all three dimensions. Spin effects and magnetism is another important field of present day research in solid state physics. The third chapter covers this physics including an article about *graphene*. The growing interest into organic materials and biological systems is reflected in a large chapter of this book with the title *Organic Materials and Water*. The last chapters of this book cover aspects which range from dynamical effects to device physics and characterization tools.

Hannover, June 2008 *Rolf J. Haug*

Contents

Part VI Characterization of Materials and Devices

Part I

Nanowires

From Ordered Arrays of Nanowires to Controlled Solid State Reactions

Margit Zacharias[1] and Hong Jin Fan[2]

[1] Institute of Microsystems Engineering, Faculty of Applied Science Albert Ludwigs University, Georges Köhler Allee 103, 79110, Freiburg, Germany
Zacharias@imtek.de

[2] Department of Earth Science, University of Cambridge, Cambridge, CB2 3EQ, UK
hfan05@esc.cam.ac.uk

Abstract. There has been increasing interest in intentional synthesis of nanowires and nanotubes based on a large variety of materials. A deeper understanding of their properties and a sufficient growth control are in the center of current research interest. Strategies for position-controlled and nano-patterned growth of nanowire arrays are demonstrated by selected examples of our work based on ZnO nanowires as well as discussed in terms of larger scale realization and future industrial prospects. The physical properties of single ZnO nanowires are presented on selected examples. Recently, we demonstrated one-dimensional free-standing spinel nanotubes which were transformed from nanowires via the Kirkendall effect and solid state reaction. The nanoscale Kirkendall effect provides a general fabrication route to hollow nanostructures, including high aspect ratio nanotubes. Such ordered arrays of spinel nanotubes may possess similar application potentials as carbon nanotubes.

1 Introduction

Approaches for nanotechnology are either realized with normal top-down processes or using bottom-up approaches. The top-down approach is based on normal microelectronic procedures reducing dimensions more and more like an artist is doing with a sculpture. However, the surfaces of such etched nanostructures are highly disturbed which drastically declines the electronic properties. On the other hand, bottom-up approaches are starting from nanosizes from the very beginning, and are trying to arrange and combine nanostructures to new functionalities. In case of such self organized nanostructures the main problem is how independently to control dimensions, positions and properties. Also the question arises how to develop methods fitting to industrial conditions or mass production, i.e. the growth of vertically arranged nanowires on full wafers. After succeeding with such a growth of millions of nanowires the task still remains to develop electrical contacts for such arrays as well as functionalities.

The one-dimensional growth of nanowires (NWs) was already demonstrated in the early sixties by Wagner et al. [1]. At that time the sizes of such structures were in the range of micrometers and the structures were denoted as "whiskers". A corresponding growth theory was already developed on the basis of the so-called "vapor-liquid-solid" (VLS) mechanism. Already in 1992, a group at Hitach [2] applied the VLS mechanism using gold droplets as catalysts for the growth of III-V NWs, e.g. GaAs and InGaAs. Already at that time the doping of wires was demonstrated which proved the general possibility for the formation of p-n junctions within the wires and, hence, for NW-based light emitting diodes. After these pioneering experiments sparse progress was reported for several years up to when the Lieber group [3] at Harvard University started new activities on the growth of Si NWs based on the Si–Au eutectic. During the last years world wide the synthesis of nanowires as well as nanotubes has been studied intensively for a wide range of materials. Such low-dimensional nanostructures are not only interesting for fundamental research due to their unique structural and physical properties relative to their bulk counterparts, but also offer fascinating potential for future technological applications. Special functionalities such as field effect devices are demonstrated most often based on nanowires chopped from the substrate and horizontally arranged on substrates by different methods including Langmuir-Blodgett techniques, nanomanipulation or statistic spread of disordered wires by spin coating.

A deeper understanding and sufficient control of the growth of nanowires are central to current research interest. Epitaxial growth of nanowires is very often depending on the selection of the correct substrate and its suitable crystal orientation, i.e. lattice constant. As demonstrated for wires based on ZnO in Fig. 1: part of the wires are showing an inclined growth on a-axis sapphire substrates whereas a perpendicular wire growth for all wires is observed on GaN (0001) substrates. The reason for the inclined wire growth is the remaining atomic steps on the polished sapphire substrates [4]. The template approach used here is quite simple: TEM grids are used as mask for the gold square arrays. During heating up the growth tube the gold film separates into small dots (an effect well known for nm thick gold films) with each dot being the starting point of a nanowire.

Vertical growth is desired for instance for field effect transistor devices but could also be useful for sensor arrays. In a recent review article we discussed the various growth and template methods [5]. Two different growth modes are discussed for nanowires (see Fig. 2). The first one is the so-called vapor-liquid-solid process (VLS). In this case the starting growth positions are metal dots on the surface of a suitable substrate where after heating up the substrates in a proper atmosphere and gas flow the nanowire growth starts due to adsorption and inclusion of vapor atoms into the melting droplets. The liquid metal droplet becomes an alloy droplet which supersaturates. At this point the

Fig. 1. Demonstration of inclined and vertical growth of ZnO nanowire arrays using the deposition of gold through TEM grids as template [4].

nanowire starts to grow at the interface of the substrate with the metal alloy droplet still at its upper head. The second process used for nanowire growth is still based on using metal dots as starting points. However the growth parameter are controlled in a way that the dots remain mainly solid. The process is then called vapor-solid growth (VS). As a result the metal catalyst remains below the growing wire at the substrate, but sometimes splits into separated small gold dots [6]. The enhanced sticking coefficient on a metal surface is used in the VS process. Hence, the growth can still be guided by a metal dot array. A schematic view of the two possible growth processes is shown in Fig. 2.

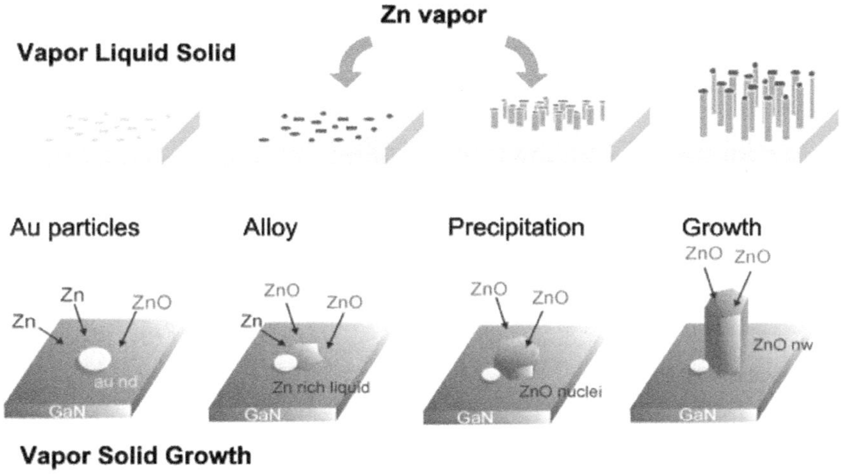

Fig. 2. Schematic demonstration of the two different nanowire growth modes, *upper image* VLS, *lower image* VS mode.

2 Methods for nano-patterning

The term “nano-patterning” refers to approaches which provide periodic arrangements with feature size and/or for a periodicity being in or below the 100 nm-range. The technology interest in the on-going search for new nano-patterning technique is based on the desire for inexpensive methods for areas in the cm^2 range and beyond. These methods should also being easier and faster in realization than conventional electron beam lithography, but sufficient reproducibility is expected. Most of the methods are still in the state of laboratory tests. The possibility of parallel sequenced writing based on such nano-patterning methods was already predicted. The realization of patterned metal arrays as catalysts for growth of semiconductor NWs is a new application of the nano-patterning techniques. In this respect, the metal catalyst arrays function as a template for the subsequently growth of NWs via VLS or VS model, so that the NWs would have the same pattern as the metal dots and the diameters of the nanowires are determined or predesigned by the size of the catalyst dots. Therefore, when growth control and position control of spatially separated NWs are desired, nano-patterning techniques become essential.

The following patterning methods are applied to bring metal dots or metal dot arrays in a controlled way on the substrate surface: photolithography or e-beam lithography, manipulation of single gold nanodots, arrangement of Au nanocrystals from suspensions, nanosphere lithography, gold deposition masks based on porous alumina templates, nanoimprint lithography, UV light interference lithography, co-block polymers for nano-lithography, and some others. These above methods are distinguished by the effort required and the possibility of large area realization. Here, we can not evaluate all of the above methods, our review published in 2006 gives more information and details [5].

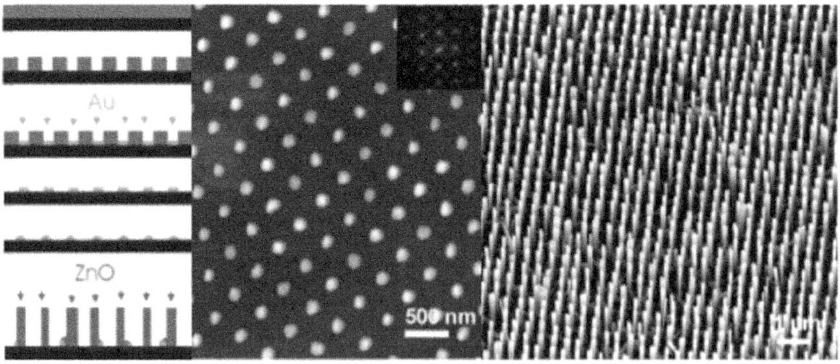

Fig. 3. Nanowire growth using templates based on interference lithography [7].

Ruling out the conventional lithographic methods due to limits in possible size and arrangement distances we selected interference lithography as one of the more promising method. The above image (Fig. 3) demonstrates the process as well as an example of the upon based nanowire arrays [7]. The interference lithography and its use for growth of ordered NW arrays are illustrated schematically in the left image of Fig. 3. After obtaining the Au nanodots an additional annealing step was applied to reduce their lateral dimension and to enhance the surface smoothness of the dots. Before the thermal annealing step, the Au NDs show a high roughness (with a mean height of 20 nm) with a diameter of 120 nm. However, after annealing the diameter and the height distributions are regular with a periodic distance of 270 nm. The mean diameter is reduced to around 80 nm and the height is increased to 30 nm. The resulting arrays of Au nanodots are ordered in a monoclinic lattice arrangement with an average density of 9 dots per μm^2. The center image shows the Au dot array after annealing, the left image the inclined view of the resulting nanowire array. The ZnO NWs have an average diameter of around 109 nm. Interestingly, the length of the ZnO NWs is quite uniform and independent of their diameter in domains of at least 10 by $10\,\mu m^2$.

3 Selected results – stimulated emission

For optical investigations we selected samples having a sufficient separation of the nanowires, an average diameter of 300 nm and a height of $1.5\,\mu m$ as seen in Fig. 4. The wide nanorods show a bulk-like excitonic photoluminescence at low excitation reflecting the high crystalline quality. In intermediate excitation regime the P-band emission is observed resulting from exciton–exciton scattering which is very pronounced in ZnO epitaxal layers and bulk samples but is weaker in our wide nanorods. We attribute this observation to the reduced phase-space for polariton–polariton scattering since the light wave coupling to the 3D exciton has a 1D confinement in the rods.

Increasing the excitation fluency in the confocal arrangement leads to the onset of stimulated emission in single wide rods (see Fig. 4) [6]. Such a transition does not occur in narrow nanorods even under global excitation. The spectral position of the stimulated recombination varies from rod to rod which is a signature for mode selection of the waveguide cavity. Together with the quenching of the P-band emission we can therefore exclude a pure excitonic origin. Instead free carriers have to be involved. The stimulated emission is present under femtosecond, confocal pumping up to 150 K. We can exclude that the observed differences in the occurrence of stimulated emission is related to sample quality. Rather, the behavior at intermediate excitation levels infers that the waveguide and resonator properties of the nanorods have a significant influence on the PL.

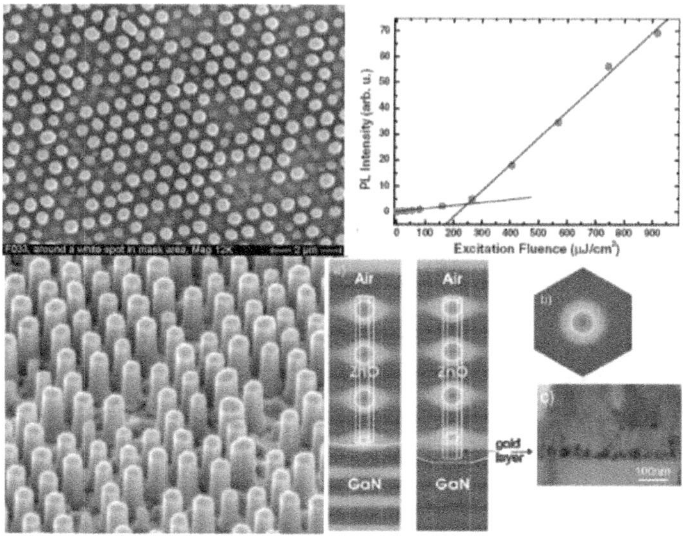

Fig. 4. Stimulated emission observed for ZnO nanowire pillars.

The fact that we have stimulated emission from the wide rods is a result of a better mode guiding but is also influenced by the presence of gold at the resonator facets. Please see the TEM image in the right part of Fig. 4. The thin gold layer influences the resonator properties. Leakage into the substrate is significantly reduced. Wide nanorods offer better overlap of the resonator eigenmodes with the gain medium. This finding can be used to improve the ZnO nanorod resonators. For more details please see Hauschild et al. [6].

4 Nanotubes based on Kirkendall diffusion and solid state reactions

Conventional treatment of the Kirkendall process only considers the bulk diffusion of growth species and vacancies. The Kirkendall effect is then a consequence of the different diffusiveness of atoms in a diffusion couple causing a supersaturation of lattice vacancies [8, 9]. This supersaturation may lead to a condensation of extra vacancies in form of so-called "Kirkendall voids" close to the interface. On the macroscopic and micrometer scale these Kirkendall voids are generally considered as a nuisance because they deteriorate the properties of the interface. In contrast, in the nanoworld the Kirkendall effect has been positively used as new fabrication route to design hollow nanoobjects. We believe that the use of the Kirkendall effect allows a rational design of nanoscale hollow objects, from metals, semiconductors to insulators, based on the proper choice of materials and different reaction properties known from thin-film diffusion couples.

In a recent paper we used $ZnAl_2O_4$ as a model system to demonstrate the transformation of nanowires into spinel nanotubes [10]. Figure 5(a) shows an example of ZnO–Al_2O_3 core-shell nanowires, which display a uniform coating of amorphous Al_2O_3 (light contrast) surrounding the single-crystal ZnO core (dark contrast). The alumina layer was uniformly deposited by atomic layer deposition (ALD), a powerful technique for precise and conformal deposition of metals and oxides over three-dimensional nanostructures. Figure 5(b–e) gives representative electron micrographs of the spinel nanotubes after thermal annealing the ZnO–Al_2O_3 core-shell nanowires. Most of the 1-D structures are hollow from one end to the other. These nanotubes are freestanding, narrow (30–40 nm in outer diameter, 10 nm in wall thickness), and smooth with excellent crystallinity.

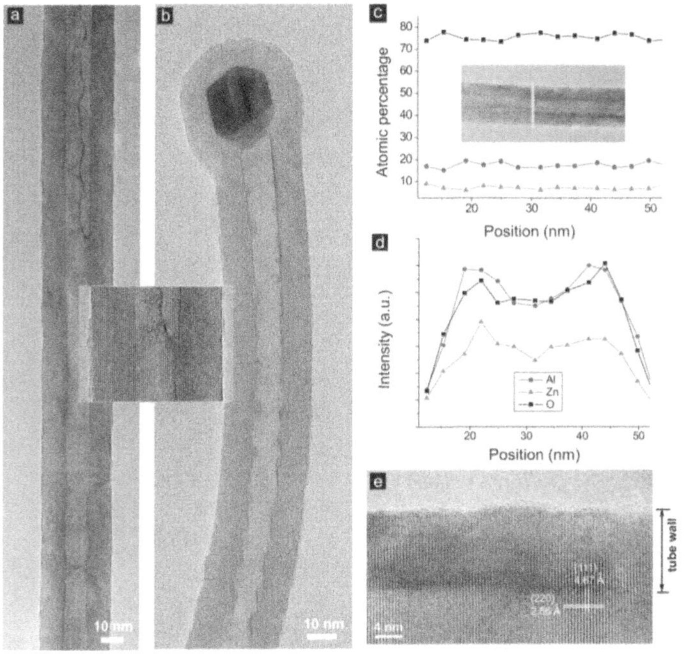

Fig. 5. Spinel nanotubes are observed after solid state reaction of a ZnO/Al_2O_3 core/shell nanowire at a temperature of 800°C [10].

In a nanoscale system, due to the finite volume and spatial confinement, a high vacancy supersaturation can readily be reached so that void formation is enhanced compared to the bulk counterparts. In this light, the voids have a high chance to touch the compound outer layer. Kirkendall voids are generated near the interface during vacancy – assisted exchange of material via bulk interdiffusion. The voids are the sinks for subsequent inward flux of vacancies and thus grow in size. The voids coalesce into bigger ones and touch the compound layer, in one sense breaking the connection for lattice diffusion and in the other sense establishing new bridges as fast transport paths for the remaining material. At this stage, diffusion of atoms of the remaining core material along the bridges, i.e., the pore surface, to the reaction front becomes the dominant material transport process, because of a much lower activation energy and higher diffusion coefficient of surface diffusion than those of bulk diffusion.

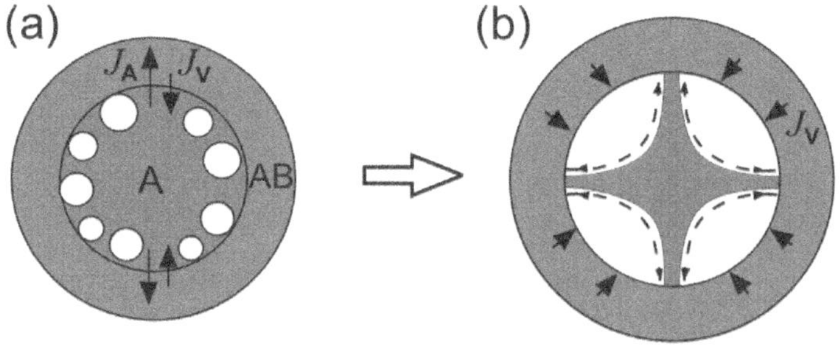

Fig. 6. Model for understanding the solid state reaction resulting in spinel nanotubes [11].

Hence, two main stages are involved in the development of the hollow interior. The respective model of the two stages is presented in Fig. 6. The initial stage is the generation of small Kirkendall voids intersecting the product interface via a bulk diffusion process (Fig. 6(a)); the second stage is dominated by surface diffusion of the core material (which is the fast diffusing species) along the pore surface (Fig. 6(a)) [11]. In this way the nanostructure is hollowed step by step. The shown results of the spinel-forming solid-state reaction of ZnO–Al_2O_3 core-shell nanowires represent a strong supporting evidence for the influence of surface diffusion on the morphology evolution. With this concept, we expect that the formation of complete hollow structures is faster than that computed by considering only bulk diffusion. Furthermore, we believe that the "Kirkendall effect in combination with surface diffusion" model should also apply to macroscopic core-shell structures, as well as to planar bilayers.

In a further experiment we investigated the influence of the core-shell ratio as well as the annealing temperature on the development of hollow nanotubes [12]. The thickness of the Al_2O_3 shell can be accurately adjusted by the number of repeated ALD cycles. We use a total number of 33, 66, 99 and 132 cycles for Al_2O_3 shells of 5.1, 10.3, 15.5 and 20.5 nm, respectively, conformal deposited onto ZnO nanowires of various diameters. Starting at 600°C the nanowire interface is intersected and irregularly shaped voids can be seen. Already in 1987 Maezawa et al. [13] reported the formation of an "aluminate-type" phase (i.e. "surface spinel") even at a synthesis temperature of 400°C. This interaction involves diffusion of Zn ions into tetrahedral sites in a surface lattice of Al_2O_3 to form spinel-like structures. In our experiments we observed scattered small voids along the interface even at lower temperatures which can be understood as consequence of the redistribution of zinc ions from the core into the alumina followed by coalescing vacancies to voids. This implies that the reaction starts at the temperature as low as 500°C. The gradually developed voids at 600°C indicate a greater diffusion of ZnO into the alumina subsurface. However, the diffusion is limited at this temperature.

Completely hollow spinel nanotubes can be formed at 700°C on condition that the ZnO core is totally consumed by the solid–solid reaction, i.e. the core/shell ration is fitting. At a higher temperature of 800°C, the unconsumed ZnO core can be further removed via diffusion, desorption and redistribution process. Therefore, all the starting core-shell nanowires with various diameters are transformed into completely hollow nanotubes. The wall thickness is approximately determined by the initial thickness of the deposited alumina layer. Reactions at 900°C lead to the formation of porous nanowires, which are derived from the collapsing of the spinel nanotubes due to their thermal instability. As the most important point, our results demonstrate that at a suitable reaction temperature, either freestanding or vertically aligned spinel nanotubes with controllable wall thickness can indeed be fabricated on large scale via the Kirkendall effect induced approach. Using the above knowledge we prepared hierarchical three-dimensional ZnO and demonstrated their shape-preserving transformation into hollow $ZnAl_2O_4$ nanostructures [14]. As shown in our latest paper, branched as well as millions of nanowires grown on planes of former Zn nanoparticles can be completely shape-preserving transferred into hollow structures.

5 Summary

During the last years the synthesis of nanowires is more and more focussed on one-dimensional growth with intentional control in structure, dimension, and spatial alignment. Nearly all existing technologies, from conventional photolithography to state of the art nanoimprint lithography, have been employed for positioning of the metal catalyst dots on suitable substrates. We found that UV interference lithography offers the possibility for wafer sized preparation of metal dot arrays combined with accuracy in size control and suitability for mass production. In our work we mainly concentrated on the realization of ordered arranged ZnO nanowires. Typical growth approaches for semiconductor NWs are the VLS and VS approaches. We focused on the VLS growth mode because it provides a technical opportunity for in situ position, definition, and size control of the nanowires. However, in special cases the inclusion of the metal catalyst at the base of the NWs might be of advantage and hinder the leakage of the optical wave into the substrate. Stimulated emission was reported for wide ZnO nanorods.

In the second part we demonstrated core-shell nanowires which were prepared by homogeneous covering of the wires by atomic layer deposition. High temperature annealing causes a diffusion of the core material ZnO into the Al_2O_3 shell. Spinel nanotubes were shown which can be realized based on the Kirkendall effect and solid state reactions using suitable temperatures and annealing times. Even complicated hierarchical three-dimensional structures can be shape-preserving transformed into hollow $ZnAl_2O_4$ nanostructures. Hence, we demonstrated that the Kirkendall and solid state reaction based process can be generally used for hollowing nanostructures of suitable diffusion couples.

6 Acknowledgement

The author (M.Z.) gratefully acknowledges funding by the German Research Foundation for the work presented here.

References

1. R.S. Wagner, W.C. Ellis: Trans. Metall. Soc. AIME **233**, 1053 (1965)
2. K. Haraguchi, T. Katsuyama, K. Hiruma, K. Ogawa: Appl. Phys. Lett. **60**, 745 (1992)
3. A. M. Morales, C.M. Lieber: Science **279**, 208 (1998)
4. H.J. Fan, F. Fleischer, W. Lee, K. Nielsch, R. Scholz, M. Zacharias, U. Gösele, A. Dadgar, A. Krost: Superlattice Microstruct. **36**, 95 (2004)
5. H.J. Fan, P. Werner, M. Zacharias: Small **2**, 700 (2006)
6. R. Hauschild, H. Lange, H. Priller, C. Klingshirn, R. Kling, A. Waag, H.J. Fan, M. Zacharias, H. Kalt: Phys. Stat. Solidi B **243**, 853 (2006)
7. D.S. Kim, R. Ji, H.J. Fan, F. Bertram, R. Dadgar, K. Nielsch, A. Krost, J. Christen, U. Gösele, M. Zacharias: Small **3**, 76 (2007)
8. E.O. Kirkendall: Trans. AIME **147**, 104 (1942)
9. A. D. Smigelskas, E. O. Kirkendall: Trans. AIME **171**, 130 (1947)
10. H.J. Fan, U. Gösele, M. Zacharias: Small **3**, 1660 (2007)
11. H.J. Fan, M. Knez, R. Scholz, D. Hesse, K. Nielsch, M. Zacharias, U. Gösele: Nano Lett. **7**, 993 (2007)
12. Y. Yang, D.S. Kim, M. Knez, R. Scholz, A. Berger, E. Pippel, D. Hesse, U. Gösele, M. Zacharias: J. Phys. Chem. **C112**, 4068 (2008)
13. A. Maezawa, Y. Okamoto, T.J. Imanaka: J. Chem. Soc., Faraday Trans. **1, 83**, 665 (1987)
14. Y. Yang, D.S. Kim, R. Scholz, M. Knez, S.M. Lee, U. Gösele, M. Zacharias: Chem. Mater. **20**, 3487 (2008)

Growth Methods and Properties of High Purity III-V Nanowires by Molecular Beam Epitaxy

D. Spirkoska *, C. Colombo *, M. Heiß *, M. Heigoldt, G. Abstreiter, and A. Fontcuberta i Morral

Walter Schottky Institut, Technische Universität München, Am Coulombwall 3, 85748 Garching, Germany
annafm@wsi.tum.de

Abstract. The synthesis and properties of catalyst-free III–V nanowires with MBE is reviewed. The two main methods are Selective Area Epitaxy and gallium-assisted synthesis. The growth mechanisms are reviewed, along with the design possibilities of each technique. Finally, the excellent structure and ultra-high purity are presented by Raman and Photoluminescence spectroscopy.

1 Introduction

Semiconductor nanowires constitute extremely promising building blocks for the XXI century electronic and optoelectronic devices. One reason is that its size will enable further down scaling of electronics [1–3]. A second reason is that nanoscale objects exhibit new properties which at the same time can be exploited into new device concepts such as high mobility transistors, thermoelectric applications and/or solar cells [4–7]. As a consequence, fundamental and applied research on nanowires has increased dramatically in the last few years.

One issue of crucial importance has been the control on the crystalline quality and impurity concentrations, as well as the reproducibility of the structures. With regards to the purity, one of the key issues has been to avoid the use of gold as nucleation and growth seed of the nanowires. Gold is a fast-diffusing metal that significantly harms the properties of semiconductors [8]. To date, synthesis without gold has been achieved by the use of alternative metals such as aluminum and titanium, or by simply avoiding the use of a catalyst [9–12]. These alternative methods involve the development of new deposition techniques, meaning that the growth mechanisms necessarily differ from the standard Au-assisted VLS/VSS growth.

Traditionally, Molecular Beam Epitaxy (MBE) has been one of the oldest techniques applied to the fabrication of high quality nanostructures. Starting

* Equal contribution

from Quantum Wells and Quantum Dots obtained by standard planar growth, the method has also been applied for the fabrication of quantum heterostructures extended in more dimensions [13]. In this paper, we study the possibilities that MBE gives us to fabricate high quality III–V As-based semiconductor nanowires, while avoiding the use of gold as seed for the nucleation. We will show how the use of MBE presents at the same time an additional interest, as this technique allows us to produce ultra-pure nanowires and quantum heterostructures on the nanowire facets with very high crystalline quality and atomically sharp interfaces. This new versatility of MBE in the growth of nanostructures opens great possibilities for the generation of novel devices with additional optical and electronic functionalities, as it has been previously shown in planar structures [14–16].

2 Experimental

The samples were grown in a high purity Gen-II MBE system. In all cases, two-inch GaAs wafers sputtered with a 10–60 nm thick silicon dioxide film were used as substrates. In the case of Selective Area Epitaxy, the oxide was patterned by combining lithography with reactive ion etching [17]. In the case of gallium assisted growth of nanowires, no patterning was realized on the surface. In order to ensure a contamination-free surface, prior to the introduction to the MBE chamber the substrates were dipped for 2 s in a buffered HF aqueous solution (10% HF). In order to desorb any remnant adsorbed molecules of the surface, the wafers were heated to 650°C for 30 min prior to growth. As a difference to standard nanowire growth, no external metal catalyst was used in any case for the growth of the nanostructures.

3 Selective area epitaxy

Selective Area Epitaxy (SAE) is one of the two techniques leading into the growth of nanowires with MBE, which at the same time avoids the use of gold as a nucleation seed and catalyst. The purpose of SAE is to restrict the incorporation of the adatoms to certain areas on a patterned substrate. Basically, the III–V substrate is masked with a patterned SiO_2 layer and growth conditions are appropriately chosen to restrict the epitaxial growth inside the apertures. This technique has both been used in Metalorganic Chemical Vapor Deposition (MOCVD) and MBE [11, 17–19]. In MOCVD the selectivity originates from a preferential decomposition of the metalorganic precursors in III–V surfaces with respect to the SiO_2. In the case of MBE, the selectivity originates from the lowering of the sticking coefficient of the species on the oxide, in comparison to the open III–V surfaces. Additionally, diffusion of adatoms from the oxide to the III–V windows plays a supplemental role.

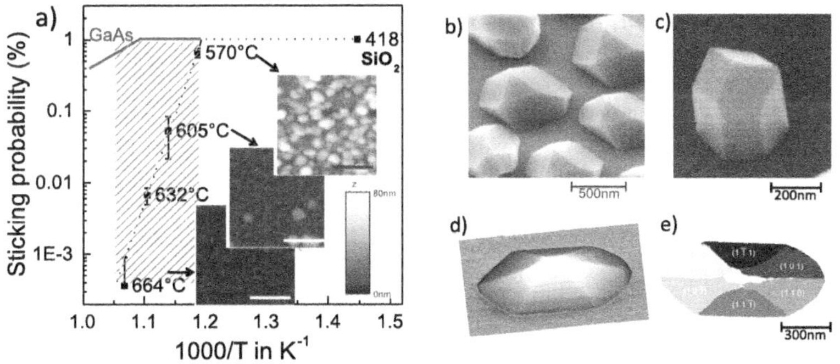

Fig. 1. (**a**) Measured sticking probability of GaAs on SiO_2 as a function of temperature, and comparison to literature values of the sticking probability on GaAs. In the inset, Atomic Force Micrographs of the surfaces after growth of 50 nm at the indicated temperatures. The scale bar stands for 400 nm. (**b,c**) Scanning Electron Micrograph of 200 nm GaAs grown by selective area epitaxy on (111)A and (111)B patterned substrates. (**d**) Atomic Force micrograph of 200 nm GaAs grown on (001) GaAs patterned substrate and (**e**) the corresponding attribution of the crystalline facets indexes.

We present first a study on the selective desorption of the Ga adatoms on surfaces by measuring the temperature dependence of the sticking coefficient, s, of GaAs on SiO_2. s is a measure of the probability of an adatom to precipitate forming a film instead of desorbing. It can be simply measured by comparing the nominal thickness with the actual thickness of material that has grown on the substrate. A nominal thickness of 50 nm GaAs was grown on SiO_2 at temperatures ranging between 418 and 664°C. The GaAs growth rate, as calibrated on GaAs at 550°C, was 0.4Å/s. The morphology of the surface after deposition was analyzed by Atomic Force Microscopy and Scanning Electron Microscopy (AFM and SEM respectively). In Fig. 1a, the results on the sticking coefficient of GaAs on SiO_2 (s_{SiO_2}) as a function of temperature are presented and compared to the literature values for GaAs on GaAs (s_{GaAs}) [20]. The AFM measurements of the surfaces after depositing nominally 50 nm of GaAs are also shown for illustration in the inset. For temperatures below 565°C , s is close to 1 in both cases meaning that 100% of the Ga adatoms precipitate on the oxide surface. For temperatures above 565°C, s_{SiO_2} starts to decrease and stays well below the value on GaAs. This means that for these temperatures, the fraction $(1 - s)$ of Ga adatoms desorb from the surface. Finally, at temperatures higher than 650°C, s on SiO_2 is very close to zero, meaning that deposition of GaAs is not possible.

We have grown different thicknesses of GaAs on patterned substrates, at 650°C. Different substrate orientation has been used, in order to compare the morphology of the obtained structures. For deposited thicknesses below 100 nm, the growth in the openings occurs in a planar way and no significant difference is observed between the substrates. However, for higher nominal thicknesses the pillars develop clearly defined facets that are always defined in terms of minimization of energy. According to our results and the existing literature, the facets with the lowest formation energy are those of the family {110} and {111}, as well as sometimes 311. In Fig. 1b–d the scanning electron microscopy (SEM) and AFM measurements of structures grown by SAE on different substrates are shown. The substrates are in all cases GaAs, the only difference is the crystalline orientation of the substrate. In Fig. 1b and 1c, the SEM of structures grown on (111)A and (111)B patterned substrates are shown. Clearly the faceting geometry is significantly different. Due to the perpendicularity between (111)B and (1–10) facets, only in the case of (111)B substrates a vertical growth in the form of nanowire is possible. In the other cases, the faceting leads into pyramidal or multipolyhedral structures. Another example is given in Fig. 1d, where an Atomic Force micrograph of 200 nm GaAs grown on (001) GaAs patterned substrate is shown. The corresponding attribution of the crystalline facets indexes is shown in Fig. 1e. It should be noted here that faceting is not a new phenomenon in MBE growth. As an example, it has been known for a long time that self assembled Stranski-Krastanov quantum dots present high index facets of the type (110), (311)... [21]. The crystallographic orientation of the facets was investigated by detailed analysis of AFM measurements. We have observed that the facets correspond generally to the plane families {110} and {111}, which are known to be the crystal facets in III–V semiconductors with the lowest energy [22, 23]. As a conclusion, minimization of surface energy and faceting is a general effect in the nearly equilibrium growth of nanostructures, faceting depends on the crystal orientation of the substrate. Here we would like to add that faceting can add degrees of freedom in the design of functional heterostructures. Heterostructures grown on faceted nanopillars will offer the possibility of in situ growing kinked quantum wells with MBE [17]. Indeed, at the interface of two or three kinked quantum wells, the existence of further confined states such as quantum wires and dots are expected, as it has been observed before by MOCVD [18]. The position of these quantum wires and dots are predetermined by the previous pattern and therefore offer many new possibilities of design for nanoscale devices.

The main advantages of SAE are the control at the monolayer level and the possible use of faceting for the exploring of additional quantum heterostructures. A drawback of SAE is the slow growth rate of the structures. For SAE to occur, high temperatures are important but also low arrival rate of the group III adatoms. This low growth rate does not allow the fabrication of high aspect ratio structures.

4 Conditions leading to group III assisted growth of nanowires

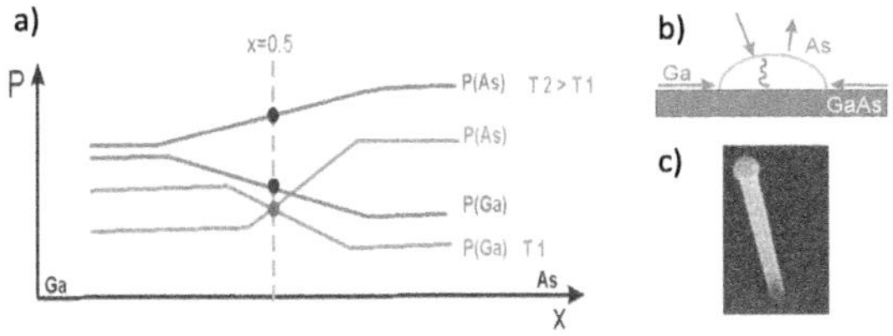

Fig. 2. (**a**) Vapor pressure schematic dependence of $Ga_{1-x}As_x$ for T1 below the congruent temperature and T2 above the congruent temperature (**b**) Schematics of the equilibrium of a Ga droplet on a GaAs substrate at Ga-rich conditions, above the congruent temperature and (**c**) Scanning electron micrograph of a 500 nm long GaAs nanowire grown under conditions of (**b**). The Ga droplet is observed on the top of the nanowire grown on a SiO_2 coated GaAs substrate.

A fundamentally different method to obtain III–V nanowires is based in the Vapor–Liquid–Solid method, in which a metal droplet is used to gather and precipitate the growth precursors. Typically, gold is a metal that works relatively well for any material. Instead, in principle it should possible to use group III metal droplets to gather group V elements and precipitate III–V nanowires underneath. In the case of GaAs nanowires, this leads to what we call "*gallium assisted growth*"; but the method can be extended to other III–V combinations such as InAs, InGaAs and AlAs .... For simplicity, we will just discuss the case of GaAs and therefore name these conditions as Ga-rich. In order to find the right conditions, it is necessary to look at the gallium and arsenic partial pressures of GaAs as a function of temperature. A schematics of this diagram is shown in Fig. 2 [24, 25]. For temperatures below 630°C the vapor pressures of Ga and As lead to a congruent evaporation of atoms, meaning that the evaporation rate of Ga and As is the same. At higher temperatures, this balance is not possible because the partial pressure of As is higher. In practice, this leads to the selective evaporation of As which in turns results in the formation of Ga droplets at the surface. This transition temperature is commonly referred as congruent temperature [24]. In order to use these Ga droplets for the gathering of As and growth of GaAs nanowires, a further element has to be considered. Indeed, in order to avoid the spreading and increase of the Ga dropet, it is necessary to avoid the wetting of the metal on the surface. For this reason, GaAs surface has to be avoided. We have proved that this purpose can be reached covering the GaAs substrate with a thin SiO_2 layer.

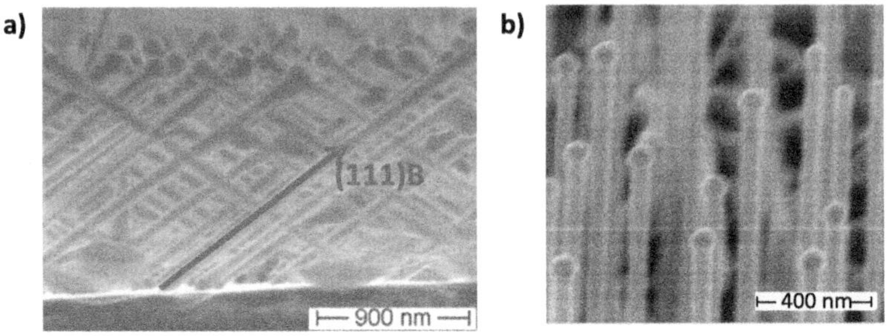

Fig. 3. (**a**) Scanning electron micrograph (SEM) of GaAs nanowires grown on a SiO_2 coated GaAs substrate. The substrate surface is (001), leading into the growth of GaAs wires on a 34° angle, which coincides with the (111)B crystalline direction of the substrate. (**b**) SEM of nanowires grown on a (111)B oriented substrate.

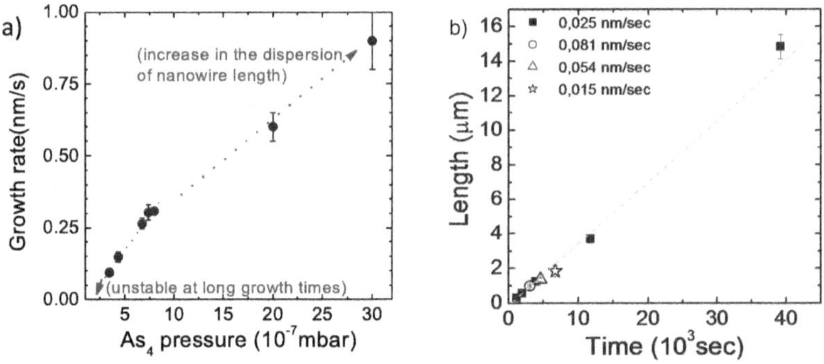

Fig. 4. (**a**) The growth rate of the GaAs nanowires tends to increase linearly as a function of the As_4 beam pressure. For pressures below $3 \cdot 10^{-7}$ mbar, the growth rate diminishes abruptly and the growth is not stable. For pressures above $8 \cdot 10^{-7}$ mbar the length dispersion from wire to wire is much higher. (**b**) The length of the GaAs nanowires as a function of time for different Ga arrival rates, for an As_4 beam pressure of $8 \cdot 10^{-7}$ mbar. The nanowire growth rate seems to be Ga rate independent.

We have observed, that when the oxide is thin enough (below 30 nm), nanowires grow following the (111)B direction of the substrate [12]. This is shown in Fig. 3, where the SEM of nanowires grown on (001) and (111)B GaAs substrates is shown. In one case, the wires grow forming a 34 ° angle with the surface, while in the other case the wires grow perpendicularly. In order to proof that the nanowire growth is governed by the arsenic, we investigated the effect of the arsenic beam flux on the growth rate of the GaAs nanowires. The results are shown in Fig. 4a. In agreement with our hypothesis, the growth rate is proportional to the arsenic beam flux in ranges from $3.5 \cdot 10^{-7}$ to $3.5 \cdot 10^{-6}$ mbar. For pressures below $3.5 \cdot 10^{-7}$ mbar, the growth rate is very low and the growth is highly unstable. Between 3.5 and $8 \cdot 10^{-7}$ mbar the dispersion in the nanowire growth rate is very small, while for higher beam pressures a large dispersion in the length exists. The reason for the increased size dispersion remains unclear, though one reason could be delayed incubation times among the nanowires. The growth rate of nanowires was also measured as a function of the Ga rate. In Fig. 4b, the length of the nanowires as a function of time is plotted for different Ga rates, with an As Beam pressure of $8 \cdot 10^{-7}$ mbar. It is clear that all points fall in the same line, indicating an identical growth rate for the Ga rates varied from 0.12 to 0.82 Å/s. This result indicates that under these conditions the growth of the nanowires is not limited by the amount of Ga adatoms arriving at the surface, as it is usually the case in epitaxial growth of GaAs thin films [26].

5 Structural and optical properties

The structural properties of the GaAs nanowires were investigated by Raman spectroscopy. Prior to the measurements, bundles of nanowires were dispersed on a silicon substrate. The Raman experiments were performed at room temperature by using the 488 nm line from an Ar^+ laser, focused with a 50x microscope objective. The measurements were realized with low excitation power (0.5 mW), in order to avoid the heating of the sample. A typical Raman spectra of the nanowires is presented in Fig. 5a. The solid black line is the recorded data while the green lines are result from a multiple Lorentzian fit. The peak positioned at 268.7 cm^{-1} can be attributed to scattering from TO phonon and the peak positioned at 292.2 cm^{-1} is due to scattering from LO phonon. The TO and the LO peaks are symmetric and have very small FWHM (around 4 cm^{-1}). The peak positions correspond exactly with the position of the TO and the LO peaks measured on bulk (111) GaAs. The measured values for the peak positions and FWHM are a good indication that the synthesized wires have excellent structural quality, free of defects and stress, which further corroborates the advantage of using MBE.

A third peak positioned at the low frequency side from the LO phonon is also clearly observed. As previous studies have shown, this peak can be attributed to scattering from surface optical phonon (SO) [27, 28]. A simple experiment further proofs the surface related nature of this mode. The nanowires were embedded in a PMMA matrix and subsequently measured by Raman. As

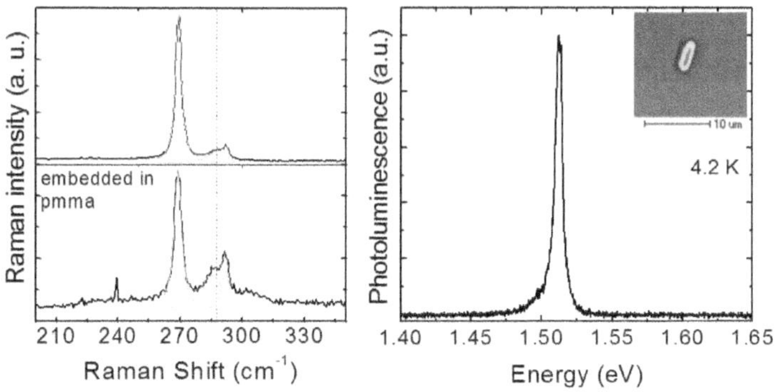

Fig. 5. (**a**) Raman spectroscopy on GaAs nanowires in air embedded in a PMMA matrix. The TO, LO and SO modes are indicated. As expected, the SO mode shifts when the wires are embedded in a matrix with a higher dielectric constant. (**b**) Photoluminescence spectroscopy on a single GaAs nanowire at 4.2 K. The small linewith corresponds well with the good crystalline quality. In the inset, an example of reflection scan realized on the sample surface to find the single nanowires.

predicted by the theory and shown in Fig. 5a, the increase in the surrounding dielectric constant leads into a shift of the SO mode, in this case of $1.8\,\mathrm{cm}^{-1}$.

In order to further assess the quality of the nanowires, photoluminescence spectroscopy (PL) on single nanowires was realized at 4.2 K, by means of a confocal microscope. The PL was excited using the 632.8 nm line of a He-Ne laser, and detected by the combination of a grating spectrometer and a silicon charge coupled device (CCD). For the measurements, the wires were dispersed on a silicon substrate. Scanning reflectivity measurements of the surfaces were realized in order to localize single nanowires. An example is shown in the inset of Fig. 5b, where the nanowire can be clearly identified in the middle of the scan. The PL emission of a single GaAs nanowire is shown in Fig. 5b. The PL spectrum is characterized by a peak centered at 1.51 eV with a full width at half-maximum of 6 meV, which corresponds well to the free exciton of undoped bulk GaAs. It should be stressed that these data are exceptional among the nanowires and further corroborate the high crystalline quality and purity of the nanowires. Here it is also important to note that the wires were not passivated, meaning that there might be very little surface states (which may be related to the fact that (110)GaAs surface has no band-gap states on the clean surface [29]). One should also note that we have observed that capped nanowires exhibit a PL higher in about a factor 100 with respect to the uncapped.

6 Potential for future structures and devices

We have shown in the previous sections, that MBE grown nanowires offer impressive structural and functional properties. At this moment it is open, the way they are going to play a significant role in the technology and basic science at the nanoscale. In this section, we would like to propose some of the future applications and perspectives of these wires.

The excellent properties of III–V materials are counterparted by their low abundancy – especially when compared with silicon – and the associated costs. For this reason, the combination of III–V semiconductors with the existing silicon-based technology has arised a great interest. This area of research has faced many challenges, mainly due to the lattice mismatch between the two materials. The problems and challenges associated with this can be overcome if the III–Vs are grown in the form of nanowire. Indeed, when the effective substrate area is reduced to the nanometer scale, the strain of the epilayer can relax laterally. The total strain energy of the system is reduced, enabling defect-free heteroepitaxy. The integration of III–V nanowires and related devices on silicon has already been shown in the past mainly by MOCVD and by using gold as a catalyst [30–32]. Catalyst-free MBE based growth of III-V nanowires on silicon still has to be demonstrated, but its achievement will bring the application of such materials much further in the technological path, as one could combine silicon technology with high performing III–V, densely packed vertical devices.

On another aspect, it is a matter of fact that the functionality of semiconductors augments significantly if dopant elements are incorporated in an efficient way. Dopants rule the conductivity of materials and allow the fabrication of devices like diodes, solar cells and transistors. Doping has been largely investigated and realized in a successful way for Si, GaN and ZnO nanowires, though theoretical works had predicted difficulties related to the diffusion of dopants towards the surface [33–37]. However, the controlled change of III–V nanowires' conductivity by doping has proven to be a difficult task. To our knowledge, highly efficient and controlled doping in the bottom-up approach of GaAs and InAs nanowires has been proved to be difficult, as dopants tend to not get incorporated during the growth process. As mentioned above, doping is key for advanced electronic and optoelectronic devices. For that, the intricate research on the synthesis, functional and structural characterization and theoretical understanding will be essential. At last, we will discuss a further advantage of MBE grown nanowires. Indeed, it is possible to switch the growth modus from nanowire-like to thin film-like. This enables the growth directly on the facets of the nanowires. This principle is indicated in Fig. 6a. Due to the directionality of the epitaxial beam, it is possible to grow on each of the facets or on the selected ones. Depending on the substrate used, the epitaxial growth will result in different geometries. For example, in the case the nanowires are grown on (111)B substrates, all layers have the same thickness. As a result, the cross-section of the layers forms a hexagon. This new design

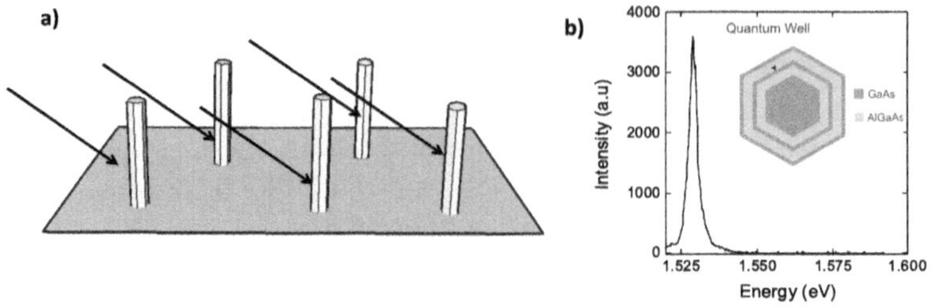

Fig. 6. (**a**) Schematics of the principle of heteroepitaxy on the facets of the nanowires. Due to the directionality of the beam, the geometry of the resulting layers results in hexagonal or asymmetric. Here, the geometry for nanowires grown on (111)B substrates is shown. (**b**) Photoluminescence of a 14 nm GaAs Quantum Well deposited on a nanowire.

possibility adds many new ways of adding functionality to the nanowires. With the purpose of proving this principle, a quantum heterostructure was grown on the {110} facets of wires grown on a (111)B substrate. The structure consisted of a quantum well (QW) of GaAs embedded in $Al_{0.3}Ga_{0.7}As$ barrier layers. The whole structure was capped with an thin layer of GaAs. The substrates were rotated at 7 rpm to ensure a uniform deposition. Thanks to the geometry of the nanowires, the deposition resulted in a prismatic configuration of the QWs, which we call p-QW. The selective growth on the {110} facets was achieved by increasing the As_4 beam flux to $5 \cdot 10^{-5}$ mbar, which is typical for 110 surfaces. Photoluminescence spectroscopy measurements were realized as a proof of principle. The spectrum is shown in Fig. 6b, where an unique emission peak centered at 1.53 eV is shown. This is consistent with a 14 nm quantum well, in agreement with the material deposited. Moreover, the same photoluminescence spectrum was obtained, when measured along the wire.

Coaxial-type nanowire heterostructures have been fabricated in the past. The function of the coating layer has been limited to the passivation of the surface states of the nanowires and has been fabricated with an isotropic radial morphology [38, 39]. Here we show that a uniform deposition on each side facets of a nanowire can be added with intrinsic functional purposes. Moreover, the lateral width of the QWs is in the order of 60 nm, meaning that the QWs pertaining to the prismatic structures can be considered as quasi one dimensional structures. The application of MBE to the fabrication of three dimensional quantum heterostructures opens a new avenue for a large variety of physical experiments and devices.

7 Conclusions

In conclusion, we have presented a novel method for growth of GaAs nanowires and prismatic quantum heterostructures by molecular beam epitaxy. We have presented and distinguished the method of selective area epitaxy and group III assisted growth. The two techniques avoid the use of gold for the nucleation and growth, solving the key issue of metal contamination. Moreover, we also show the advantages of using MBE which are: (1) a high purity leading into excellent structural and optical properties and (2) the selective growth of quantum heterostructures on the facets of the wires, providing a large new range of functionalities and applications.

8 Acknowledgements

The authors kindly thank D. Grundler, R. Gross, M. Stutzmann, J. R. Morante, M. Bichler and B. Laumer for experimental support and discussions. The financial support from Marie Curie Excellence Grant SENFED, Nanosystems Initiative Munich (NIM) and SFB 631 is greatly acknowledged.

References

1. M. T. Björk, O. Hayden, H. Schmid, H. Riel, W. Riess: Vertical surround-gated silicon nanowire impact ionization field-effect transistors, Applied Physics Letters **90**, 142110 (2007) doi:10.1063/1.2720640
URL `http://link.aip.org/link/?APL/90/142110/1`
2. R. Chau, S. Datta, A. Majumdar: Opportunities and challenges of iii-v nano-electronics for future high-speed, low-power logic applications, Compound Semiconductor Integrated Circuit Symposium, 2005. CSIC '05. IEEE SP-4 pp. (30 Oct.–2 Nov. 2005) doi:10.1109/CSICS.2005.1531740
3. J. Knoch, W. Riess, J. Appenzeller: Outperforming the conventional scaling rules in the quantum-capacitance limit, Electron Device Letters, IEEE **29**, 372–374 (April 2008) doi:10.1109/LED.2008.917816
4. W. Lu, J. Xiang, B. P. Timko, Y. Wu, C. M. Lieber: One-dimensional hole gas in germanium/silicon nanowire heterostructures, Proceedings of the National Academy of Sciences **102**, 10046–10051 (2005) doi:10.1073/pnas.0504581102
URL `http://www.pnas.org/cgi/content/abstract/102/29/10046`
5. A. Guichard, D. Barsic, S. Sharma, T. Kamins, M. Brongersma: Tunable light emission from quantum-confined excitons in tisi2-catalyzed silicon nanowires, Nano Letters **6**, 2140–2144 (2006)
URL `http://pubs3.acs.org/acs/journals/doilookup?in$\_$doi=10.1021/nl061287m`
6. L. Shi, D. Li, C. Yu, W. Jang, D. Kim, Z. Yao, P. Kim, A. Majumdar: Measuring thermal and thermoelectric properties of one-dimensional nanostructures using a microfabricated device, Journal of Heat Transfer **125**, 881–888 (2003) doi:10.1115/1.1597619
URL `http://link.aip.org/link/?JHR/125/881/1`

7. J. Maiolo, B. Kayes, M. Filler, M. Putnam, M. Kelzenberg, H. Atwater, N. Lewis: High aspect ratio silicon wire array photoelectrochemical cells, Journal of the American Chemical Society **129**, 12346–12347 (2007)
URL `http://pubs3.acs.org/acs/journals/doilookup?in$\_$doi=10.1021/ja074897c`
8. S. D. Brotherton, J. E. Lowther: Electron and hole capture at au and pt centers in silicon, Physical Review Letters **44**, 606–609 (1980) doi:10.1103/PhysRevLett.44.606
9. Y. Wang, V. Schmidt, S. Senz, U. Gosele: Epitaxial growth of silicon nanowires using an aluminium catalyst, Nature Nanotechnology **1**, 186–189 (2006)
URL `http://dx.doi.org/10.1038/nnano.2006.133`
10. T. I. Kamins, R. S. Williams, D. P. Basile, T. Hesjedal, J. S. Harris: Ti-catalyzed si nanowires by chemical vapor deposition: Microscopy and growth mechanisms, Journal of Applied Physics **89**, 1008–1016 (2001) doi:10.1063/1.1335640
URL `http://link.aip.org/link/?JAP/89/1008/1`
11. B. Mandl, J. Stangl, T. Martensson, A. Mikkelsen, J. Eriksson, L. Karlsson, G. Bauer, L. Samuelson, W. Seifert: Au-free epitaxial growth of inas nanowires, Nano Letters **6**, 1817–1821 (2006)
URL `http://pubs3.acs.org/acs/journals/doilookup?in$\_$doi=10.1021/nl060452v`
12. A. F. i Morral, C. Colombo, G. Abstreiter, J. Arbiol, J. R. Morante: Nucleation mechanism of gallium-assisted molecular beam epitaxy growth of gallium arsenide nanowires, Applied Physics Letters **92**, 063112 (2008) doi:10.1063/1.2837191
URL `http://link.aip.org/link/?APL/92/063112/1`
13. G. Schedelbeck, W. Wegscheider, M. Bichler, G. Abstreiter: Coupled quantum dots fabricated by cleaved edge overgrowth: From artificial atoms to molecules, Science **278**, 1792–1795 (1997) doi:10.1126/science.278.5344.1792
URL `http://www.sciencemag.org/cgi/content/abstract/278/5344/1792`
14. K. Brunner, G. Abstreiter, G. Böhm, G. Tränkle, G. Weimann: Sharp-line photoluminescence and two-photon absorption of zero-dimensional biexcitons in a gaas/algaas structure, Physical Review Letters **73**, 1138–1141 (1994) doi:10.1103/PhysRevLett.73.1138
15. S. F. Fischer, G. Apetrii, U. Kunze, D. Schuh, G. Abstreiter: Energy spectroscopy of controlled coupled quantum-wire states, Nature Physics **2**, 91–96 (2006)
URL `http://dx.doi.org/10.1038/nphys205`
16. T. Egeler, G. Abstreiter, G. Weimann, T. Demel, D. Heitmann, P. Grambow, W. Schlapp: Anisotropic plasmon dispersion in a lateral quantum-wire superlattice, Physical Review Letters **65**, 1804–1807 (1990) doi:10.1103/PhysRevLett.65.1804
17. M. Heiß, E. Riedlberger, D. Spirkoska, M. Bichler, G. Abstreiter, A. F. i. Morral: Growth mechanisms and optical properties of gaas-based semiconductor microstructures by selective area epitaxy, Journal of Crystal Growth **310**, 1049–1056 (2008)
URL `http://www.sciencedirect.com/science/article/B6TJ6-4RHFVM6-1/1/b4df6ccca6568fdb6a0605ad3db318ab`

18. J. M. Hong, S. Wang, T. Sands, J. Washburn, J. D. Flood, J. L. Merz, T. Low: Selective-area epitaxy of gaas through silicon dioxide windows by molecular beam epitaxy, Applied Physics Letters **48**, 142–144 (1986) doi:10.1063/1.96977 URL `http://link.aip.org/link/?APL/48/142/1`
19. P. Mohan, J. Motohisa, T. Fukui: Controlled growth of highly uniform, axial/radial direction-defined, individually addressable inp nanowire arrays, Nanotechnology **16**, 2903–2907 (2005)
URL `http://stacks.iop.org/0957-4484/16/2903`
20. R. Fisher, J. Klem, J. T. Drummond, E. R. Thorne, W. Kopp, H. Morkoc, Y. A. Cho, *J Appl. Phys.* **54**, 2058–2510 (1983)
21. J. Márquez, L. Geelhaar, K. Jacobi: Atomically resolved structure of inas quantum dots, Applied Physics Letters **78**, 2309–2311 (2001) doi:10.1063/1.1365101 URL `http://link.aip.org/link/?APL/78/2309/1`
22. N. Moll, A. Kley, E. Pehlke, M. Scheffler: Gaas equilibrium crystal shape from first principles, Physical Review B **54**, 8844–8855 (1996) doi:10.1103/PhysRevB.54.8844
23. J. Márquez, P. Kratzer, L. Geelhaar, K. Jacobi, M. Scheffler: Atomic structure of the stoichiometric gaas(114) surface, Physical Review Letters **86**, 115–118 (2001) doi:10.1103/PhysRevLett.86.115
24. C. Chatillon, D. Chatain: Congruent vaporization of gaas(s) and stability of ga(l) droplets at the gaas(s) surface, Journal of Crystal Growth **151**, 91–101 (1995)
URL `http://www.sciencedirect.com/science/article/B6TJ6-3Y5MNBT-CW/1/618e13e48368370025ebddc5476f2a8a`
25. J.-y. Shen, C. Chatillon: Thermodynamic calculations of congruent vaporization in iii-v systems; applications to the in-as, ga-as and ga-in-as systems, Journal of Crystal Growth **106**, 543–552 (1990)
URL `http://www.sciencedirect.com/science/article/B6TJ6-46D26H6-2C/1/61e6f54efa66285c41035fc80d35848d`
26. C. Colombo, D. Spirkoska, M. Frimmer, G. Abstreiter, A. F. i Morral: Ga-assisted catalyst-free growth mechanism of gaas nanowires by molecular beam epitaxy, Physical Review B (Condensed Matter and Materials Physics) **77**, 155326 (2008) doi:10.1103/PhysRevB.77.155326
URL `http://link.aps.org/abstract/PRB/v77/e155326`
27. M. Watt, C. M. S. Torres, H. E. G. Arnot, S. P. Beaumont: Surface phonons in gaas cylinders, Semiconductor Science and Technology **5**, 285–290 (1990)
URL `http://stacks.iop.org/0268-1242/5/285`
28. D. Spirkoska, G. Abstreiter, A. F. i Morral: Size dependence of the bulk and surface phonon modes of gallium arsenide nanowires as measured by raman spectroscopy, Nanotechnology **19**, 435704 (2008)
29. J. Van Laar, J. J. Scheer: Influence of volume dope on fermi level position at gallium arsenide surfaces, Surface Science **8**, 342–356 (1967)
URL `http://www.sciencedirect.com/science/article/B6TVX-46T3C5Y-29B/1/2da1fa525f4de8a02322ee95cea78ee0`
30. T. Martensson, C. Svensson, B. Wacaser, M. Larsson, W. Seifert, K. Deppert, A. Gustafsson, L. Wallenberg, L. Samuelson: Epitaxial iii-v nanowires on silicon, Nano Letters **4**, 1987–1990 (2004)
URL `http://pubs3.acs.org/acs/journals/doilookup?in$\_$doi=10.1021/nl0487267`

31. M. A. Sanchez-Garcia, E. Calleja, E. Monroy, F. J. Sanchez, F. Calle, E. Muñoz, R. Beresford: The effect of the iii/v ratio and substrate temperature on the morphology and properties of gan- and aln-layers grown by molecular beam epitaxy on si(1 1 1), Journal of Crystal Growth **183**, 23–30 (1998)
URL `http://www.sciencedirect.com/science/article/B6TJ6-3W8STD4-4/1/ceb4d2577504cc8815a0de8507ed2712`
32. L. Cerutti, J. Ristić, S. Fernández-Garrido, E. Calleja, A. Trampert, K. H. Ploog, S. Lazic, J. M. Calleja: Wurtzite gan nanocolumns grown on si(001) by molecular beam epitaxy, Applied Physics Letters **88**, 213114 (2006) doi:10.1063/1.2204836
URL `http://link.aip.org/link/?APL/88/213114/1`
33. Y. Cui, X. Duan, J. Hu, C. Lieber: Doping and electrical transport in silicon nanowires, Journal of Physical Chemistry B **104**, 5213–5216 (2000)
URL `http://pubs3.acs.org/acs/journals/doilookup?in$\_$doi=10.1021/jp0009305`
34. K.-K. Lew, L. Pan, T. E. Bogart, S. M. Dilts, E. C. Dickey, J. M. Redwing, Y. Wang, M. Cabassi, T. S. Mayer, S. W. Novak: Structural and electrical properties of trimethylboron-doped silicon nanowires, Applied Physics Letters **85**, 3101–3103 (2004) doi:10.1063/1.1792800
URL `http://link.aip.org/link/?APL/85/3101/1`
35. S. Bae, C. Na, J. Kang, J. Park: Comparative structure and optical properties of ga-, in-, and sn-doped zno nanowires synthesized via thermal evaporation, Journal of Physical Chemistry B **109**, 2526–2531 (2005)
URL `http://pubs3.acs.org/acs/journals/doilookup?in$\_$doi=10.1021/jp0458708`
36. G. Cheng, A. Kolmakov, Y. Zhang, M. Moskovits, R. Munden, M. A. Reed, G. Wang, D. Moses, J. Zhang: Current rectification in a single gan nanowire with a well-defined p–n junction, Applied Physics Letters **83**, 1578–1580 (2003) doi:10.1063/1.1604190
URL `http://link.aip.org/link/?APL/83/1578/1`
37. M. V. Fernández-Serra, C. Adessi, X. Blase: Surface segregation and backscattering in doped silicon nanowires, Physical Review Letters **96**, 166805 (2006) doi:10.1103/PhysRevLett.96.166805
URL `http://link.aps.org/abstract/PRL/v96/e166805`
38. N. Skold, L. Karlsson, M. Larsson, M.-E. Pistol, W. Seifert, J. Tragardh, L. Samuelson: Growth and optical properties of strained gaas-gaxin1-xp core-shell nanowires, Nano Letters **5**, 1943–1947 (2005)
URL `http://pubs3.acs.org/acs/journals/doilookup?in$\_$doi=10.1021/nl051304s`
39. N. Skold, J. Wagner, G. Karlsson, T. Hernan, W. Seifert, M.-E. Pistol, L. Samuelson: Phase segregation in alinp shells on gaas nanowires, Nano Letters **6**, 2743–2747 (2006)
URL `http://pubs3.acs.org/acs/journals/doilookup?in$\_$doi=10.1021/nl061692d`

Simple Ways to Complex Nanowires and Their Application

Mady Elbahri, Seid Jebril, Sebastian Wille, and Rainer Adelung

Institute for Materials Science, Christian-Albrechts-Universität, Kaiserstr. 2, 24143 Kiel, Germany
me@tf.uni-kiel.de for nanowires: ra@tf.uni-kiel.de

Abstract. Many syntheses exist for nanowire fabrication. This development was driven by the promise of novel applications. In order to realize these applications, simple and rapid ways have to be found for organized nanowire fabrication, because applications often require large area coverage with nanostructures or low production costs for mass fabrication. Furthermore, contact formation to nanowires is often difficult to achieve. One solution is the fabrication of nanowires within thin film cracks. Two examples for this method will be shown, one for flexible electronics, the other for the integration in microchips. The first is discussing about the change in properties of polymer foils by covering them with a large amount of parallel and crossed nanowires, leading to a variation of the optical and electrical conductivity behaviour. The other is showing the fabrication of ZnO nanowires and their integration into microchips, it is demonstrated that they can be used as field effect transistors for sensor applications.

1 Introduction

Recently, nanowires were determined as second "hottest topic" in physics [1]. Nanowires play an essential role as building blocks for electronics, mechanics, and sensing technology [2–5]. Beside quantum mechanical effects, the interesting features of nanowires are due to the relatively high surface to volume ratio. Nanowires are mechanically more robust than their macroscopic counterparts. This is not only due to anisotropy effects in the grain structure of metal nanowires, but also simply due to the small diameter of the wires. Compressive and tensile stress at the sides of the wires are in simplest Euler-Bernoulli beam equation case proportional to the wire diameter. This simple fact gives nanowires superior resistance against momentum against their axis. Even single crystalline nanowires can be bent to a much larger curvature than their macroscopic counterparts [6]. This makes them a good candidate for flexible electronics, e.g., combining them with flexible materials like polymer foils. Among other advanced technological applications, the optical properties of the metal nanowires based on characteristic phenomena of surface plasmon polariton (SPP) represent an exotic one. The concentration and channeling of light using metal nanowires in a wave guide like fashion by, e.g., plasmon polariton tunneling, is discussed to be promising for a generation of photonic devices [7–10]. Moreover, periodic wire arrays are used to fabricate photonic materials and efficient Bragg reflectors [11] as well as left-hand material with a negative permeability and permittivity [12, 13]. Other applications like sensors do not rely so much on the mechanical stability, an integration in a flexible surrounding is often not necessary. Instead, for using them in devices like portable electronics, e.g., in cell phones, a basic requirement is to integrate them into silicon technology. Meanwhile, very precise techniques for the nanoscale structuring on silicon do exist, but in most cases, the techniques used are either relatively slow, expensive (like electron-beam [14] or dip-pen lithography [15]) or material specific and are partially limited to distinct deposition techniques. On the other hand, the cost effective self-organized techniques often suffer from a lack of controllability, leading to problems to connect nanowires in a desired way and integrate them into standard microfabrication processes. Demonstrators for both, metal and semiconducting nanowires have shown the great potential of nanowires as sensors integrated into electronics. Examples for sensors based on metal nanowires are made by Paladium [5, 16] to detect hydrogen or Silver to detect NH_3 or semiconducting nanowires forming pH-based sensors [17].

2 Experimental

The production of the nanowires is based on the fracture approach which can be effectively used either on rigid or flexible substrates. The basic steps for the nanowire formation are illustrated in Fig. 1 and [18, 19]. In general, the fabrication of the nanowire contains 4 steps. First, a thin film that acts as a mask is deposited on a subtrate. Here, in both cases (rigid and flexible), it will be a photoresist. Second this film is exposed to stress, as described later on for the different cases, resulting in thin film cracks. Third the material that will form nanowires is deposited, it will be deposited into the cracks and on top of the thin film. As fourth step the thin film will be removed together with the on top deposited material, while the material in the cracks, which is in contact with the substrate will stay.

To demonstrate the use of the fracture approach for flexible electronics, and to compare 0-dimensional with the 1-dimensional nanostructures, clusters as well as nanowires were produced from sputter deposited Au and Ag on a transparent polymer foil. Magnetron sputtering was carried out in an ultra high vacuum chamber for all the samples using an Ar ions at a pressure 2.5×10^{-3} mbar with a sputtering rate 0.5 Å/s. Thickness was determined by a quartz microbalance and AFM (atomic force microscope, Park Autoprobe) in order to achieve an equivalent film thickness of 2 nm for all samples. While the clusters form directly on the substrate, the meso- and nano-wires were

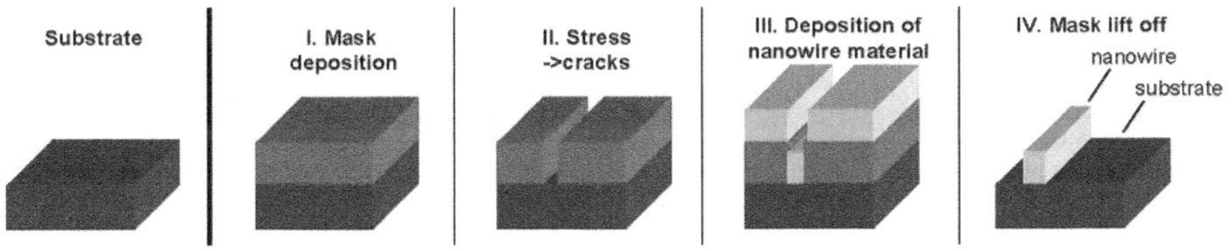

Fig. 1. General fabrication steps to form nanowires based on thin film fracture.

created by using a simultaneous fractured [18] and partially delaminated [20] thin film as a template on top of a substrate, as sketched in Fig. 2a–d. Here, we want to apply customary, widely available materials to form the nanowires. Standard inkjet printer foil (Canson, France) was coated with commercial photoresist (POSITIV 20, Contact Chemie) and dried in air. In principle, the use of the photoresist allows a simultaneous micro structuring of the thin film, e.g., to create macroscopic contacts for the nanowires as demonstrated further below to create a field effect transistor. The photoresist thickness was determined using a profilometer to be in the range of 1 μm, after deposition it is exposed to UV light for 15 h to effect an embrittlement. Bending of the elastic substrate foil (width 200 μm) leads due to the above mentioned "bending beam" setup for enough tensile stress in the brittle photoresist thin film, see Fig. 2a and 2b. The stress is uniaxial and leads to cracks that can be down to the nanoscale with a well defined periodicity, as known for the case of fracture under constrain [21, 22]. Each crack under constrain will relax only a limited area, which explains that neighboring cracks keep a well defined distance between each other. The obtained cracks are mainly delaminated as observed by AFM. This occurs if the stress at the bottom of a crack is large enough that it can overcome the adhesion of the thin film with substrate, which is advantageous in nanowire production, as shown in [20]. After metal deposition, Fig. 2c, the photoresist was removed by a subsequent mask lift off with acetone as a solvent. In this way, only the material which was deposited directly on the substrate in the cracks is remaining on the surface, resulting in nanowires, see Fig. 2d. Based on this bending idea and by controlling the direction of the applied stress, parallel or disordered metallic (Au, Ag) wire arrays can be designed and fabricated in photoresist on a polymer foil. Disordered wires can be formed by a second bending step into another direction.

For the integration in micro-lithographic steps, described above, the procedure is modified in the following way: First, photoresist, Shipley 1813, is deposited on a silicon samples by spin coating in a various thickness. Second, as additional step compared to the procedure above, photolithography is used to microstructure the photoresist. These lithography based photoresist microstructure patterning was done on silicon substrates (76 mm-diameter, p-doped, 1–10 Ω/cm resistivity, 380 μm thick, <100> oriented and coated with

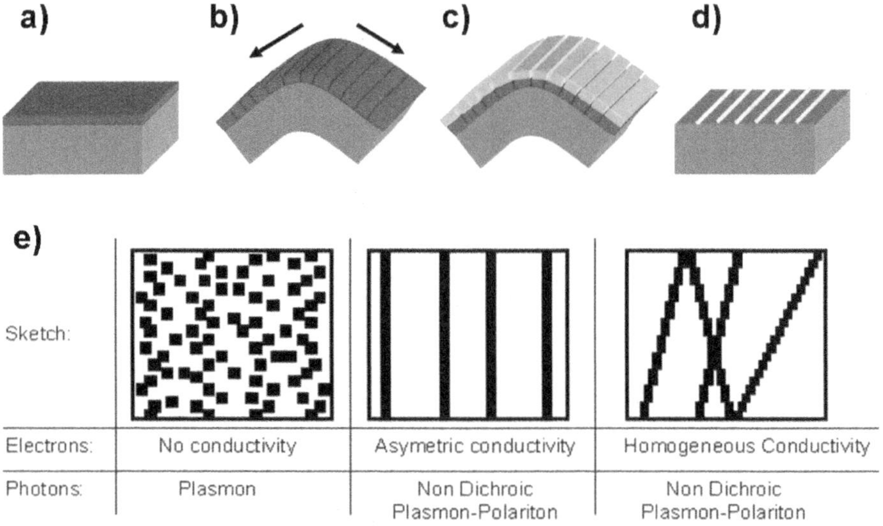

Fig. 2. Fabrication of different nanostructure patterns containing the same amount of material (**a**) A photoresist thin film on a polymer substrate gets exposed to tensile stress. (**b**) by bending, resulting in parallel cracks. (**c**) Those cracks get filled with metal by sputtering (**d**) A subsequent mask lift removes additional deposited material and leaves parallel nanowires on a substrate (**e**) Sketch of clusters, parallel and disordered wires showing different electrical and optical properties.

100 nm thick thermally grown SiO_2. For stress generation, as silicon wafers are not flexible, bending can not be applied to create tesile stress. Therefore in a third step, the samples are exposed to a thermal cycling down to cryogenic temperatures, e.g., dipping into a bath of liquid nitrogen which has a temperature about 77 K. This drastic change of temperature on the resist film generates thermal stress which results a mechanical failure of a resist film and leads to formation of nanoscopic cracks. The fracturing of thin films can be determined by strain fields which depends on the design of the film that can be chosen intentionally to create predefined crack patterns. One way to achieve this is by employing a photo- or e-beam lithography based micro-structuring of the photoresist (to be discussed later). If films are too thin (below 400 nm), the strain developed on the film may not be enough to induce the fracture in the second step and can yield almost no cracks on the resist, whereas if films are too thick (above 1,000 nm), the risk for uncontrolled delamination is very high and also the cracks may not able to reach down to the substrate. The fourth step contains the deposition of ZnO by magnetron sputter deposition, deposition was done at ultrahigh vacuum (UHV) at a base pressure of 10^{-8} mbar. The final and crucial step (fifth step) is separating the superfluous ZnO on the photoresist from the microstructured contacts (formed

in the microstructured areas) and the nanosized wires formed in the surface cracks. The electrical measurements described below are carried out by using a computer controlled setup including a Keithley pico-amperemeter.

3 Results and discussion

In order to demonstrate the feasibility of the fracture approach for flexible electronics, nanoclusters, parallel ordered and disordered meso- and nanowires having a width in the range of 100–300 nm and length in the mm scale were fabricated. Note that the metal content is in all cases equivalent to a nominal average film thickness of only about 2 nm, enabling the comparison of the properties arising only from the different arrangement of material. The different structures which are examined here are sketched in the Fig. 2e. The global optical properties were investigated by UV-Vis. spectroscopy as well as the electrical properties were examined by performing conductivity measurements.

Even metal nanowires with dimensions far below the light wavelength are typical objects with a weak optical contrast. Similar to others, also our metal wires start to differ in their conductivity properties from a critical size of around 70 nm, strongly depending on the metal [23]. Those wires exhibit anomalous temperature behavior and are very sensitive to the attachments on their surface [24]. Such size dependent electrical properties are similar to the general trend of decreasing the plasmonic propagation length with the miniaturization width of the wires [25]. In other words, typically, wider nano- or mesowires are therefore a better choice for the use of electrical conductors as well as optical guiding materials, if no sensing should be effected.

The anisotropic electrical properties of the nanowire arrays can be directly observed. A reasonable conductivity can be measured by evaporating contact pads with a distance of around 500 μm perpendicular to the parallel wires. This reveals typical resistivity values in the range of 1–10 MΩ/cm, strongly depending on the wire density and diameter. No conductivity can be found across the wire arrays. In order to visualize the conductivity, we used the charging that occurs in a Philips XL 30 scanning electron microscope (SEM). Figure 3a shows parallel nanowires, covered with a conductive carbon film. This standard procedure for non-conductive SEM samples ensures that the scanning electron beam will not be deflected by an insulator like our polymer substrate. Without covering, the polymer will charge up and the e-beam will be partially deflected, causing an intensity increase. This can be observed in Fig. 3b. Between two wires, a broad area of enhanced intensity shows the charging and thus the isolation between the wires. Directly next to the wire, no charging can be observed, meaning a sufficient amount of electrons will be able to reach the wire before they charge the material significantly. In contrast, Fig. 3c shows a grid of unorganized, crossing wires, which are able to provide conductivity in both directions and contain insulated islands instead of insulated stripes that act as barriers. As expected, a cluster area consisting of the same amount of material doesn't show any conductivity.

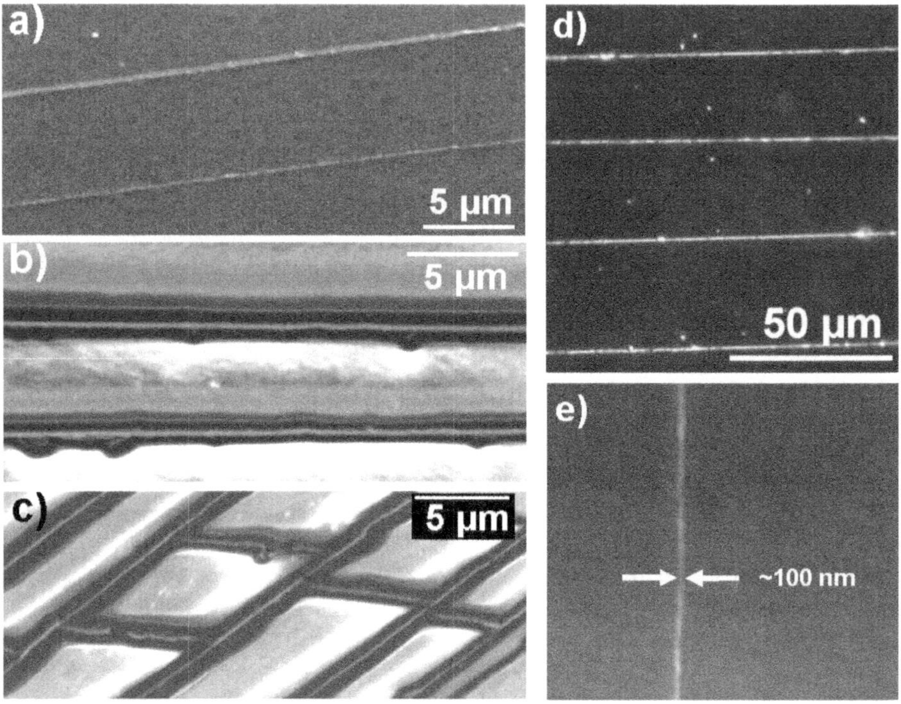

Fig. 3. SEM images of Au wires on the surface of the polymer foil. (**a**) Coated with carbon (**b**) without coating. the "structures" between the wires represent the charging, showing the insulating character perpendicular to the wires. (**c**) Part of a criss-cross wire network, showing biaxial conductivity. The intensity occurring from charging appears as islands between the network meshes. (**d**) Optical dark field image shows enough contrast from light scattering, (**e**) even though the nanowires are 100 nm or less in diameter, as seen from the carbon coated wire in this SEM image.

For comparison, Fig. 3d shows a dark field optical image of the wires from the sample in Fig. 3a, showing that even though wires are 100 nm or less, there is sufficient contrast by scattering. In contrast to metal clusters showing a Mie resonance [26] the optical properties of nanowires are governed by its geometry and by the plasma-like response of a free electron gas inside it. Such kind of a surface wave is known as surface plasmon polariton (SPP) [27] which can propagate along a metal-dielectric interface of a long wire. The interaction between the electromagnetic field and gold clusters and wires, which is studied in a wavelength of 300–2000 nm using a Perkin Elmer Lambda

900 spectrometer, is shown in Fig. 3. The absorption of the substrate as well as Au/polymer before mask lift off, (here, the intensity is divided by 10 to be comparable with the other spectra) is shown. Where the gold nanoclusters show a broad well known localized plasmon resonance located at 697 nm, which is known to depend on several factors like size, shape, dielectric constant of the surrounding etc., a single extinction peak for the parallel array of Au wires at 558 nm is observed. The extinction of gold clusters is also divided by two to be comparable with the extinction spectra of the wires reflecting the higher absorption/scattering coefficients of the clusters and the transparency gain using wires. The plasmon peak of the wires in the present case appears to be shifted towards the asymptotic value in the range of 510–530 nm and do not show any dichroic behavior in contrary to the earlier reports [28], with microwires. The surface plasmons of nanowires can show a dichroic behavior, according to the transversal and longitudinal oscillation of the free electrons [29, 30]. However such behavior is expected to be absent as the aspect ratio becomes high enough, where the resonances merges into a single peak [31–33]. As one can see in Fig. 4, using different metal wires like silver, there is only one peak at 370 nm which is also shifted to the asymptotic value representing a finger print for polaritonic behavior. This is in good agreement with the observed SPP value for high aspect ratio nanowires [34, 35] as well as the expected behavior from long wires. The guiding possibility in the direction of the long axis of our wires is currently under investigation.

This analysis shows how the rearrangement of a 2 nm thin film can vary its properties from insulating and localized plasmons in the case of nanoclusters to an uni- and bi-axial conductive and plasmonic mode behavior in the case of parallel and randomly aligned nanowire arrays. The asymmetric features of the parallel arrays of nanowires showed by different techniques are promising candidates for a multi-channel directional transport in the opto-electronic field. Even for plain electrical and optical applications, a parallel alignment of nanowires is useful. An extreme difference in transport occurs perpendicular and parallel to the wires. This allows in principle the fabrication of a broadband cable just by clipping multi-conductor plugs to both ends of the wires arrays: If all wires are well insulated against each other, an individual contact on one plug should just lead to only one at the plug on the other side. be more effective as a multi wave guiding channels alleviate the possibility for a plasmonic optical fibers and hybrid photonic integrated circuits. Such a direction dependence of electronic-photonic tunneling could be a step towards a directional opto-electronic device. Furthermore realizing such parallel wires with an extremely low filling factor on a transparent and flexible substrate like a polymer foil forms a type of transparent conductor. For instance applications like transparent electrodes require two properties namely, transparency and conductivity. In this context the metal/polymer composite can be an interesting candidate. However the interplay between the conductivity and transparency using metal nanocluster seems to be problematic. For instance, a higher transparency requires an extremely low metal filling factor,

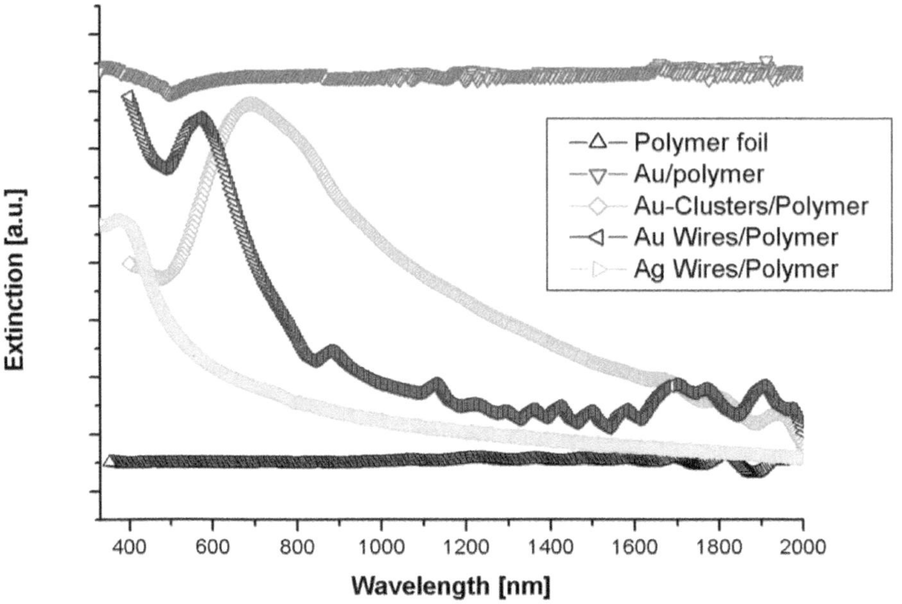

Fig. 4. Extinction spectra of the foils coated in different ways.

which means sacrificing conductivity. Otherwise increasing the conductivity by increasing the filling factor will diminish the transparency . A compromise can be found if the nanoculsters with extremely low filling factor are organized in a wire like fashion. such arrangement will cause a fundamental change in the electrical as well as in the optical properties.

In contrast to the requirements of nanowires on flexible substrates for optical and electronic applications, nanowires for sensor applications require other properties. Not a high conductivity is required, but a large conductivity difference between the "on" and "off" state of the sensor, if the sensor is exposed to the detected substance or not. Typically, sensitivity can be achieved by a large surface area compared to the volume that is transporting the current. As gas molecules typically attach to the surface and thus modify the surface layer, often with respect to the conductivity. It is not only the surface to volume ratio, that forms a good basis for the sensitivity, but also its seriality in conduction. While a thin film has a much higher surface to volume ratio, it often has less sensitivity. Simply, because if only one small part of a nanowire is covered and thus its conductivity is lowered at this point, the conductivity through this point determines the conductivity through the hole wire as a serial circuit. If this principle is applied in the right manner, nanowires have a higher sensitivity and a faster response time. Resistive wires have the advantage that they can be further controlled by additional fields like in a field effect transistor setup. Especially the conductivity of semiconductors depends strongly from the applied electrical field. For a sensor, a field effect transistor setup is desired [17].

The here chosen setup is illustrated in Fig. 5. Figure 5a shows zig zag types of cracks, that form on a 10 µm thin rectangular slab of photoresist with a length of 200 µm [19]. In order to produce nanowires in the most effective way, the simplest and plausible pattern should be employed as a form for the cracks. The examination of several simple shapes of microstructure patterns revealed that the here shown simple rectangular stripe of photoresist are very effective to create a well ordered fracture pattern. An advantage of this structure is that it needs only a minimal requirement to be formed by a microstructuring process. The dimension of the strip appears to be a guiding factor for the design of the pattern at which the crack follows. For a 10 µm stripe width, for example, the cracks are very well defined, equidistantly distributed (about 20 cracks in a zig zag shape). The parallel connection of the nanowires is a further advantage of the setup, as it guarantees a certain defect tolerance: if one single wire fails, the others still work, changing the overall performance in a tolerable way. As explained above, these cracks and the surrounding contacts were exposed to ZnO resulting, after mask lift off, in connected nanowires as seen in Fig. 5b. The schematic of the device is depicted in Fig. 5c. First field effect transistor curves are shown in Fig. 5d. A gate control can be achieved, enabling to drive the transistor at an optimal current for sensing. Please note that the transistor is not optimized at in order to be proof of principle. This is the first transistor demonstrated with the facture approach on a microstructured silicon chip.

Currently, the setup is tested as an ozone sensor in a collaboration with Volker Cimalla at the Fraunhofer-Institut für Angewandte Festkörperphysik IAF in Freiburg. First results point towards the fact that the nanowires have a much faster response time than conventional thin films.

4 Conclusion

It was demonstrated that nanowires can be formed by the fracture approach on flexible substrates as well as rigid ones where nanowires can be integrated onto microchips. It was shown that metallic as well as semiconducting nanowires can be made and contacted in a relatively simple manner, meaning that either "household" type of materials are used or machines were employed that are common everywhere microstructure processing or fabrication. It was not necessary to do any other nanoscale processing. On flexible polymer substrates it is showed how the rearrangement of an equivalent amount of a 2 nm metal thin film can vary its properties. These range from insulating and localized plasmons in the case of nanoculsters to an uni- and bi-axial conductive and plasmonic mode behavior in the case of parallel and randomly aligned nanowire arrays. The asymmetric features of the parallel arrays of nanowires showed by

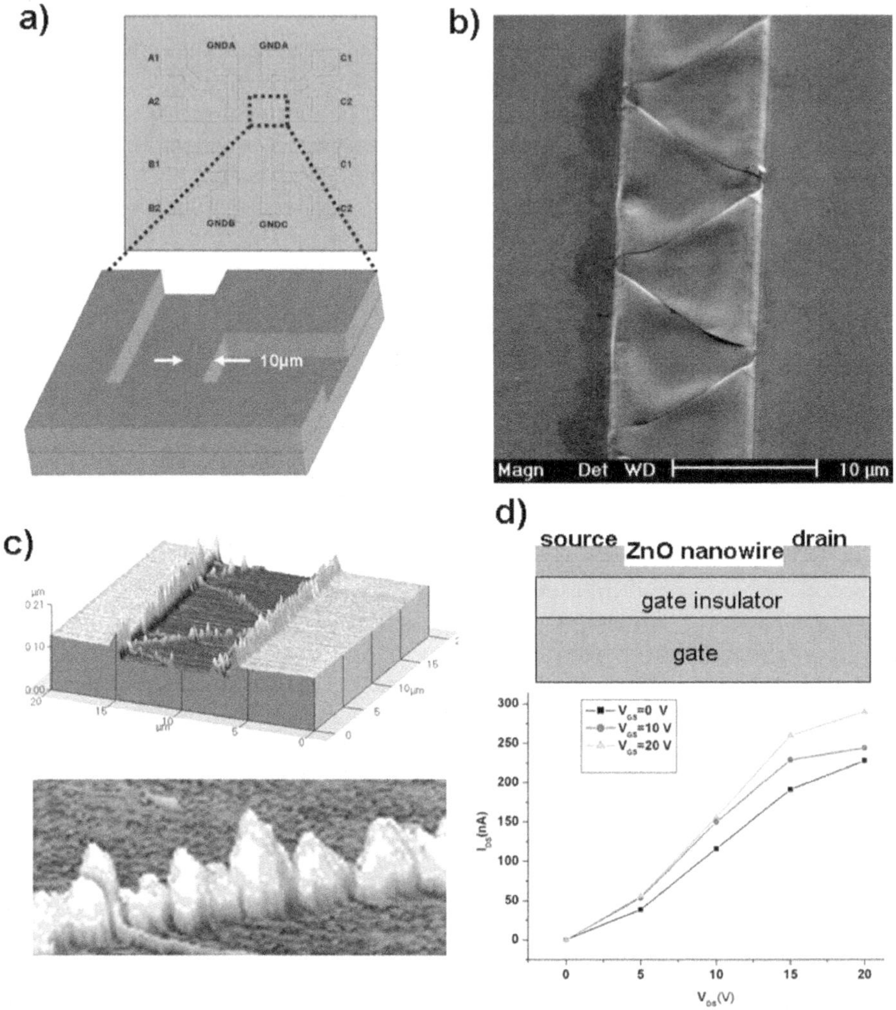

Fig. 5. ZnO Nanowire field effect transistor. (**a**) sketch of the microchip layout. Please note in the magnification the bar of photoresist (10 µm) (**b**) after the thermo cycling, fracture is induced resulting in a zig zag pattern along the photoresist bar. (**c**) Channel after mask lift of, zig zag nanowires connect two larger contact pads, all obtained from sputtered ZnO. The magnification shows that the wire consists of individual grains which can be less than 20 nm at the position where they touch each other (**d**) sketch of the field effect transistor setup and current voltage curves with different gate voltages.

different techniques are promising candidates for a multi-channel directional transport in the opto-electronic field. Even for plain electrical and optical applications, a parallel alignment of nanowires is useful. An extreme difference in transport occurs perpendicular and parallel to the wires. This allows in principle the fabrication of a broadband cable just by clipping multi-conductor plugs to both ends of the wires arrays: If all wires are well insulated against each other, an individual contact on one plug should just lead to only one at the plug on the other side. A semiconducting electronic application was realized by using sputter deposited ZnO nanowires on a silicon chip in a field effect transistor setup, proving the compatibility of the fracture approach with micro electronic fabrication. This makes the here chosen fabrication pathway an ideal candidate for the integration of sensor properties in microchips.

5 Acknowledgement

We are grateful to acknowledge the German Science Foundation (DFG) for financial support under grant SPP 1165 (AD 183/2), and one of the authors (R. A.) gratefully acknowledges a Heisenberg professorship. The authors are also thankful to Mrs. Heike Felsmann for photoresist spin-coating and lithographic patterning and S. Rheders for technical assistance.

References

1. J. Giles, Nature **441**, 265 (2006)
2. Y. Xia, P. Yang, Y. Sun, Y.Wu, B. Mayers, B. Gates, Y. Yin, F. Kim, and H. Yan: Adv. Mater. **15**, 353–389 (2003)
3. T. Hassenkam, K. Moth-Poulsen, N. Stuhr-Hansen, K. Nørgaard, M. S. Kabir, and T. Bjørnholm: Nano Lett. **4**, 19 (2004)
4. H. He and N. Tao: Adv. Mater. **2**, 161–164 (2002)
5. F. Favier, E. C. Walter, M. P. Zach, T. Benter, and R. M. Penner: Science **293**, 2227–2230 (2001)
6. S. Hoffmann, I. Utke, B. Moser, J. Michler, S. Christiansen, V. Schmidt, S. Senz, P. Werner, U. Gösele, and C. Ballif: Nano Lett. **6**, 622 (2006)
7. W. L. Barnes, A. Dereux, and T. W. Ebbesen: Nature (London) **424**, 824 (2003)
8. E. Ozbay: Science **311**, 189 (2006)
9. R. Zia, J. A. Schuller, A. Chandran, and M. L. Brongersma: Mater. Today **9**, 20 (2006)
10. H. Ditlbacher, J. R. Krenn, G. Schider, A. Leitner, and F. R. Aussenegg: Appl. Phys. Lett. **81**, 1762–1764 (2002)
11. J. R. Krenn, H. Ditlbacher, G. Schider, A. Hohenau, A. Leitner, and F. R. Aussenegg: J. Microscopy **209**, 167–172 (2002)
12. J. B. Pendry, A. J. Holden, W. J. Stewart, and I. Youngs: Phys. Rev. Lett. **76**, 4773 (1996)
13. J. Wood, Mater. Today **9**, 18 (2006)

14. A. M. Haghiri-Gosnet, C. Vieu, G. Simon, M. Majias, F. Carcenac, and H. Launois: J. Phys. IV **9**, pr2-133 (1999)
15. R. D. Piner, J. Zhu, F. Xu, S. Hong, C. A. Mirkin: J. Phys. IV **283**, 661 (1999)
16. B. J. Murray, E. C. Walter, and R. M. Penner: Nano Lett. **4**, 283 (2004)
17. Y. Cui, Q. Wei, H. Park, and C. M. Lieber: Science **293**, 1289–1292 (2001)
18. R. Adelung, O. C. Aktas, J. Franc, A. Biswas, R. Kunz, M. Elbahri, J. Kanzow, U. Schürmann, and F. Faupel: Nat. Mater. **3**, 375 (2004)
19. S. Jebril, M. Elbahri, G. Titazu, K. Subannajui, S. Essa, F. Niebelschütz, C. C. Röhlig, V. Cimalla, O. Ambacher, B. Schmidt, D. Kabiraj, D. Avasti and R. Adelung: Small (submitted)
20. M. Elbahri, S. K. Rudra, S. Wille, S. Jebril, M. Scharnberg, D. Paretkar, R. Kunz, H. Rui, A. Biswas, R. Adelung: Adv. Mater. **18**, 375 (2006)
21. Z. Suo: In Encyclopedia of Materials: Science and Technology, 2nd ed., (2001)
22. R. F. Cook, Z. Suo: MRS Bull. **27**, 45 (2002)
23. Y. M. Lin, O. Rabin, S. B. Cronin, J. Y. Ying, and M. S. Dresselhaus: Appl. Phys. Lett. **81**, 2403 (2002)
24. H. He and N. Tao: Adv. Mater. **14**, 161 (2002)
25. B. Lamprecht, J. R. Krenn, G. Schider, H. Ditlbacher, M. Salerno, N. Felidj, A. Leitner, F. R. Aussenegg, and J. C. Weeber: Appl. Phys. Lett. **79**, 51 (2001)
26. U. Kreibig and M. Vollmer: Optical Properties of Metal Cluster, Springer, Berlin (1995)
27. H. Raether: Surface Plasmons, Springer, Berlin (1988)
28. A. Biswas, R. Kunz, O. C. Aktas, M. Elbahri, R. Adelung, H. Takele, U. Saeed, U. Schürmann, V. Zaporojtchenko: Appl. Phys. Lett. **85**, 1952 (2004)
29. S. Link and M. A. El-Sayed: J. Phys. Chem. B **103**, 8410 (1999)
30. S. Link, M. B. Mohammed, and M. A. El-Sayed: J. Phys. Chem. B **103**, 3073 (1999)
31. Y. Sun, Y. Yin, B. T. Mayers, T. Herricks, and Y. Xia: Chem. Mater. **14**, 4736 (2002)
32. K. K. Caswell, C. M. Bender, and C. J. Murphy: Nano Lett. **3**, 667 (2003)
33. J. Jhu, S. Liu, O. Palchik, Y. Koltypin, and A. Gedanken: Langumir **16**, 6396 (2000)
34. Y. Sun, Y. Yin, B. Gates, B. T. Mayers and Y. Xia: Nano Lett. **2**, 165 (2002)
35. C. S. Ah, S. D. Hong, and D. J. Jang: J. Phys. Chem. B **103**, 3073 (1999)

ZnO Nanostructures: Optical Resonators and Lasing

Klaus Thonke[1], Anton Reiser[1], Martin Schirra[1], Martin Feneberg[1], Günther M. Prinz[1], Tobias Röder[1], Rolf Sauer[1], Johannes Fallert[2], Felix Stelzl[2], Heinz Kalt[2], Stefan Gsell[3], Matthias Schreck[3], and Bernd Stritzker[3]

[1] Institut für Halbleiterphysik, Universität Ulm, 89069 Ulm, Germany
klaus.thonke@uni-ulm.de

[2] Angewandte Physik, Universität Karlsruhe,
Gaedestrasse, 76128 Karlsruhe, Germany

[3] Experimentalphysik IV, Universität Augsburg,
Universitätsstraße 1, 86135 Augsburg, Germany

Abstract. We review briefly reports on lasing in different kinds of ZnO nanostructures ranging from clusters to pillars, tetrapods, and belts. Then detailed studies on individually in-situ accessible, well-faceted upright pillars grown on a metal interlayer on silicon substrate are presented. We observe in cathodoluminescence directly ultraviolet light standing wave resonator modes, and under pulsed excitation lasing activity based on (e,h) plasma recombination.

1 Introduction

ZnO allows to grow a wide variety of nanostructures: Simple nano-clusters, nano-wool, nano-ribbons, comb- and tree-like structures, tetrapods, pillars, nanopropellors, nanonails etc. [1]. All kinds of growth methods have been used to obtain such different species of nanostructures: metal-organic vapor phase epitaxy (MOVPE), molecular beam epitaxy (MBE), magnetron sputtering, pulsed laser deposition, vapor deposition with or without catalysts, electrodeposition, spray pyrolysis, or even simple wet chemistry methods. For many of the nanostructures obtained lasing was reported, but quite often that part of the object under investigation giving rise to the stimulated emission remained unidentified with respect to its relevant dimensions and shape, or simply ensembles of lasing structures with a distribution of sizes and shapes were investigated. Much more specific information can be obtained from studies on individual objects; to this end investigations were carried out mostly on single pillars suspended on supporting substrates.

Out of the big zoo of available nanostructures, nano-pillars are of special interest for applications as sensors and optoelectronic devices, since they allow – at least in principle – individual addressing. These nanopillars can be produced in excellent structural quality and purity using high-temperature growth processes working above or around 900°C. In the present article, examples of rather well-faceted hexagonal pillars grown on sapphire and silicon/iridium (fcc) substrates are presented. We managed to grow such pillars with large distances of around 1 µm which can be individually addressed and retrieved as-grown in a controlled way by optical methods like confocal microscopy.

Performing cathodoluminescence (CL) with high spatial resolution on as-grown single pillars, we find ultraviolet (UV) light standing waves in pillars with appropriate diameters. Under high excitation, these pillars show spectrally well-resolved competing laser modes. Time-resolved measurements reveal the transition from spontaneous to stimulated emission and allow to study the lasing dynamics from electron-hole recombination in a high-density plasma (EHP) in more detail.

2 Lasing in nanostructures

Lasing in ZnO bulk material was reported as early as 1966 by Nicoll [2]. He excited a sample in the form of a cleaved, c-oriented platelet by a pulsed electron beam with 15 kV energy and a current density of 5 A/cm^2 at a temperature of 77 K and found a super-linear increase of the emission. Using also e-beam excitation, Hvam [3] identified exciton-exciton collisions (the "P band") as the origin of stimulated emission at 3.327 eV for 10 K.

Optically pumped lasing from ZnO epilayers grown by MBE on sapphire substrates was reported by Bagnall et al. in 1997 [4]. They observe in their laser cavities, with dimensions of 5 mm × (300 – 1000) µm and cut by cleaving, for pumping levels above ≈ 660 kW/cm^2 rather broad (≈ 200 meV) emission with unresolved mode structure at 3.06 eV, which they assign to the "P_2" exciton–exciton collision process, where one of two interacting excitons is scattered into its first excited ($n = 2$) state [5]. For still higher excitation above 1 MW/cm^2, they find at room temperature an EHP emitting a band at ≈ 3.1 eV dominating the spectra [6].

Room-temperature lasing in ZnO nanocrystalline material was reported in 1997 by Zu et al. [7] in thin microcrystalline MBE ZnO epilayers grown on sapphire substrates, which consisted of c-oriented grains of ≈ 55 nm size. A peak observed at 390 nm for pumping levels above ≈ 24 kW/cm^2 was interpreted as stimulated emission fed by exciton-exciton collisions (commonly labeled "P_∞ band"). This peak was reported to be only observable for grain sizes ≤ 55 nm, from which the authors concluded that the reduced size enforced higher overlap of the electron and hole, resulting in an increased matrix element for the recombination. For pumping levels above ≈ 50 kW/cm^2 a peak at lower energy (labeled "N band") become dominant, which red-shifted under further increase of the pumping intensity, and was ascribed to the formation of electron-hole pairs yielding enough gain for lasing activity.

In the following years, lasing was reported for ZnO nanostructures of virtually any shape and produced by almost every growth method applicable. A lot of rumor was created by the paper of Huang et al. [8], who reported room-temperature lasing from ZnO nanorods grown by a simple carbo-thermal chemical vapor deposition (CVD) process on Au catalyst films deposited on a-plane sapphire substrates. Whereas in their original report the authors presented a relatively noisy spectrum recorded under 100 kW/cm^2 excitation and found an increase in intensity of just a factor of two when increasing the pump power over approximately two orders of magnitude (see Fig. 3B in Ref. [8]), they could more convincingly demonstrate the laser action in their later reports. For more detailed investigations in a scanning near-field optical microscope (SNOM) arrangement, they prepared a suspension of wires in ethanol and drip-coated them onto a quartz substrate. Under pulsed sub-picosecond excitation, above a threshold of 120 kW/cm^2 a relatively narrow peak at 380 nm with 1–3 nm halfwidth was observed for a wire with ≈ 200 nm diameter and a few μm length. In a similar study of the same group, time-resolved emission spectra were presented and a bimodal decay with a fast component ($\tau = 9\,\mathrm{ps}$) and a lasing threshold of $1\,\mu\mathrm{W/cm}^2$ was reported [9]. Further details were presented by Sirbuly et al. [10], who tentatively assigned the typically observed 4–6 spectral modes to HE_{11}, TE and TM resonator modes. They reported an extremely low lasing threshold of some 80 nJ/cm^2 – three orders of magnitude less than they obtained for similar single GaN nanowires, where $I_{\mathrm{th}} \approx 150\,\mu\mathrm{J/cm}^2$.

So-called nanoribbons, i.e. flat, long structures of ZnO with more or less rectangular cross sections of $\approx 1\,\mu\mathrm{m} \times 200$ nm were either grown by a vapor-liquid–solid approach using Au as catalyst at 900°C, or by a catalyst-free process of the vapor-solid type at higher temperatures around 1350°C. Suspended on substrates, these nanoribbons were found to show laser action with a similar threshold of around $3\,\mu\mathrm{W/cm}^2$ under 200 femtosecond pulsed excitation at room temperature [11]. Complicated mode structures were observed at higher pump levels, with mode spacings of some 0.5 nm. Since the geometry of the wires was not sufficiently well known, no theoretical modeling and assignment of the modes observed was possible.

In polycrystalline laser-MBE grown ZnO films, 50–200 nm thick and consisting of hexagonal pillars with ≈ 50 nm diameter and multiple terraces, Tang et al. [12] found for higher pumping levels at room temperature spectra consisting of sharp, close lying lines in the range of the "P" band. From their spacings, the authors assumed a lateral Fabry-Pérot resonator defined in its length by the size of the stripe-shaped pumping beam to be responsible for the mode pattern. A threshold of $40\,\mathrm{kW/cm}^2$ ($\simeq 0.6\,\mu\mathrm{J/cm}^2$) for this excitonic lasing process was determined, which is about one order of magnitude lower than the values reported for bulk ZnO [4]. Besides the "P" band, at lower energy the "N" band emerged originating from EHP decay for higher pumping levels.

Similarly, Cao et al. [13] found room temperature lasing in a polycrystalline ZnO film deposited by laser ablation on fused silica substrates. Only for pulsed excitation of larger spots they obtained broad spectra at around $\lambda = 388$ nm with superimposed sharp lines. The emission patterns looked like distorted rings reminiscent of phase-distorted images of point-like sources. For an explanation the authors argued that only for large enough laser spots some accidentally closed paths along multiple scattering microcrystallites exist. These remain unlikely for too small spot sizes. The lasing threshold quoted in this work was 400 $\mathrm{kW/cm^2} \times 15$ ps $= 6\,\mu\mathrm{J/cm^2}$.

Yu et al. [14] observed room temperature "random lasing" in a relatively complicated structure grown by MOVPE, where hexagonally shaped ZnO pillars were embedded in a ZnO epilayer serving as a waveguide, which in turn was placed on a MgO buffer layer. Some sharp peaks with linewidths around 0.4 nm superimposed on a broad 385 nm-peaked band were observed under pulsed Nd:YAG excitation tripled to 353 nm. The threshold was quoted to be between 0.8 $\mathrm{MW/cm^2}$ ($\simeq 5$ $\mathrm{mJ/cm^2}$) and 1.5 $\mathrm{MW/cm^2}$, depending on the polarization direction. For the size of the pillars the authors quote an average diameter of 70 nm only, which in the light of simulation calculations presented later in this article appears questionable since too narrow to hold a UV resonator made.

Even ZnO pillars grown on silicon substrate with a wet chemistry low-temperature process at 95°C showed lasing activity, although their shape was not really well defined, and transmission electron microscopy (TEM) investigations revealed numerous defects in these samples [15]. At pumping levels above 70 $\mathrm{kW/cm^2}$, a super-linear increase in the emitted intensity as a function of pumping power was detected and assigned to exciton–exciton scattering processes. For still higher pumping with 200 $\mathrm{kW/cm^2}$, sharp emission peaks superimposed on a broad band at 386 nm were found and ascribed to EHP-driven stimulated emission. Surprisingly, the PL lifetime was reported to *increase* by some 20% for higher pumping levels, in our opinion contradicting real laser activity.

Also simple pellets pressed from ZnO powder can serve as a laser active medium, as reported by Sun et al. [16]. They identified the gain mechanism to be exciton–exciton scattering at moderate pumping intensities (i.e. above $I_{\mathrm{th}} = 230$ $\mathrm{kW/cm^2}$), and at high intensity (i.e. $I > 8 \times I_{\mathrm{th}}$) as recombination in an EHP.

A better defined lasing system apart from nanopillars are single ZnO tetrapods grown with low density on sapphire or silicon substrates by a chemical vapor transport and condensation (CVTC) process [10]. Again, PL bands around 385 nm with multiple superimposed sharp lines were found, and the maximum emitted intensity was found in the center of the tetrapods. The lasing threshold ($\approx 430\,\mu\mathrm{J/cm^2}$) was higher for three arms in touch with the substrate as compared to the situation with only one arm in touch. Leung et al. [17] produced ZnO tetrapods with less clearly defined shapes by evaporation of Zn in a N_2 gas flow at 950°C. Stimulated emission with a threshold of $\approx 100\,\mu J/cm^2$ was observed, and the linewidths of individual modes were in the range of 0.7–0.9 nm. In their time-resolved spectra, they unexpectedly observe the onset of PL related to exciton–exciton scattering delayed by some 5 ps after the end of the excitation pulse, whereas the EHP is virtually immediately formed. The same group of authors reports similar results, especially the same rather unusual delay of the "P" band emission [18] from investigations on "highly faceted" ZnO nanorods grown on silicon substrate.

A general problem when investigating nanostructures suspended on substrates is the leakage of the optical modes into the substrate. When sapphire substrates are used, the difference in the refractive index n relative to ZnO is rather low. Theoretical modeling of the complete arrangement is thus difficult, and any simple high-symmetry resonator models are generally inapplicable. Therefore it is desirable to investigate single pillars in situ, i.e. in a preferentially upright position as-grown on a substrate. These are the structures lending themselves most obviously to applications, anyhow.

Sirbuly et al. [10] have reported such a study, but the single pillar investigated had no really well defined shape looking rather conical with a larger bottom diameter, and the spectra recorded under pulsed excitation were noisy. Thus no theoretical interpretation of the data was possible.

Using a confocal microscope, Hauschild et al. [19] managed to record low-temperature time-resolved PL spectra of a few narrow nanorods or single wider rods. Unlike Zu et al. [7], they found under pulsed excitation at intermediate levels decreasing intensity of the "P" band for decreasing ZnO pillar diameters, casting doubts on the argument that the exciton–exciton scattering matrix element might be larger for smaller diameters. Instead it was concluded that the reduction of phase-space for polariton-polariton scattering might be important [19]. Hauschild et al. found stimulated emission in the "P" band range for single ZnO pillars with larger diameters, when the reflectivity at the bottom end was increased by some remains of Au catalyst particles residing at the interface to the sapphire substrate, as confirmed by TEM micrographs. A super-linear increase of the emitted intensity was found for excitation densities above $250\,\mu J/cm^2$, but no resonator modes could be resolved. The "P" band itself was quenched above the threshold, and the stimulated emission varied from rod to rod. From this fact the authors excluded a pure excitonic origin of the laser gain, and instead suggested that free carriers should be involved. Model calculations confirmed the importance of the reflectivity increase at the bottom end, and predicted a minimum diameter of the ZnO pillars of $\approx$ 120 nm to obtain high enough modal gain in the pillars [20]. A similar value for the minimum diameter necessary for lasing had been estimated before by Johnson et al. [21].

Recently, Zhou et al. [22] have reported lasing from ordered arrays of ZnO nanorods with $\approx$ 200 nm diameter and $4.7\,\mu m$ length, grown on GaN/sapphire substrate. Since the pillars obviously have a very narrow diameter distribution, they showed well resolved modal structure, although always several pillars were excited simultaneously. The average distance of 500 nm did not allow to address and identify single pillars.

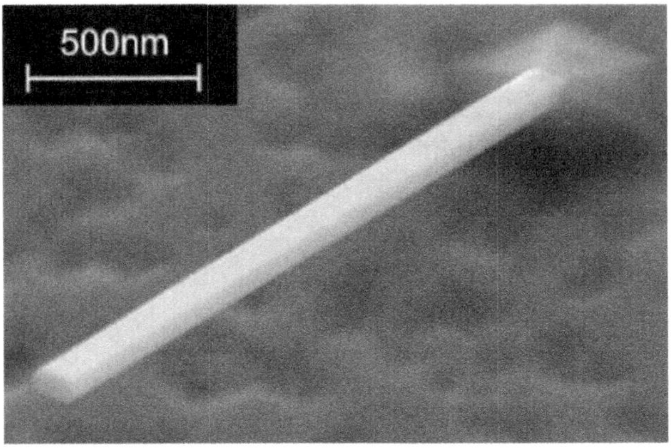

Fig. 1. High-resolution SEM micrograph of a well-faceted single ZnO pillar grown on (111) silicon with an (fcc) Ir interlayer. This pillar is similar to the one who yielded the spectra depicted in Fig. 8 under high excitation.

3 Single ZnO pillars as nano-resonators

In order to get insight into the modal structure of ZnO nano-resonators and their dependence on geometrical and optical parameters, investigations on individual nanorods are required. A growth method was developed in our group which allows growth of well-faceted ZnO pillars with hexagonal cross section and average distances above 1 μm [23]. On such samples, markers can be written e.g. by the focused ion beam (FIB) technique, and individual pillars can then be retrieved both in optical experiments in a confocal microscope and in an electron microscope for CL measurements and precise structural characterization. The diameters of the pillars used here are in the range of 100 nm–200 nm, and their lengths are several μm (see Fig. 1). They were grown on a 150 nm fcc-crystalline iridium/40 nm yttria-stabilized zirconia layer sequence deposited on a Si (111) substrate [23, 24]. The iridium layer has the main function of preventing alloying of the ZnO with the silicon substrate [25] as a necessary condition for the interface serving as an optical mirror. Growth on silicon substrates has several advantages: The substrate is conductive as required for future electric connections, it is inexpensive, and the difference in the refractive index relative to that of ZnO is larger as compared to sapphire substrates. This way, a higher reflectivity for nano-resonators at the bottom end can be achieved.

The resulting pillars have first been characterized by scanning electron microscopy (SEM). We find them tilted from the surface normal along three preferential directions, each rotated by 120°C relative to each other in top view. The individual facets are very well developed and the pillars are free of crystalline defects. This is confirmed in low-temperature macroscopic PL measurements, where we find exclusively shallow excitonic luminescence in the form of In- and Al-related donor-bound excitons (at 369 nm corresponding to 3.36 eV) in the near-bandgap range. The full width at half maximum (FWHM) of the In-related (D^0, X) line is around 1 meV at 4.2 K (see Fig. 2).

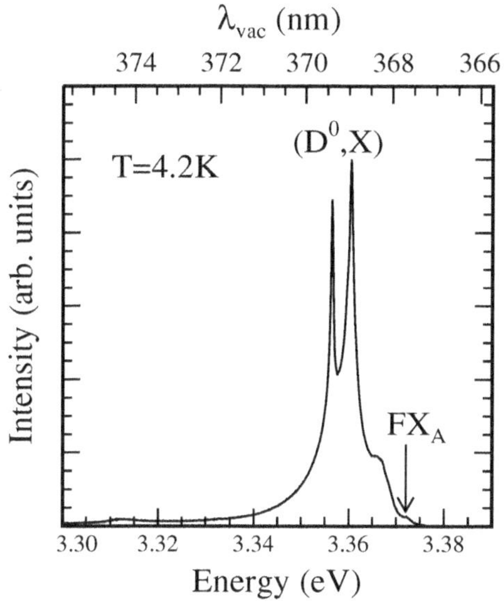

Fig. 2. Low-temperature PL spectrum of ZnO pillars grown on (111) silicon with an (fcc) Ir interlayer. The spectrum is dominated by sharp donor-bound exciton peaks.

To study the emission properties of single pillars, we mounted the sample on the cold finger of a home-built special CL setup. Since high spatial resolution in the CL measurements is needed, a low acceleration voltage around 2 kV has to be chosen. Numerical simulations predict an electron scattering range of $\approx$ 40 nm in ZnO [26], which – besides the exciton diffusion length - limits the obtainable spatial resolution. For such low electron voltages, the working distance has to be as small as 2–3 mm, i.e. the sample has to be brought as close as 2–3 mm to the last electron lens. Such small distances do not allow mounting of a spacious mirror to couple the e-beam generated light out. For this purpose, we installed a UV transparent glass fiber with 100 μm diameter at a distance of 50 μm to the sample, providing efficient large-angle CL light collection. This method has the further advantage, that still perfect SEM pictures can be recorded even when using the backscattering detector, since secondary electrons are not shielded by bulky mirrors. For light detection and dispersion, either a $f = 90$ cm monochromator with a UV optimized cooled charge coupled device (CCD) camera was used, or for CL intensity mappings at individual emission wavelength ranges, a $f = 1/4$ m monochromator with a photomultiplier tube and current-amplifier were used. The e-beam current was typically in the range of 0.1 nA.

When exciting in a CL scan an area of some 6 μm × 6 μm, we obtain practically the same spectrum as recorded in the macro-PL experiment, with the minor disadvantage of a slightly reduced spectral resolution. Again, the spectrum is dominated by donor-bound exciton (D^0, X) features. In the next experiment, the detected light is restricted to this (D^0, X) range by setting the 1/4 m monochromator to the corresponding central wavelength and adjusting the slit width appropriately. The attached photomultiplier yields an intensity map reflecting the local excitation efficiency of the (D^0, X) emission. This recording can be viewed as spatially resolved "CL-excitation-spectroscopy", see Fig. 3.

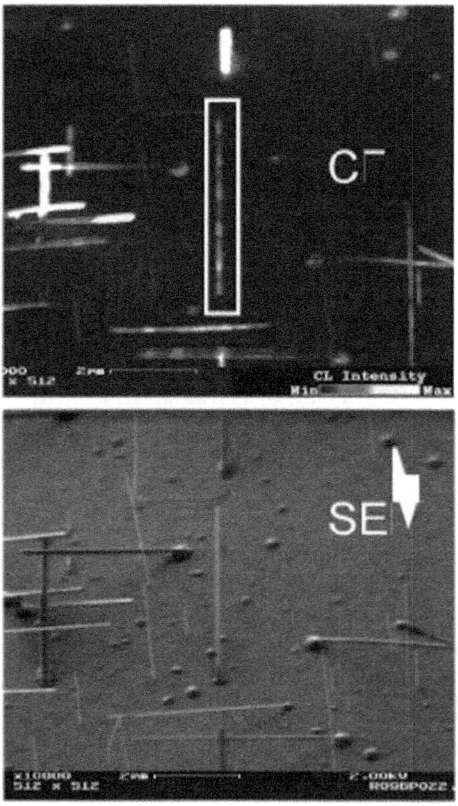

Fig. 3. *Top*: CL excitation efficiency map of the donor-bound exciton emission range for ZnO pillars grown on (111) silicon. A vertically aligned ZnO pillar in the middle of the CL image exhibits periodic light intensity modulations to be discussed in the text. *Bottom*: SEM image of the same sample area.

(D^0, X) emission is found from the pillars, whereas their environment remains dark. Hence, the CL map is a true counterpart of the SEM image, as nicely seen in Fig. 3. The highest signal is obviously generated at spots, where relatively thick pillars are grown like on the middle left and top of the scanned region. The most remarkable feature in this scan is a periodic intensity modulation along some of the thinner pillars, which is most pronounced in the vertical pillar in the middle of the CL image and framed by a white box: Along a length of $\approx 5\,\mu m$, (the pillar is tilted from the surface normal, and the angle of view is 45° relative to the surface), the modulation of the intensity comprises six maxima and minima in-between. A line scan of the intensity distribution along this pillar is depicted in Fig. 4 obviously reflecting the lobes and nodes of the electric field distribution of a standing wave in the nano-resonator. At first glance, the rather large mode spacing is puzzling: For purely longitudinal modes in a Fabry-Pérot resonator one would expect a spacing of order $\lambda_{vac}/2n$ where λ_{vac} is the wavelength in vacuum, and n the refractive index of ZnO, resulting in a spacing of $\approx 370\ \text{nm}/(2 \times 2.85) = 65$ nm far below our experimental value of $\approx 1\,\mu m$.

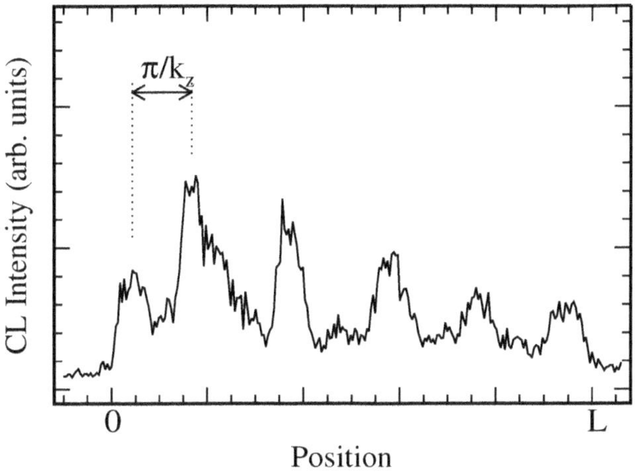

Fig. 4. CL intensity recorded as a linescan along the pillar positioned vertically in the middle of Fig. 3. The light intensity is modulated with a period of $\approx 1\,\mu m$.

This unexpected modulation pattern can be explained in terms of a 3-dimensional nano-resonator for which the transverse dimension is comparable to the wavelength of the radiation. The electric field properties are described by the Helmholtz equation

$$\Delta \boldsymbol{E} + \boldsymbol{k}^2 \boldsymbol{E} = 0, \tag{1}$$

where $\boldsymbol{E}$ is the electric field vector and $\boldsymbol{k}^2 = (nk_0)^2$. Assuming as the simplest model a square-shaped resonator cavity with metal walls and filled with dielectric material, this equation can be separated yielding different k_i components in different directions with

$$k_x^2 + k_y^2 + k_z^2 = \boldsymbol{k}^2. \qquad (2)$$

All electromagnetic waves are here assumed to be of transverse type, i.e. the electric and magnetic field components are perpendicular to each other and any anisotropy of the medium is neglected. Observed in the experiment is the intensity distribution $I \propto |\boldsymbol{E}|^2$, and the longitudinal modulation period is given by π/k_z.

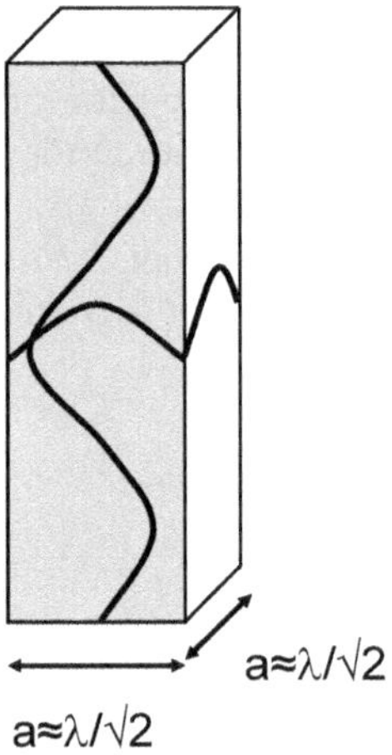

Fig. 5. Simple model resonator: a *square-shaped column* with conductive walls. If the transverse dimensions approach $\lambda/\sqrt{2}$, the *standing wave* is stretched along the z-direction and is finally "squeezed out" of the resonator.

In the following we make the natural assumption that in the two transverse directions the electric field is in its fundamental eigenmode. This is equivalent to choosing the dimension of the transverse directions to $\lambda/\sqrt{2} = \sqrt{2}\pi/k_x = \sqrt{2}\pi/k_y$. Then the sum $k_x^2 + k_y^2$ becomes almost equal to $\boldsymbol{k}^2$, and the remaining k_z^2 is rather small. Solutions for Eq. (2) do only exist, as long as $(k_x^2 + k_y^2) \leq \boldsymbol{k}^2$. Illustrated graphically, this means that the standing wave is stretched more and more along the z direction, and finally is "squeezed out" of the resonator when a decreases (see Fig. 5). For the values given in our example here, this should happen for a thickness of ≈ 94 nm of such a square-shaped resonator consistent with the actual pillar diameter of ≈ 100 nm.

A further model to approach our real hexagonal dielectric resonator is based on the theory developed for dielectric cylindrical waveguides. Therefore, one has to calculate solutions of the Helmholtz equation in cylinder coordinates. The fraction of intensity confined in the waveguide is given by

$$\eta = 1 - (2.405 \exp{(1 - V)})^2 V^{-3} \qquad (3)$$

with $V = kR\sqrt{n^2 - 1}$, $k = 2\pi/\lambda_n$, and R the cylinder radius.

The number of bound transverse modes M_{bm} in a multimode cylindrical optical cavity can be approximated by:

$$M_{\rm bm} = 2(\pi R/\lambda)^2(n^2 - 1) \tag{4}$$

For our case, just one bound mode exists for the minimum diameter (i.e. the "cutoff diameter") $2R \approx 130$ nm, a number close to the diameter of $\approx$ 110 nm as determined from the SEM micrograph Fig. 3. If finally the diameter is too small, no confined mode exists in the core of the resonator.

A realistic calculation for the true hexagonal pillar cross section (instead of a rectangular or circular one) and a finite difference of the refractive indices of the resonator and the surrounding vacuum cannot be obtained by analytical methods, but must be calculated numerically. For the correct inclusion of boundary conditions, the so-called "boundary elements method" [27] can be used. As a preliminary result, a 2-dimensional calculation applying the "finite elements method" is represented here. This approach does account for the shape of the resonator, but does not for the step in the refractive index.

Therefore, only the hexagonal 2-dimensional resonator is calculated in the first instance. The intensity distribution $|\boldsymbol{E}|^2$ is shown for the first 15 modes in Fig. 6. Unity edge length hexagons were assumed. The corresponding eigenvalues of these modes are summarized in Table 1.

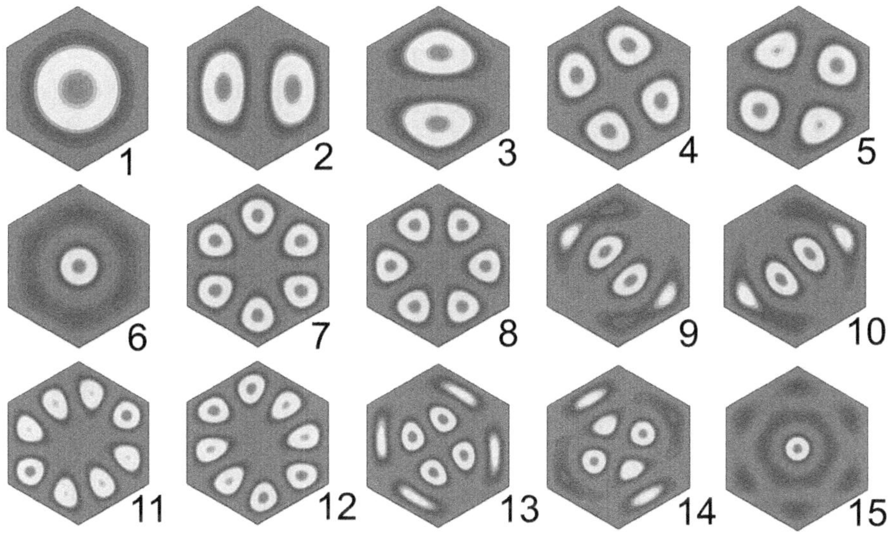

Fig. 6. Intensity distributions of the first 15 modes in the hexagonal 2-dimensional resonator with unity edge length as calculated by a finite element method [28].

These eigenvalues have to be scaled to the real problem before entering the calculation sketched above: The eigenvalue $k_{\rm hex}$ has to be divided by the measured edge length of the hexagon which is half the diameter. Reformulating Eq. (2) results in

Table 1. Eigenvalues of the first 15 modes in the hexagonal 2-dimensional resonator with unity edge length as calculated by a finite element method.

mode #	k_{hex}	mode #	k_{hex}	mode #	k_{hex}
1	2.67495	6	6.12304	11	8.37507
2	4.25814	7	6.90144	12	8.37508
3	4.25814	8	7.25524	13	9.35599
4	5.69666	9	7.75281	14	9.35601
5	5.69666	10	7.75281	15	9.49007

$$k_{\text{hex}}^2 + k_z^2 = \boldsymbol{k}^2, \tag{5}$$

where k_{hex} is the scaled eigenvalue taken from Table 1. For a given wavelength, e.g. $\lambda_{\text{vac}} = 370$ nm, the lobe distance in the z-direction of a nanopillar can easily be calculated (see Fig. 7). Each line represents a different eigenvalue k_{hex}. By decreasing the diameter of the pillar, the lobe distance along the z-direction of the pillar is increased until finally the mode is “squeezed out” of the resonator. The smallest diameter of a nanopillar allowing standing waves with $\lambda_{\text{vac}} = 370$ nm is therefore $d = 115$ nm.

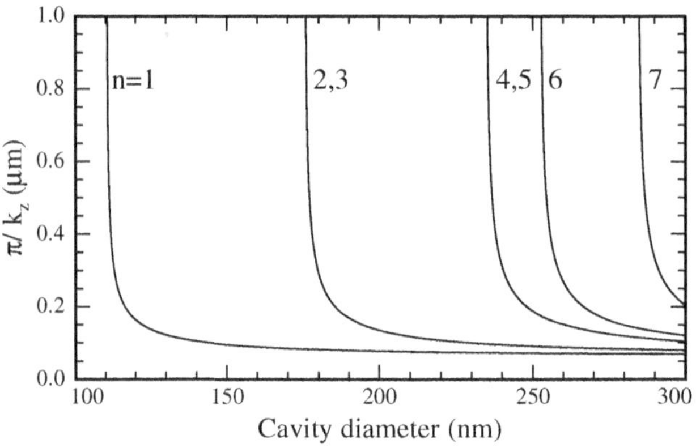

Fig. 7. Lobe distance along the z-direction of a nanopillar as a function of the diameter for $\lambda_{\text{vac}} = 370$ nm. The cutoff-behavior of the modes n (same labeling as in Fig. 6) manifests itself in the steep increase in the lobe distance for decreasing diameter.

The diameter of the pillar visible in Fig. 3 can now be estimated by the measured lobe distance π/k_z. In our case this distance is about 1 μm resulting in a pillar diameter of 110 nm (see line for $n = 1$ in Fig. 7). This value coincides well with the micrograph in Fig. 3 yielding a diameter of about 120 nm. Note, that only the fundamental mode can resonate in this specific pillar. Thinner nanopillars cannot show any resonance for $\lambda_{\text{vac}} = 370$ nm. Indeed, in our experiments we do not find standing waves in thinner pillars. For thicker diameters of the nanopillars, higher order modes are allowed which would obscure the well-defined intensity distribution along the z-direction of the pillar because they have a different lobe distances for a given pillar diameter. Furthermore, if the lobe distance becomes smaller than the spatial resolution of our instrument, the intensity distribution also would become unresolvable. Our spatial resolution of 40 nm would allow observation of a second order mode in a cavity of 175 nm diameter.

4 Laser-activity of single ZnO pillars

Time-resolved PL measurements were performed to study the lasing dynamics of single ZnO nanorods at low temperature. The sample was mounted on the cold finger of a microscope cryostat allowing for temperature variation in the range from 5 K to room temperature. In a confocal arrangement the laser was focused to a single nanorod. Markers written on the sample substrate by a focused ion beam system (FIB) allowed to retrieve specific pillars both in the optical experiments and in electron microscopy. The pumping system used for excitation consists of a frequency-doubled Ti:sapphire laser to obtain an above-bandgap excitation with $\lambda = 355$ nm and a pulse length of 150 fsec. The excitation power is regulated by a combination of a Pockels cell and a linear polarizer. A Streak camera mounted to a $f = 0.46$ m monochromator allows to record energy- and time-resolved spectra with a resolution of $\Delta E = 1$ meV and a time resolution of 5 ps. More experimental details can be found in [29].

In time-integrated spectra (Fig. 8) recorded at low temperature (≈ 5 K) under mid-level pulsed excitation we find a broad, structureless emission band centered around 3.355 eV, which generally is ascribed to the recombination of biexcitons [30–32]. Upon increase of the excitation density, periodic modulations appear on the low-energy side of this band representing individual laser modes. They become more and more pronounced at further increased pumping. For the highest laser powers, the number of active modes decreases and sequentially, lower-energy modes become dominant. The underlying low-energy shift of the gain spectrum indicates bandgap-renormalization due to electron-hole exchange interaction in the high-density plasma. The individual modes are upshifted in energy in subsequent spectra related to a decrease of the refractive index, i.e. a decrease of the effective cavity length. Under low excitation, the refractive index in this energy range is augmented due to the resonance of the exciton-polariton. For high excitation leading to carrier densities above the Mott density a transition from excitonic states to a Coulomb correlated EHP occurs and thus the excitonic contribution to the refractive index is reduced [33]. These findings are summarized in Fig. 9.

Let us consider now more closely how the emission spectra develop shortly after the exciting laser pulse (see Fig. 10). For relatively low excitation powers (upper left picture, 1 mJ/cm^2) the broad, unstructured band of spontaneous emission from biexcitons ("M band") is observed, which decays with a time constant of around 100 ps. During the decay, no shift of the emission maximum is visible. For 1.8 mJ/cm^2 excitation power, 6-7 distinct laser modes develop superimposed on the broad band seen before. The PL lifetime of these modes of the stimulated emission drops to $\approx$ 8 ps and is then much shorter than the

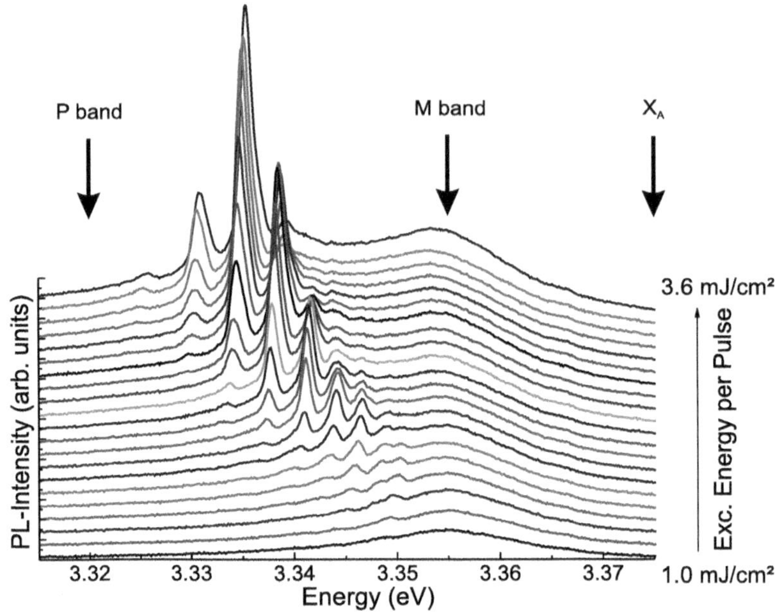

Fig. 8. Sequence of time-integrated low-temperature PL spectra recorded for increasing pumping power. Above a threshold of ≈ 2 mJ/cm^2 laser activity sets on, and the modes sequentially switch to lower energy. X_A marks the energy position of the lowest free exciton state, "M" is the position of the biexciton, and "P" of decay processes involving exciton–exciton scattering.

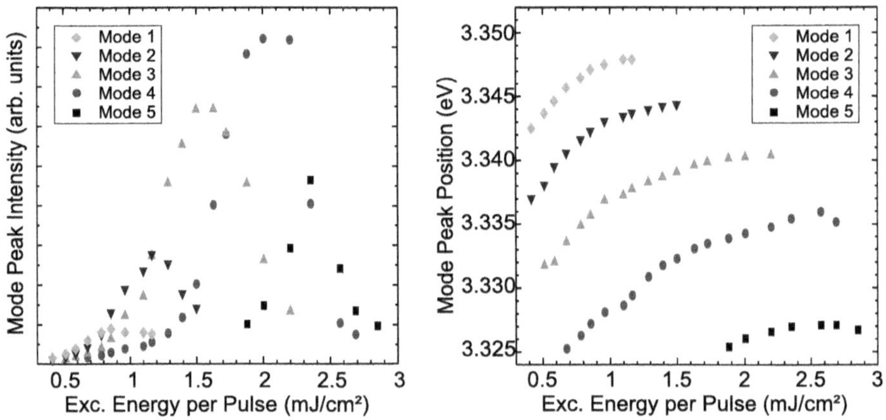

Fig. 9. *Left*: Intensity of the competing laser modes observed for increasing pump power. *Right*: Energy shift of the modes for varying pump power. For the specific pillar investigated here, the dimensions are (3.1 ± 0.1) µm for the length, and 145 nm for the diameter.

biexciton lifetime. Furthermore, during the decay all lines shift towards lower energy by some meV. This shift is due to the change in the refractive index caused by the decreasing carrier density. Further increase of the pumping power leads to a reduction of the number of competing modes – similar to standard edge-emitting semiconductor lasers – and to a further drop in their decay time. The initial line shift immediately after the exciting pulse towards longer wavelengths becomes more and more pronounced, as the maximum carrier density increases and with it the maximum change in the refractive index occurs.

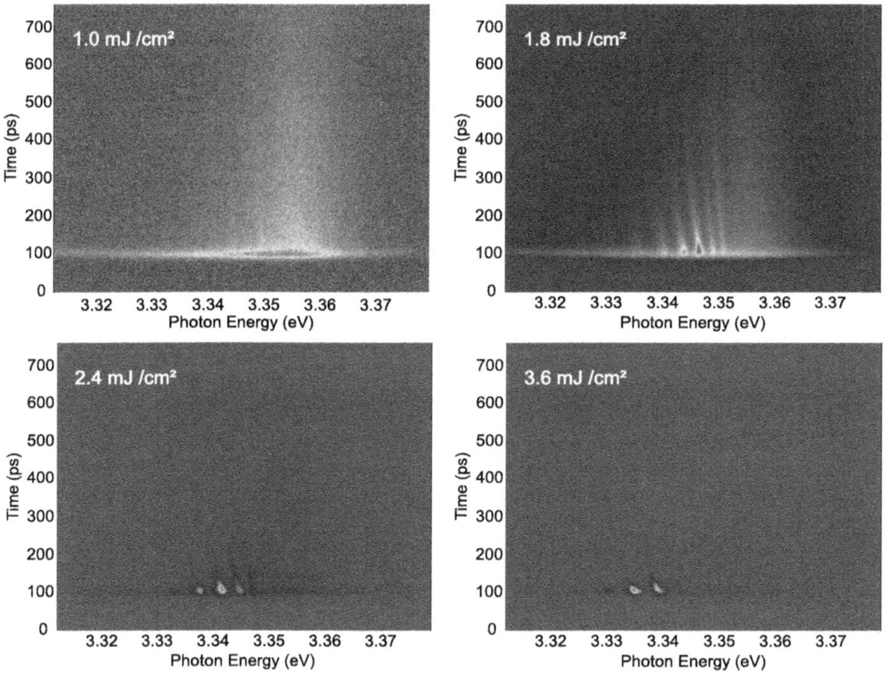

Fig. 10. Time- and energy-resolved spectra of a single nanorod recorded with a Streak camera.

Full analysis of the temporal and spectral properties of the emission spectra were carried out on numerous pillars of this sample, and always rather similar results were found. For the pillars discussed in detail here, the gain mechanism leading to laser emission is clearly recombination of electron-hole pairs in a plasma, where all excitons are ionized. Based on the argument, that the exciton binding energy in ZnO is rather high ($\approx$ 60 meV), most previous studies have assumed an exciton–exciton scattering recombination mechanism ("P" band) as the source of the optical gain [10]. In our samples, the P-band and M-band (biexciton recombination) are clearly separated from the lasing wavelengths and do not contribute to stimulated emission. A theoretical simulation of the gain profile expected under the present excitation conditions can be found in Ref. [29].

A rough estimate for the required threshold gain is obtained from the following consideration: The threshold gain for a Fabry-Pérot-resonator is given by [34]

$$g_{\text{th}} = -\frac{1}{2L} \times \ln\sqrt{R_1 R_2} \tag{6}$$

We assume the same reflectivities at both ends, $R_1 = R_1 \approx 0.2$ as expected for the interface ZnO/air. For a resonator length of 10 µm, the threshold gain is in the range of 800 cm^{-1}.

The resulting line width for the Fabry-Pérot-resonator is

$$\Delta\nu = -c/(4\pi L n)\ln\left[R_1 R_2 (1-\alpha_i)^2\right] \tag{7}$$

with α_i the internal losses and L the length of the resonator. The values $R_i = 0.2, L = 10\,\mu\text{m}$, $n = 2.85$, and $\alpha_i = 0.1$ yield a line width of ≈ 1 meV, similar to the value found in our experiments.

5 Summary and outlook

Controlled growth procedures have allowed us to grow hexagonally shaped, well-faceted nano-resonators of ZnO. Excellent resonator properties can be realized without sophisticated sequences of Bragg-mirrors. The resonator quality is manifested in two types of experiments on single pillars: In CL measurements, UV light standing waves are observed, and under high optical pulse excitation multimode lasing spectra are measured.

The intermediate (111) fcc Ir interlayer on the (111) silicon substrate suppresses alloying of the Zn with Si, which is an important issue for undisturbed epitaxial growth, and enhances the reflectivity at the bottom end. The major improvement of the bottom reflectance comes presumably from the silicon substrate replacing the usually used sapphire substrate: the refractive index of Si at $h\nu \approx 3.36$ eV ($\lambda \approx 369$ nm) is $n \approx 7$ and thus much higher than that of ZnO ($n \approx 2.85$) and sapphire ($n \approx 1.8$).

The present pillars should be good candidates for coupled emitter-resonator systems in the weak or strong coupling limit, i.e. for the observation of the Purcell spontaneous emission enhancement effect, or of Rabi splittings [35].

The Purcell spontaneous emission enhancement factor is given by

$$P = \frac{3}{4\pi^2}\left(\frac{\lambda}{n}\right)^3 \frac{Q}{V} \tag{8}$$

with V the resonator volume and Q the resonator quality factor. We estimate a quality factor $Q = E/\Delta E \approx 3\,\text{eV}/1\,\text{meV} = 3000$ from the laser spectra, with a halfwidth ΔE of the emitted lines of around 1 meV and an emitted photon energy of $\approx$ 3 eV. Due to their small diameters, the resonators presented here have a rather low volume of only $\approx 0.2\,\mu m^3$ – much lower than micro-resonators etched from AlGaAs Bragg stacks. With these numbers a Purcell factor of 10–20 should result. For strong coupling, the Q factor must be further improved by increasing the reflectivity of both end faces, and possibly by decreasing the leakage of the optical field through the side facets.

Van Vugt et al. [36] investigated the angular emission characteristics of exciton polaritons in suspended ZnO nanowires at room temperature. They derived indirectly by modelling an extremely strong coupling which they considered characteristic for ZnO, and quoted the Rabi splitting energy to be more than 100 meV. This would be much more than obtained in any resonator based on sophisticated Bragg mirrors or photonic crystals, opening a wide field for new applications of ZnO nanostructures.

6 Acknowledgements

We thank Prof. D. Gerthsen and her group for high-resolution SEM micrographs and the FIB processes to define marks. The financial support by the Kompetenznetzwerk "Funktionelle Nanostrukturen" within the Landesstiftung program of the Government of the Federal State of Baden-Württemberg is gratefully acknowledged.

References

1. Z.L. Wang, Mater. Today **7**, 26 (2004)
2. F.H. Nicoll, Appl. Phys. Lett. **9**, 13 (1966)
3. J.M. Hvam, Sol. Stat. Comm. **12**, 95 (1973)
4. D.M. Bagnall, Y.F. Chen, Z. Zhu, T. Yao, S. Koyama, M.Y. Shen, and T. Goto, Appl. Phys. Lett. **70**, 2230 (1997)
5. C. Klingshirn, Phys. Stat. Sol. B **244**, 3027 (2007)
6. D.M. Bagnall, Y.F. Chen, Z. Zhu, T. Yao, M.Y. Shen, and T. Goto, Appl. Phys. Lett. **73**, 1073 (1998)
7. P. Zu, Z.K. Tang, G.K.L. Wong, M. Kawasaki, A. Ohtomo, H. Koinuma, and Y. Segawa, Sol. Stat. Comm. **103** 459 (1997)
8. M.H. Huang, S. Mao, H. Feick, H. Yan, Y. Wu, H. Kind, E. Weber, R. Russo, and P. Yang, Science **292**, 1897 (2001)
9. J. Johnson, K. Knutsen, H. Yan, M. Law, Y. Zhang, P. Yang, and R. Saykally, Nano Lett. **4**, 197 (2004)
10. D. Sirbuly, M. Law, H. Yan, and P. Yang, J. Phys. Chem. B **109**, 15190 (2005)
11. H. Yan, J. Johnson, M. Law, R. He, K. Knutsen, J. McKinney, J. Pham, R. Saykally, and P. Yang, Adv. Mater. **15**, 1907 (2003)
12. Z.K. Tang, G.K.L. Wong, P. Yu, M. Kawasaki, A. Ohtomo, H. Koinuma, and Y. Segawa, Appl. Phys. Lett. **72**, 3270 (1998)

13. H. Cao, Y.G. Zhao, H.C. Ong, S.T. Ho, J.Y. Dai, J.Y. Wu, and R.P.H. Chang, Appl. Phys. Lett. **73**, 3656 (1998)
14. S.F. Yu, C. Yuen, and S.P. Lau, Appl. Phys. Lett. **84**, 3241 (2004)
15. J. Choy, E. Jang, J. Won, J. Chung, D. Jang, and Y. Kim, Adv. Mater. **15**, 1911 (2003)
16. Y. Sun, J.B. Ketterson, and G.K.L. Wong, Appl. Phys. Lett. **77**, 2322 (2000)
17. Y.H. Leung, W.M. Kwok, A.B. Djurisic, D.L. Phillips, and W.K. Chan, Nanotechnology **16**, 579 (2005)
18. W. Kwok, A. Djurisic, Y. Leung, W. Chan, D. Phillips, H. Chen, C. Wu, S. Gwo, and M. Xie, Chem. Phys. Lett. **412**, 141 (2005)
19. R. Hauschild, H. Lange, H. Priller, C. Klingshirn, R. Kling, A. Waag, H.J. Fan, M. Zacharias, and H. Kalt, Phys. Stat. Sol. B **243**, 853 (2006)
20. R. Hauschild and H. Kalt, Appl. Phys. Lett. **89**, 123107 (2006)
21. J. Johnson, H. Yan, P. Yang, and R. Saykally, J. Phys. Chem. B **107**, 8816 (2003)
22. H. Zhou, M. Wissinger, J. Fallert, R. Hauschild, F. Stelzl, C. Klingshirn, and H. Kalt, Appl. Phys. Lett. **91**, 181112 (2007)
23. G.M. Prinz, A. Reiser, T. Röder, M. Schirra, M. Feneberg, U. Röder, R. Sauer, K. Thonke, S. Gsell, M. Schreck, and B. Stritzker, Appl. Phys. Lett. **90**, 233115 (2007)
24. S. Gsell, T. Bauer, J. Goldfuß, M. Schreck, and B. Stritzker, Appl. Phys. Lett. **84**, 4541 (2004).
25. Y. Li, M. Feneberg, A. Reiser, M. Schirra, R. Enchelmaier, A. Ladenburger, A. Langlois, R. Sauer, K. Thonke, J. Cai, and H. Rauscher, J. Appl. Phys. **99**, 054307 (2006)
26. M. Schirra, A. Reiser, G.M. Prinz, A. Ladenburger, K. Thonke, and R. Sauer, J. Appl. Phys. **101**, 113509 (2007)
27. J. Wiersig, Phys. Rev. A **67**, 023807 (2003)
28. M. Feneberg et al., to be published
29. J. Fallert, F. Stelzl, H. Zhou, A. Reiser, K. Thonke, R. Sauer, C. Klingshirn, and H. Kalt, Opt. Express **16**, 1125 (2008)
30. J.M. Hvam, G. Blattner, M. Reuscher, and C. Klingshirn, Phys. Stat. Sol. B **118**, 179 (1983)
31. A. Yamamoto, K. Miyajima, T. Goto, H.J. Ko, and T. Yao, J. Appl. Phys. **90**, 4973 (2001)
32. C. Klingshirn, Chem. Phys. Chem. **8**, 782 (2007)
33. C. Klingshirn, R. Hauschild, J. Fallert, and H. Kalt, Phys. Rev. B **75**, 115203 (2007)
34. K. Domen, K. Kondo, A. Kuramata, and T. Tanahashi, Appl. Phys. Lett. **69**, 94 (1996)
35. M. Bayer, A. Forchel, T. Reinecke, P. Knipp, and S. Rudin, Phys. Stat. Sol. A **191**, 3 (2002)
36. L.K. van Vugt, S. Ruhle, P. Ravindran, H.C. Gerritsen, L. Kuipers, and D. Vanmaekelbergh, Phys. Rev. Lett. **97**, 147401 (2006)

Waveguiding and Optical Coupling in ZnO Nanowires and Tapered Silica Fibers

Tobias Voss

Institute of Solid State Physics, University of Bremen, Otto-Hahn-Allee 1, 28359 Bremen, Germany
voss@ifp.uni-bremen.de

Abstract. Tapered silica fibers are shown to be efficient and easy to use tools to controllably inject laser light into ZnO nanowires with diameters of about 250 nm to study their waveguiding properties. The excitation of high-order waveguide modes in ZnO nanowires is observed in the experiments. Numerical simulations based on the finite-difference time domain technique reproduce the experimental observations and allow for a study of the mode profiles and the coupling efficiencies for different angles between the silica and ZnO nanowire waveguides.

1 Introduction

Due to their uniform diameter $d \lesssim 300\,\mathrm{nm}$, their low absorption throughout the visible spectral region, and their large refractive index $n \gtrsim 2$ ZnO nanowires [1–7] are interesting candidates for sub-wavelength optical waveguides. In such waveguides, light is propagating through structures whose lateral dimensions are smaller than the corresponding vacuum wavelength. In addition, in the visible spectral region even wide-bandgap semiconductor nanowires exhibit a large index contrast to the surrounding air, $\Delta n \gtrsim 1$, which is about 2 orders of magnitude larger than that of conventional optical fibers [8]. In pioneering works the waveguiding properties of semiconductor nanowires and nanoribbons have been demonstrated [9–11].

To study the waveguiding properties of single ZnO nanowires, an efficient and controllable means of coupling external light into the waveguide modes of the nanowire is required. Silica nanowires offer a convenient solution to this problem: they can be produced by a taper-drawing technique from standard silica optical fibers to which they are still naturally attached after the drawing process [12]. Therefore, light can be coupled into the silica optical fiber with conventional fiber-optical components, and it will be guided through the taper region into the silica nanowire. The silica nanowire is mounted onto a micro-positioning stage so that its end can be brought close to an individual ZnO nanowire to study the coupling of the optical modes between the two waveguides.

In this paper, we will review our previous results on waveguiding in ZnO nanowires [13] and present a study of the mode profiles of the optical field in the nanowires. We present additional results of numerical simulations of the coupling efficiency between silica tapered fibers and ZnO nanowires.

2 Experiment

Single-crystalline ZnO nanowires synthesized by a vapor-liquid-solid (VLS) growth process in a horizontal tube furnace [14] were mechanically dispersed onto thin glass substrates coated with an 800-nm thick mesoporous silica film. This technique typically yielded 20-μm to 80-μm long ZnO nanowires that can be individually addressed in a conventional optical microscope. The mesoporous films have monodisperse, 8-nm wide pores and were deposited using a dip-coating process [15, 16]. The resulting films have a refractive index of about $n = 1.185$ throughout the visible spectral region [15], preventing parasitic coupling into the substrates. Combined with their extremely high homogeneity and flatness, these layers also help minimize losses and noise due to scattering at the output sides of the silica and ZnO nanowires, which is crucial for investigating coupling and waveguiding processes. Because the refractive index of mesoporous silica is much lower than that of both the silica and ZnO nanowires, the nanowires are treated as freestanding, air-clad waveguides in the numerical simulations.

Silica nanowires were fabricated with a conventional fiber tapering technique to produce low-loss submicrometer diameter wires that remain attached to a standard fiber on one side. Details about the taper drawing process have been previously published [12, 13].

The coupling and waveguiding properties of the ZnO nanowires were studied by launching continuous wave laser light ($\lambda = 532\,\text{nm}$, $P = 1\,\text{mW}$) into a silica optical fiber mounted on a micropositioning stage. The silica nanowires were cut off close to the tapered region, using the end of the tapered region to couple light into the ZnO nanowires. The coupling was observed using an inverted microscope with a 100x oil-immersion objective (numerical aperture 1.4) with attached micropositioning stages to bring the end of the silica fiber close to an individual ZnO nanowire.

3 Results and discussion

Figure 1 shows an optical image of the coupling process between a silica tapered fiber and a ZnO nanowire. The taper region was brought close to one end of the wire so that the green laser light couples into the waveguide mode of the nanowire. While significant scattering is observed in the coupling region, no additional waveguiding losses of the ZnO nanowire are observed. The light is guided to the opposite end of the wire where it is emitted. The angle

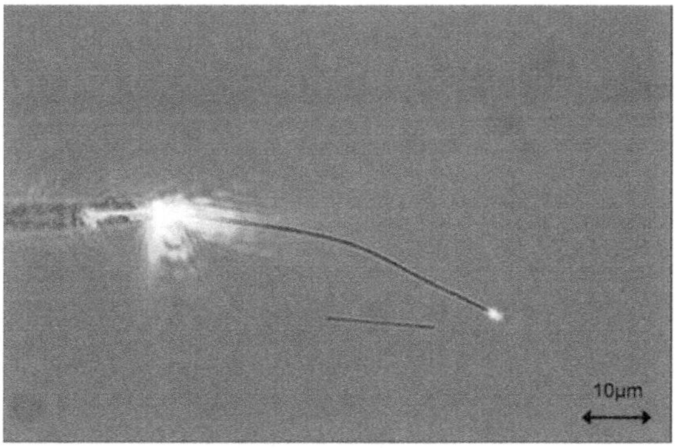

Fig. 1. Coupling of green laser light from a silica tapered fiber into a ZnO nanowire.

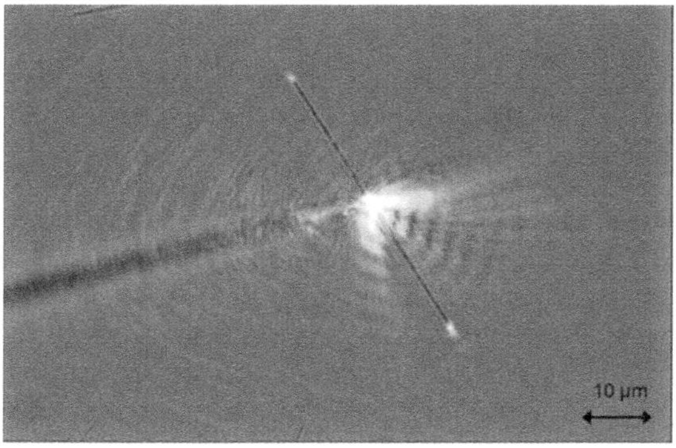

Fig. 2. Coupling of green laser light from a silica tapered fiber into a ZnO nanowire at the middle of the nanowire.

of emission from a nanowire with a typical diameter of about $d = 200\,\text{nm}$ was found to be approximately $\alpha = 90°$ [13]. This result was confirmed in numerical simulations based on the finite-difference time domain technique [13].

Coupling is not only possible at the nanowire ends but can also be achieved if the tapered fiber is brought close to the center region of the nanowire such that the two waveguides form an angle of about 90° (see Fig. 2). In this case, the light is guided to both ends of the ZnO nanowire from where it is emitted. Considering the alignment between the silica fiber and the ZnO nanowire, it is expected that the light should preferentially be coupled into the waveguide mode propagating to the lower right side of the ZnO nanowire in Fig. 2. The corresponding end of the nanowire is indeed the one from which a higher light intensity is emitted. Since part of the light is reflected when it reaches one of the ends of the ZnO nanowires, reflected modes further contribute to the emitted intensity observed at both ends of the nanowire.

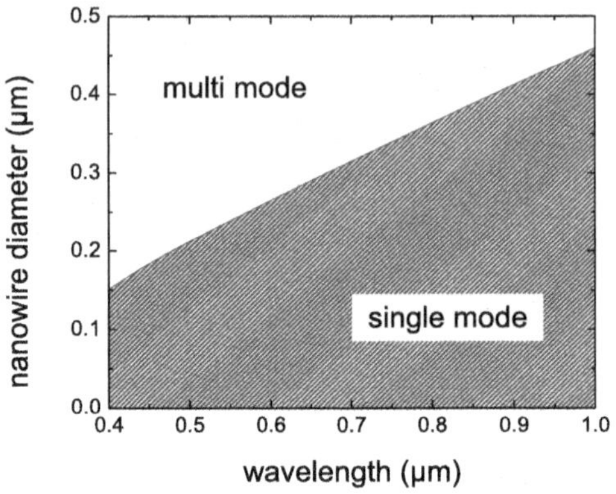

Fig. 3. Conditions for single- and multi-mode waveguiding in ZnO nanowires as a function of the wire diameter and the vacuum wavelength.

The diameter of the analyzed VLS-grown ZnO nanowires typically varies between 100 and 400 nm. For cylindrical waveguides in this diameter range, both single- and multi-mode waveguiding is expected from waveguide theory [8]. The single-mode cutoff limit can be estimated with the V number defined as $V = \frac{2\pi}{\lambda_0} d \sqrt{n_{\mathrm{ZnO}}^2 - n_{\mathrm{air}}^2}$ (λ_0: vacuum wavelength, n: refractive index) [8]. If the V number is smaller than roughly 2.4 the corresponding waveguide supports only a single mode [8, 17]. Taking into account the ZnO refractive index approximated by a Sellmeier type equation for wavelengths above the exciton resonance [18], we get the results shown in Fig. 3 for the regions of single- and multi-mode waveguiding in ZnO nanowires as a function of the vacuum wavelength and the nanowire diameter. The results show that for light in the visible spectral region the transition from multi-mode to single-mode waveguiding occurs for ZnO nanowire diameters between 150 and 300 nm. This is the typical diameter range for VLS-grown ZnO nanowires which thus may act as either single-mode or multi-mode waveguides depending on their actual diameter.

Multi-mode waveguiding in ZnO nanowires with diameters of about 250 – 300 nm can be studied by exciting the corresponding waveguide modes using the silica tapered fibers. Figure 4(a) and (b) demonstrate that a slight change in the alignment between the silica tapered fiber and the ZnO nanowire waveguide leads to significantly different waveguide modes: in (a) low-order modes are excited only whereas in (b) light is guided also in high-order waveguide modes. The high-order modes possess additional evanescent-field contributions which exponentially decrease with the radial distance from the nanowire surface. These evanescent components are very sensitive to roughness and

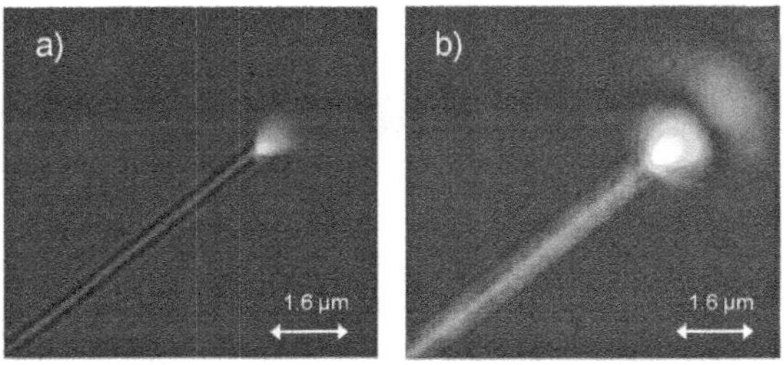

Fig. 4. ZnO nanowire carrying (**a**) low order modes with no visible evanescent-field contributions and (**b**) high order modes with substantial evanescent-field contributions. The change from (**a**) to (**b**) is achieved by slightly varying the alignment between the silica tapered fiber and the ZnO nanowire. The true width of the ZnO nanowire (around 300 nm) cannot be resolved with the optical microscope.

impurities in the substrate and thus experience noticeable scattering leading to the green appearance of the ZnO nanowire surface in Fig. 4b.

Additional investigations with a white-light source instead of the green laser have demonstrated that the high-order waveguide modes with strong evanescent contributions experience wavelength dependent losses. This is expected because the penetration depth of the evanescent mode into the surrounding medium increases with the wavelength. The white-light waveguiding measurements therefore support the interpretation of the experimentally observed effects as being due to high-order waveguide modes with significant evanescent-field contributions.

The results discussed so far demonstrate that ZnO nanowires with diameters around 200 nm can serve as multimode waveguides for visible light. In order to analyze the achievable coupling efficiency and to understand the origin and properties of the high-order waveguide modes, we numerically simulated the waveguiding and coupling between silica and ZnO nanowires using the finite-difference time domain (FDTD) technique [19]. The results of the FDTD simulations that were performed with the software package MEEP [19] are scalable if the wavelength and the lateral dimensions of the nanowires are multiplied by the same number, and as long as the refractive indices of the involved materials are constant. For the two-dimensional simulations presented here, we assume constant refractive indices of $n_{\mathrm{ZnO}} = 2$, $n_{\mathrm{silica}} = 1.46$, and $n_{\mathrm{air}} = 1$, which is a good approximation for the wavelength range between 400 and 800 nm. In all simulations the vector of the electric field is perpendicular to the plane of the nanowires. We simulated the propagation of a pulse from a silica to a ZnO nanowire for various angles between the symmetry axes of the two nanowires. A light pulse uniformly illuminates the left end of the silica fiber, exciting a low-order mode that travels to the right. At the interface between the silica and the ZnO nanowire, part of the pulse is reflected, a second part is scattered, and a third part is coupled into the waveguide modes of the ZnO nanowire. The two-dimensional refractive-index profile used for the simulations together with the mode profiles excited in the ZnO nanowire for an angle $\alpha = 10°$ between the two waveguides and the wavelengths (a) $\lambda_0 = 2.8d_{\mathrm{ZnO}}$ and (b) $\lambda_0 = 2.2d_{\mathrm{ZnO}}$ is shown in Fig. 5. The results of the numerical simulations demonstrate that in the multi-mode waveguiding regime (small wavelengths as in (b)) high-order modes actually are excited in the ZnO nanowire during the coupling process. Comparing the mode profiles in Fig. 5(a) and (b), the mode profile in (b) exhibits additional intensity minima perpendicular to the propagation direction and has a significant evanescent contribution propagating outside of the ZnO nanowire. The simulations therefore confirm the above given interpretation of the results presented in figure 4 as being due to the excitation of either low- or high-order waveguide modes.

Fig. 5. Mode profiles of a light pulse after the coupling process from a silica fiber (*left*) into a ZnO nanowire (*right*) for two different wavelengths (**a**) $\lambda_0 = 2.8d_{\mathrm{ZnO}}$ and (**b**) $\lambda_0 = 2.2d_{\mathrm{ZnO}}$.

Figure 6 shows the power of the light field that is actually coupled from the silica tapered fiber into the ZnO nanowire (transmission) as a function of the angle between the two waveguides for three different wavelengths (vacuum wavelengths, given in units relative to the diameter of the ZnO nanowire). For angles below 60° the upper limit for the transmission is determined to vary between about 50 and 75%. For larger angles between the two waveguides, the transmitted power rapidly drops to very low values of only a few percent. Still, the rather high refractive index of the ZnO nanowires implies a correspondingly high acceptance angle for the waveguide modes which results in the nanowires being rather insensitive to the precise coupling conditions, as long as they are kept in a moderate range. Although the calculated coupling efficiencies will rarely be achieved in the experiments, silica tapered fibers offer a rather robust and easy-to-use means of controllably injecting external light from macroscopic light sources into semiconductor nanowire waveguides.

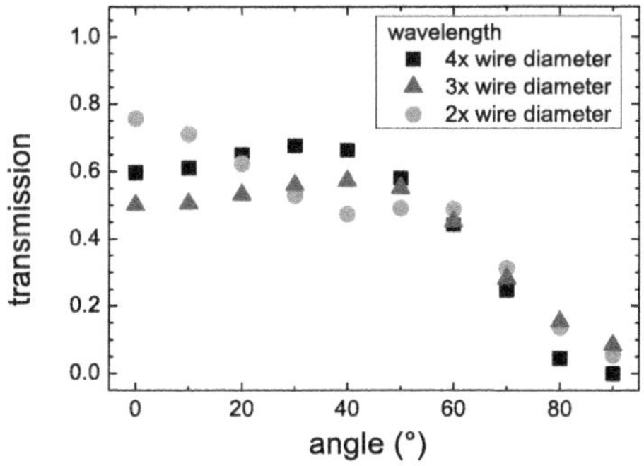

Fig. 6. Simulated transmission (coupling efficiency) from a silica tapered fiber into a ZnO nanowire as a function of the angle between the two waveguides for three different wavelengths.

4 Conclusions

In summary, we have demonstrated that silica tapered fibers can be used as efficient tools to couple external light into the waveguide modes of individual ZnO nanowires. The silica fibers can be fabricated by an easy taper-drawing process. With this tool it is possible to study the different single- and multi-mode waveguiding conditions in sub-wavelength semiconductor nanowires. For light in the visible spectral region, we have calculated the single- and multi-mode waveguiding conditions for ZnO nanowires. We have shown that for typical wires grown by a vapor-liquid-solid process both conditions may apply, depending on the actual nanowire diameter. In experiments and numerical simulations, we have demonstrated that the high-order waveguide modes may carry a significant intensity as evanescent fields outside the wire which experience wavelength-dependent losses. We have also analyzed the coupling efficiency between the tapered fibers and the ZnO nanowires and found a weak dependence on the precise alignment between the waveguides, at least under moderate conditions.

5 Acknowledgements

Several people contributed to the work described in this paper. T. Voss conceived the basic idea for this work, designed the experiment, carried out the experiments, and performed the FDTD simulations. G. T. Svacha (Harvard University, USA) fabricated the silica tapered fibers. C. Ronning and S. Müller (University of Göttingen, C. R. now University of Jena, Germany) fabricated and characterized the ZnO nanowires. D. Konjhodzic and F. Marlow fabricated and characterized the mesoporous silica substrates. E. Mazur (Harvard University, USA) supervised the research. The author wishes to thank Jürgen Gutowski, University of Bremen, for critically reading the manuscript. He further acknowledges funding by the German Research Foundation through the grants VO1265/3 and VO1265/4-1.

References

1. Z. L. Wang and J. Song, Science **312**, 242 (2006)
2. Y. Li, F. Qian, J. Xiang, and C. M. Lieber, Mater. Today **9**, 18 (2006)
3. Y. Huang, X. Duan, and C. M. Lieber, Small **1**, 142 (2005)
4. J. Bao, M. A. Zimmler, and F. Capasso, Nano Lett. **6**, 1719 (2006)
5. P. J. Pauzauskie and P. Yang, Mater. Today **9**, 36 (2006)
6. Z. Ye, X. Hu, M. Li, K.-M. Ho, and P. Yang, Appl. Phys. Lett. **89**, 241108 (2006)
7. T. Voss and L. Wischmeier, J. Nanosci. Nanotechnol. **8**, 228 (2008)
8. K. Okamoto: *Fundamentals of Optical Waveguides.* (Academic Press, 2nd Edition, 2006).
9. R. Hauschild and H. Kalt, Appl. Phys. Lett. **89**, 123107 (2006)
10. M. Law, D. J. Sirbuly, J. C. Johnson, J. Goldberger, R. J. Saykally, and P. Yang, Science **305**, 1269 (2004)
11. A. V. Maslov and C. Z. Ning, Appl. Phys. Lett. **83**, 1237 (2003)
12. L. M. Tong, R. R. Gattass, J. B. Ashcom, S. L. He, J. Y. Lou, M. Y. Shen, I. Maxwell, and E. Mazur, Nature **426**, 816 (2003)
13. T. Voss, G. T. Svacha, S. Müller, C. Ronning, D. Konjhodzic, F. Marlow, and E. Mazur, Nano Lett. **7**, 3675 (2007)
14. C. Borchers, S. Müller, D. Stichtenoth, D. Schwen, and C. Ronning, J. Phys. Chem. B **110**, 1656 (2006)
15. D. Konjhodzic, H. Bretinger, U. Wilczok, A. Dreier, A. Ladenburger, M. Schmidt, M. Eich, and F. Marlow, Appl. Phys. A **81**, 425 (2005)
16. M. Schmidt, G. Boettger, M. Eich, W. Morgenroth, U. Huebner, H. G. Meyer, D. Konjhodzic, H. Bretinger, and F. Marlow, Appl. Phys. Lett. **85**, 16 (2004)
17. L. Tong, J. Lou, and E. Mazur, Opt. Exp. **12**, 1025 (2004)
18. H. Yoshikawa and S. Adachi, Jpn. J. Appl. Phys. **36**, 6237 (1997)
19. A. Farjadpour, D. Roundy, A. Rodriguez, M. Ibanescu, P. Bermel, J. D. Joannopoulos, S. G. Johnson, and G. Burr, Opt. Lett. **31**, 2972 (2006)

Part II

Quantum Dots and Nanoparticles

Electrically Driven Single Quantum Dot Emitter Operating at Room Temperature

Tilmar Kümmell[1], Robert Arians[1], Arne Gust[2], Carsten Kruse[2], Sergey Zaitsev[1], Detlef Hommel[2], and Gerd Bacher[1]

[1] Werkstoffe der Elektrotechnik and CeNIDE, Universität Duisburg-Essen, Bismarckstraße 81, 47057 Duisburg, Germany
`tilmar.kuemmell@uni-due.de`

[2] Institut für Festkörperphysik, Universität Bremen, Otto-Hahn-Allee, 28359 Bremen, Germany

Abstract. We present both optically and electrically driven room temperature emission from single CdSe quantum dots, realized by self-organized epitaxial growth. A structure design that embeds the CdSe quantum dots into ZnSSe/MgS barriers results in high carrier confinement and exceptionally large quantum efficiencies at room temperature. Microphotoluminescence with a spatial resolution of 200 nm exhibits single dot emission that remains visible up to 300 K. When integrating these quantum dots into p-i-n diode structures, an electrically driven single dot emitter with pronounced room temperature emission is realized. The linewidth of the single dot emission increases with temperature due to exciton-phonon interaction and reaches 26 meV at 300 K. This value is only slightly larger than the biexcitonic binding energy, opening a way to solid state single photon emitters operating at elevated temperatures.

1 Introduction

One fundamental challenge for establishing semiconductor single quantum dots (SQDs) in commercial applications is the need for room temperature operation of SQD devices. SQDs can be the base for devices, where single electrons or single photons act as information carriers. A promising application field that has been pushed forward during the last years encompasses light sources emitting photons on demand. Great effort has been made to implement concepts for such single photon sources [1–7]. For a future application, compact and stable electrically driven devices are most attractive, and this has fostered the development of structures based on epitaxially grown single quantum dots. However, the operation of the great majority of these SQD emitters is limited to temperatures below 100 K. Moreover, most prototypes are optically pumped; only few electrically driven single photon sources have been realized up to now [8–10].

Several attempts towards room temperature (RT) emission of SQDs have been performed in case of optical excitation. SQD emission at 300 K has been reported in $In_xGa_{1-x}As$ quantum dots (QDs), but obviously with weak signals and long integration times [11, 12]. By enhancing the band gap of the barrier material, the luminescence efficiency for InAs-based QD ensembles could be improved substantially at room temperature [13], however, no data for SQD emission has been reported.

Wide band gap materials like II–VI or III-nitride semiconductors seem to have a high potential for room temperature applications. RT single photon emission from CdSe/ZnS nanocrystals was observed after optical excitation already a few years ago [14]. Publications on *epitaxially grown* CdSe-QDs also have stated a high carrier confinement and an enhanced quantum efficiency at elevated temperatures in this material system [15] and indeed single photon emission up to 200 K was obtained for optically pumped SQDs [16]. Comparable results have been reported recently also for other material systems: Single photon emission at 200 K was found in GaN-based SQDs [17] and for a single color center in diamond [18] actually RT operation could be reached. While first successful attempts have been made to achieve RT electroluminescence from single CdSe/ZnS nanocrystals [19], where, however, strong line broadening and a pronounced background emission was observed, electrically driven emitters based on epitaxially grown SQDs are limited to quite low temperatures up to now.

In order to develop an electrically driven SQD emitter for operation at elevated temperatures we will concentrate on epitaxially grown CdSe/ZnSe QDs embedded into a p-i-n device structure. In this contribution, we demonstrate optically and electrically driven SQD emission at room temperature, that is achieved by optimizing the confining barriers with respect to both, quantum yield and current injection.

2 Quantum dots optimized for RT emission

2.1 Enhanced carrier confinement in CdSe/ZnSSe/MgS structures

A key building block for attaining RT emission are quantum dots with strong carrier confinement, which prevents a thermally induced escape into the wetting layer or the barrier causing a quenching of quantum efficiency with rising temperature. We realized this on the basis of self-assembled CdSe/ZnSSe quantum dots using additional MgS barriers for carrier confinement. MgS exhibits one of the highest band gaps known for II–VI semiconductors ($E_{gap} = 5.4\,eV$), on the other hand the lattice mismatch allows for an epitaxial growth on ZnSe [20, 21]. This opens a way to embed the active quantum dot area between MgS barriers.

To analyse the influence of MgS barriers on the high temperature quantum efficiency, three structures with different active regions were fabricated

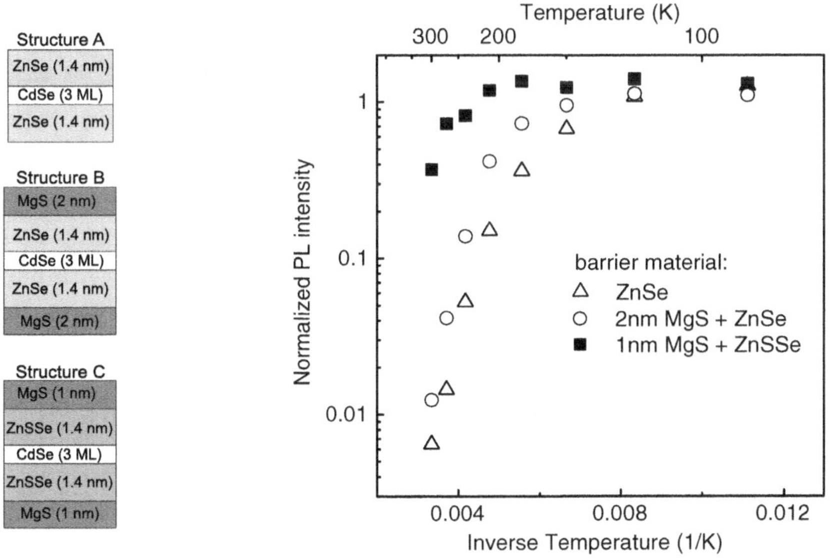

Fig. 1. Spectrally integrated intensities for QD structures with different confinement layers in dependence on the temperature. The intensity is normalized to the value found at $T = 4\,\mathrm{K}$ for each sample. In the *left part* of the figure, the nominal layer sequence of the different structures studied is schematically shown.

by molecular beam epitaxy. These structures are designed for optical spectroscopy and therefore undoped. A 50 nm thick ZnSe buffer was grown on a GaAs (001) oriented substrate. The quantum dot region of the reference sample consists of the sequence ZnSe/CdSe/ZnSe with the nominal thicknesses 1.4 nm/3 ML/1.4 nm. The upper ZnSe layer covering the CdSe is essential for the formation of QD structures by a Zn induced Cd reorganization. These quantum dots typically have sizes in a range of 1.2–2 nm (height) and 5–10 nm (diameter) and the average QD density is typically between 10^{10} and $10^{11}\,\mathrm{cm}^{-2}$ [22]. In the first sample (A) that serves as a reference, no additional barriers were grown and the structure was directly capped by 25 nm ZnSe. In the second sample (B), the ZnSe/CdSe/ZnSe QD area was sandwiched between two layers of 2 nm thick MgS and capped again by 25 nm ZnSe. Sample (C) is similar to sample B, except that the quantum dot area consists of $ZnS_{0.4}Se_{0.6}/CdSe/ZnS_{0.4}Se_{0.6}$ and the MgS layer thicknesses are only 1 nm. The sample structures are schematically summarized in Fig. 1, left.

In order to address single CdSe quantum dots, metal apertures with diameters between 100 μm and 200 nm were prepared on the sample surface. The structures were defined by electron beam lithography using a highly sensitive negative resist. After developing the resist, we evaporate a metal mask consisting of 50 nm chromium and 25 nm of gold. The processing is finished by a lift-off step. This metal sequence suppresses both the exciting laser beam and the sample luminescence outside the apertures effectively. Furthermore, it can be used as a top contact, which will be of importance for the electrically driven devices. For photoluminescence (PL) measurements in a temperature range between 4 and 300 K, the samples were mounted in a He–flow cryostat and excited by a continuous wave Ar^+ laser with an excitation wavelength of 457 nm. The PL signal was dispersed by a 0.55 m monochromator with a 1800/mm grating and collected by a liquid nitrogen cooled charged coupled device camera.

In the first experiment the variation of the quantum efficiency with temperature is analyzed for the different structures by macro–PL–measurements. The defocussed laser beam excites the 100 μm aperture with an excitation density of 200 W/cm^2, while the temperature was varied between 4 and 300 K. The spectrally integrated signal intensities are plotted in Fig. 1. The intensities are normalized to the values at 4 K for the respective samples. The range below 90 K, where no significant intensity change is observed, is not shown. For temperatures above 90 K, clear differences become evident: The intensity of reference sample A significantly drops for temperatures above 150 K; at 300 K the quantum efficiency has decreased more than two orders of magnitude compared with 4 K. Sample B, with the active area embedded between MgS barriers, reveals a clear improvement of the quantum efficiency at temperatures around 200 K. However, the intensity decrease between 4 K and room temperature is still relatively high. For sample C, in contrast, no significant decrease of the PL intensity is detectable for temperatures below 240 K and at room temperature still 35% of the intensity at 4 K can be observed. Obviously, the ZnSSe/MgS matrix has a considerable impact on the carrier confinement. It is interesting to discuss our findings briefly in the light of literature data. On one hand, single dot emission of CdSe/ZnSSe-QDs could be observed in earlier work up to more than 200 K [16], while on the other hand, macro-PL measurements performed on CdSe QDs embedded directly into MgS barriers have shown a significant quenching of the PL intensity above 50 K [23]. Apparently by combining these two approaches a distinct enhancement of the room temperature PL efficiency can be achieved. One reason could be that adding S to the barrier results in a significant increase of the valence band offset as compared to pure CdSe/ZnSe structures [24]. In combination with the MgS barriers, this might account for the remarkable quantum efficiency at high temperatures of sample C.

It is instructive to study the recombination dynamics in the high barrier quantum dots in comparison with conventional CdSe/ZnSe samples. A frequency doubled mode-locked Ti-sapphire laser with a wavelength of 456 nm was used for excitation and the transient PL spectra were recorded by a two-dimensional (2D) synchroscan streak camera with an overall time-resolution of 4 ps and a spectral resolution of 0.8 meV. The spectrally integrated temporal decay curves of samples A and C are depicted in Fig. 2 for various temperatures. For sample A, a significant dependence of the decay times on the temperature is found. While the excitonic lifetime amounts to 270 ps at $T = 4$ K,

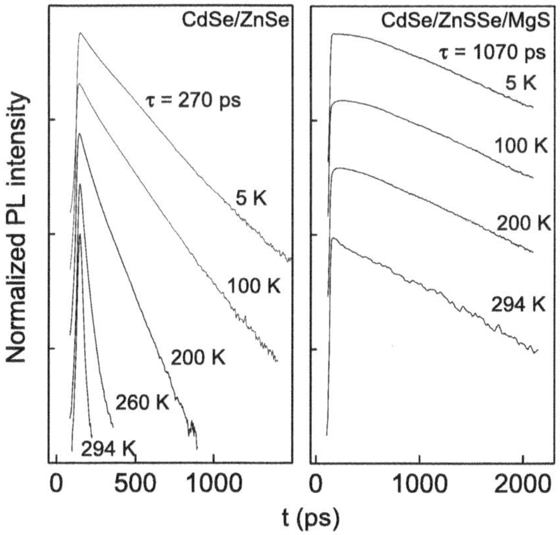

Fig. 2. Temperature dependent decay curves of CdSe/ZnSe QSs (*left*) and of CdSe/ZnSSe/MgS QDs (*right*).

the lifetime decreases drastically with rising temperature. At room temperature, the lifetime is only 15 ps, mainly reduced due to nonradiative processes like carrier escape into the barrier states. The time-resolved study of sample C reveals a completely different behavior: First, the radiative lifetime is longer at low temperatures (about 1 ns) and secondly, it remains nearly constant over the whole temperature range up to room temperature. These findings nicely state the 0D character of the excitons [25] that can be observed even at 300 K. Moreover, the data confirm that the nonradiative decay channels due to thermal escape do not play a significant role in the CdSe/ZnSSe/MgS QDs even at room temperature.

2.2 Single quantum dot photoluminescence spectroscopy

We get access to the SQD emission of this promising class of CdSe quantum dots by micro-PL measurements at structures masked with nanoapertures. The laser light was focused by an achromatic microscope objective with a numerical aperture of 0.55 to a spot diameter below 2 μm. The PL signal was collected by the same microscope objective. Measuring the signal from an aperture with a diameter of 300 nm at an excitation power of 60 μW reveals a number of discrete sharp lines, typical for the emission of individual QDs with a three-dimensional carrier confinement. We record temperature-dependent micro-PL spectra of a single line at the low energy tail of the QD ensemble PL emission (Fig. 3). At 4 K, the emission line at E=2.388 eV exhibits a full width at half maximum (FWHM) of 340 μeV. This linewidth is above the spectral resolution of our setup and can most probably ascribed to spectral diffusion [26]; other groups have reported values between 70 and 500 μeV for CdSe-SQDs at low temperatures [16, 27, 28]. The most remarkable fact in

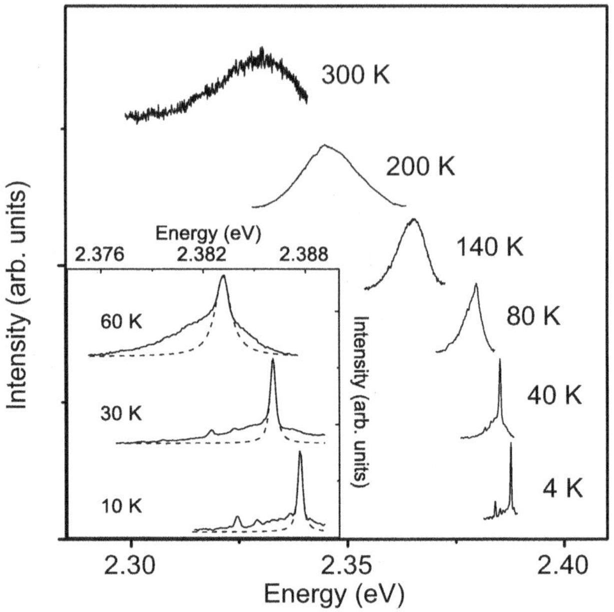

Fig. 3. Optically pumped SQD emission for temperatures between 4 K and 300 K. In the inset, selected low temperature PL spectra are enlarged. The *dashed lines* indicate the Lorentzian lineshape of the zero phonon contribution to the PL spectrum.

Fig. 3 is the observability of the SQD emission up to 300 K that has not been found for epitaxially fabricated II–VI-semiconductor QDs before. For comparison: In sample B (CdSe/ZnSe/MgS), we detect SQD emission up to 160 K [29] (not shown here) and in CdSe/ZnSSe-structures, SQDs are visible up to 200 K [16]. Thus, the combination of the ZnSSe-Matrix surrounding the CdSe QDs and of the MgS-Barriers seems to be essential again to expand the operating point successfully up to room temperature.

With rising temperature the SQD line shifts to lower energy and broadens strongly. Although this is principally well-known for semiconducting quantum dots, it is worth to have a closer look at the line profile. For low temperatures between 4 and 60 K, the emission line can be separated into two components [30, 31] (see inset of Fig. 3). The central, spectrally narrow line, called zero-phonon line (ZPL), is assigned to the excitonic zero-phonon transition, and is nearly constant in width for $T < 40\,\mathrm{K}$. The coupling of the excitons to the acoustic phonons gives rise to additional shoulders, so-called sidebands. These sidebands are much more pronounced in II–VI systems than in InAs quantum dots, because the II–VI QDs are significantly smaller and the excitons can couple with a broader band of acoustic phonons [30]. The sidebands rise in intensity with increasing temperature, but for temperatures up to 80 K the

ZPL is still separable. For $T > 80\,\mathrm{K}$, the sidebands eventually overlap the ZPL, merging into one inseparable line.

The FWHM of this line is further increasing with T, reaching 15 meV at 200 K, which is comparable to values reported for CdSe/ZnSSe QDs [16]. At room temperature, we end up with a linewidth of 25 meV [32]. We attribute the broadening at high temperatures to the interaction of excitons with optical phonons. Generally, the line broadening is larger than in InAs-SQDs [11, 12]. However, for the QDs' possible applicability for single photon sources the linewidth has to be set into a context with the biexciton binding energy; this point will be addressed later.

3 Electrically driven single quantum dot emitter

Electrically driven devices based on the new CdSe/ZnSSe/MgS QDs require an integration of the active region into doped (p-i-n) structures and the development of appropriate ohmic contacts. A schematic setup of the p-i-n structures is shown in Fig. 4, left side. The active region is surrounded symmetrically by $ZnS_{0.06}Se_{0.94}$ layers (doped with Cl and N for the n and the p side, respectively), and the structure is capped by a ZnSe:N spacer (100 nm) and a 45 nm thick ZnSe/ZnTe:N multi quantum well p-contact layer on top. To achieve a good p-contact, the nanoapertures were defined in a 50 nm Pd/25 nm Au mask, and additional Au pads with 200 nm thickness were evaporated in a second lithographic step to ensure high stability of the contacting bond wires. The n-contact consists of 200 nm AuGe.

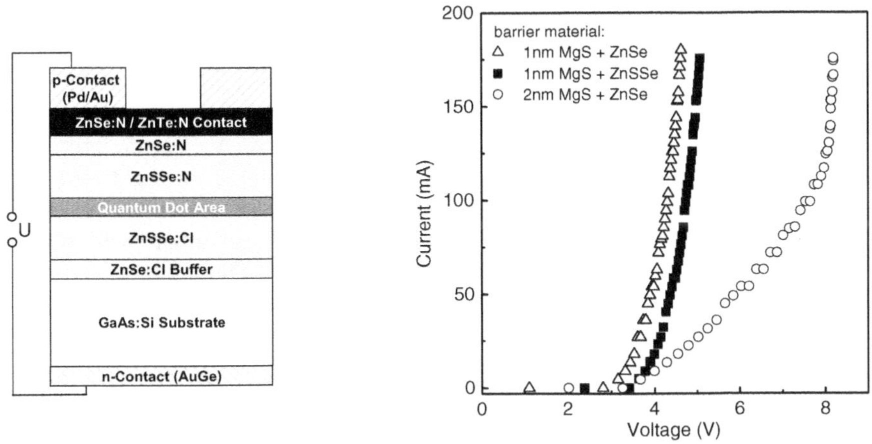

Fig. 4. *Left*: Basic structure for electrically driven devices. *Right*: I–U curves for diode structures with different active areas.

Figure 4 shows I–U curves for p-i-n diode structures with different active layers, detected under ambient conditions at room temperature. We focussed our study on optimizing the MgS layer thickness with respect to current injection and quantum yield. As can be seen directly, the sample with 2 nm thick MgS barriers reveals a relatively slow and irregular increase of the current and a large turn-on voltage, whereas structures with 1 nm MgS barriers exhibit a clear diode-like behavior with turn-on voltages between 3.5 V and 4.5 V for both CdSe QDs embedded into ZnSe and into ZnSSe barriers. Obviously, the current injection is not hampered by these very thin MgS barriers and the structures can be operated as a light emitting device [33].

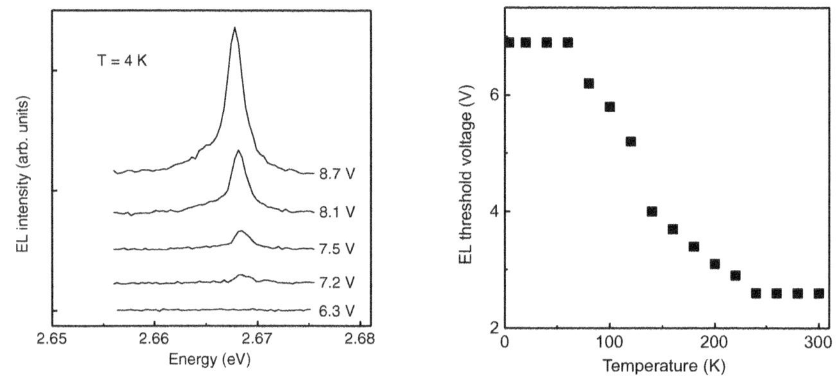

Fig. 5. *Left*: Electrically driven *single dot* emission. *Right*: Dependence of the luminescence threshold voltage on the temperature.

Access to single quantum dots is gained as before by detecting the emission from a nanoaperture. The samples are driven by a pulsed voltage source with a pulse length of 300 ns and a duty cycle of 0.3 in order to avoid heating. At low temperatures, single lines start to emit at voltages of 7 V (Fig. 5, left). The intensity is increasing distinctly with bias, thus confirming the light emitting diode-like behavior. The slight red shift seen with rising voltage is observed for many quantum dots with different magnitude and can most probably be attributed to the Stark effect that depends on the local QD geometry and, in part, to local heating due to the increasing current.

Indications for a biexcitonic emission are visible in some of our measurements (not shown here). An additional line at the low energy side of the SQD emission, separated by 20 meV is found, that increases much faster with bias than the other lines. This line emerges for a current density exceeding $j = 0.4\,\mathrm{A/cm^2}$. This value is quite low, corresponding to approximately 10 pA (10^8 e^-/s) per quantum dot and indicates a non homogeneous population of the QDs within the device during current injection.

In Fig. 5, right, the EL threshold voltage, defined here as the voltage where the EL signal clearly becomes distinguishable from the signal background (signal/noise ratio 3:1), for one specific SQD is plotted versus temperature. It can be seen that the threshold voltage of the EL signal strongly decreases with increasing temperature, i.e. when the doping in the diode becomes active, and remains nearly constant at a value of 2.6 V for the temperature range between 240 K and room temperature. It should be noted that the absolute values of this voltage are varying between individual SQDs, but the general behavior is the same for all quantum dots and confirms a high performance of the device at high temperatures. The single quantum dot emission for the

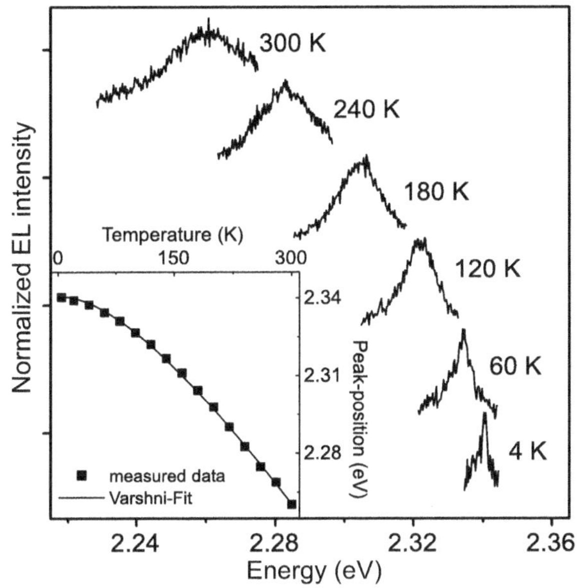

Fig. 6. Electrically driven *single quantum dot* emission between 4 K and room temperature. Inset: SQD peak position in dependence on T, fitted by Varshni's law with the parameters of ZnSe.

complete temperature range between 4 K and 300 K is depicted in Fig. 6. The EL emission of this SQD at 300 K is detected at a bias of only 2.6 V, which is remarkably low considering the emission energy of 2.26 V. In analogy to the optically pumped SQD spectra, the electrically driven emission shows a redshift and a distinct broadening with temperature. The energetic position of the emission line follows Varshni's law $E(T) = E_0 - \frac{\alpha T^2}{T+\beta}$ with the values $\alpha = 5 \cdot 10^{-4}\,\mathrm{eV/K}$ and $\beta = 250\,\mathrm{K}$; these parameters are in good agreement with literature data for ZnSe [34].

The FWHM exhibits a continuous broadening with rising temperature and eventually reaches 26 meV at room temperature as can be seen in Fig. 7. As a reference, the data obtained for the optically pumped samples are depicted additionally in the figure. While we find slightly larger linewidths at low temperature for the electrically pumped devices, maybe caused by a stronger spectral diffusion, at higher temperatures the FWHM is very similar to the optically excited samples: Obviously, the SQD in the electrically driven device is subject to the same phonon interactions as discussed before.

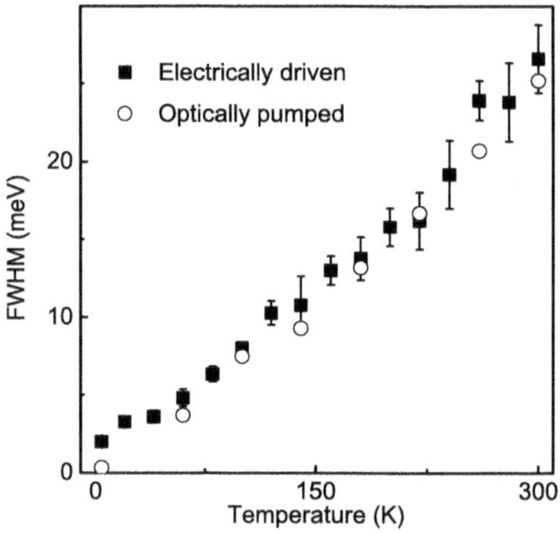

Fig. 7. FWHM of the single quantum dot electroluminescence in dependence on the temperature (filled symbols). Reference values for the optically pumped SQD emitter are depicted with open symbols.

This point is of practical importance when discussing the application of our single quantum dot device as a high-temperature single photon emitter. Because the cascaded biexciton–exciton emission is discussed as a key element for highly efficient single photon sources [35], biexcitonic and excitonic recombination must be spectrally separable; otherwise the reliability of a single photon emission process would be spoiled. One great advantage of the used II–VI material system based on selenides and sulfides is the high biexciton binding energy E_{XX} found therein for QDs. Values between 18 and 30 meV have been reported [20, 36, 37]. Therefore, even distinctly broadened emission lines at elevated temperatures due to phonon interaction are much less critical than in e.g. InAs QDs, where E_{XX} is only in the range between 2 and 4 meV [38, 39].

4 Summary

In conclusion, we presented single CdSe QDs as highly attractive candidates for an electrically driven single photon emitter operating at room temperature. Due to a combination of ZnSSe matrix and additional MgS barriers, a strong confinement could be achieved that is evidenced in both high luminescence intensity at room temperature and nearly constant radiative lifetimes of the excitons over the whole temperature range from 5 to 300 K. Although the barriers provide good confinement, an efficient electrical injection of carriers can be realized in p-i-n diode structures. Thus, single dot luminescence at room temperature is observable for both optically pumped and in particular electrically driven structures. The linewidth at room temperature amounts to about 25 meV and is in the range of the biexcitonic binding energy; therefore a spectral separation of excitonic and biexcitonic recombination, necessary for single photon devices, seems to be feasible.

5 Acknowledgements

The authors gratefully acknowledge the financial support of the Deutsche Forschungsgemeinschaft under Contract Nos. Ba1422/5 and Ho1388/28.

References

1. P. Michler, A. Kiraz, C. Becher, W. V. Schoenfeld, P. M. Petroff, L. Zhang, E.Hu, and A. Imamoğlu, Science **290**, 2282 (2000)
2. E. Waks, K. Inoue, C. Santori, D. Fattal, J. Vuckovic, G. S. Solomon, and Y. Yamamoto, Nature **420**, 762 (2002)
3. M. H. Baier, E. Pelucchi, E. Kapon, S. Varoutsis, M. Gallart, I. Robert-Philip, and I. Abram, Appl. Phys. Lett. **84**, 648 (2004)
4. M. B. Ward, O. Z. Karimov, D. C. Unitt, Z. L. Yuan, P. See, D. G. Gevaux, A. J. Shields, P. Atkinson, and D. A. Ritchie, Appl. Phys. Lett. **86**, 201111 (2005)
5. S. Kiravittaya, M. Benyoucef, R. Zapf-Gottwick, A. Rastelli, and O. G. Schmidt, Appl. Phys. Lett. **89**, 233102 (2006)
6. P. Ester, L. Lackmann, S. M. de Vasconcellos, M. C. Hübner, A. Zrenner, M. Bichler, Appl. Phys. Lett. **91**, 111110 (2007)
7. S. Strauf, N. G. Stoltz, M. T. Rakher, L. A. Coldren, P. M. Petroff, and D. Bouwmeester, Nature Phot. **1**, 704 (2007)
8. Z. Yuan, B. E. Kardynal, R. M. Stevenson, A. J. Shields, C. J. Lobo, K. Cooper, N. S. Beattie, D. A. Ritchie, and M. Pepper, Science **295**, 102 (2002)
9. R. Schmidt, U. Scholz, M. Vitzethum, R. Fix, C. Metzner, P. Kailuweit, D. Reuter, A. Wieck, M. C. Hübner, S. Stufler, A. Zrenner, S. Malzer, and G. H. Döhler, Appl. Phys. Lett. **88**, 121115 (2006)
10. A. Lochmann, E. Stock, O. Schulz, F. Hopfer, D. Bimberg, V. A. Haisler, A. I. Toropov, A. K. Bakarov, and A. K. Kalagin, Electron. Lett. **42**, 774 (2006)
11. K. Matsuda, K. Ikeda, T. Saiki, H. Tsuchiya, H. Saito, and K. Nishi, Phys. Rev. B **63**, 121304 (2001)
12. M. Bayer and A. Forchel, Phys. Rev. B **65**, 041308 (2002)
13. A. Polimeni, A. Patanè, M. Henini, L. Eaves, and P. C. Main, Phys. Rev. B **59**, 5064 (1999)
14. P. Michler, A. Imamoğlu, M. D. Mason, P. J. Carson, G. F. Strouse, and S. K. Buratto, Nature **406**, 968, (2000)

15. F. Gindele, K. Hild, W. Langbein, and U. Woggon, Phys. Rev. B **60**, R2157 (1999)
16. K. Sebald, P. Michler, T. Passow, D. Hommel, G. Bacher, and A. Forchel, Appl. Phys. Lett. **81**, 2920 (2002)
17. S. Kako, C. Santori, K. Hoshino, S. Götzinger, Y. Yamamoto, and Y. Arakawa, Nat. Mater. **5**, 887 (2006)
18. E. Wu, J. R. Rabeau, G. Roger, F. Treussart, H. Zeng, P. Grangier, S. Prawer and J.-F. Roch, New J. Phys. **9**, 434 (2007)
19. H. Huang, A. Dorn, V. Bulovica, and M. G. Bawendi, Appl. Phys. Lett. **90**, 023110 (2007)
20. C. Bradford, B. Urbaszek, M. Funato, A. Balocchi, T. C. M. Graham, E. J. McGhee, R. J. Warburton, K. A. Prior, B. Cavenett, J. Cryst. Growth **251**, 581 (2003)
21. A. Gust, C. Kruse, K. Sebald, H. Lohmeyer, J. Gutowski, and D. Hommel, Phys. Stat. Sol. C **3**, 767 (2006)
22. T. Passow, H. Heinke, T. Schmidt, J. Falta, A. Stockmann, H. Selke, P. L. Ryder, K. Leonardi, and D. Hommel, Phys. Rev. B **64**, 193311 (2001)
23. M. Funato, K. Omae, Y. Kawakami, Sg. Fujita, C. Bradford, A. Balocchi, K. A. Prior, and B. C. Cavenett, Phys. Rev. B **73**, 245308 (2006)
24. K. Shahzad, D. J. Olego, and C. G. Van de Walle, Phys. Rev. B **38**, 1417 (1988)
25. J. Feldmann, G. Peter, E. O. Göbel, P. Dawson, K. Moore, C. Foxon, and R. J. Elliott, Phys. Rev. Lett. **59**, 2337 (1987)
26. J. Seufert, R. Weigand, G. Bacher, T. Kümmell, A. Forchel, K. Leonardi, and D. Hommel, Appl. Phys. Lett. **76**, 1872 (2000)
27. J. C. Kim, H. Rho, L. M. Smith, H. E. Jackson, S. Lee, M. Dobrowolska, and J. K. Furdyna, Appl. Phys. Lett. **75**, 214 (1999)
28. T. Flissikowski, A. Hundt, M. Lowisch, M. Rabe, and H. Henneberger, Phys. Rev. Lett. **86**, 3172 (2001)
29. R. Arians, T. Kümmell, G. Bacher, A. Gust, C. Kruse, and D. Hommel, Physica E **40**, 1938 (2008)
30. L. Besombes, K. Kheng, L. Marsal, and H. Mariette, Phys. Rev. B **63**, 155307 (2001)
31. S. Moehl, F. Tinjod, K. Kheng, and H. Mariette, Phys. Rev. B **69**, 245318 (2004)
32. R. Arians, T. Kümmell, G. Bacher, A. Gust, C. Kruse, and D. Hommel, Appl. Phys. Lett. **90**, 101114 (2007)
33. A. Gust, C. Kruse, and D. Hommel, J. Cryst. Growth **301–302**, 789 (2007)
34. L. Malikova, W. Krystek, F. H. Pollak, N. Dai, A. Cavus, and M. C. Tamargo, Phys. Rev. B **54**, 1819 (1996)
35. O. Benson, C. Santori, M. Pelton, and Y. Yamamoto, Phys. Rev. Lett. **84**, 2513 (2000)
36. V. D. Kulakovskii, G. Bacher, R. Weigand, T. Kümmell, A. Forchel, E. Borovitskaya, K. Leonardi, and D. Hommel, Phys. Rev. Lett. **82**, 1780 (1999)
37. G. Bacher, R. Weigand, J. Seufert, V. D. Kulakovskii, N. A. Gippius, A. Forchel, K. Leonardi, and D. Hommel, Phys. Rev. Lett. **83**, 4417 (1999)
38. G. Chen, T. H. Stievater, E. T. Batteh, X. Li, D. G. Steel, D. Gammon, D. S. Katzer, D. Park, and L. J. Sham, Phys. Rev. Lett. **88**, 117901 (2002)
39. M. Bayer, T. Gutbrod, A. Forchel, V. D. Kulakovskii, A. Gorbunov, M. Michel, R. Steffen, and K. H. Wang, Phys. Rev. B **58**, 4740 (1998).

Silicon Nanoparticles: Excitonic Fine Structure and Oscillator Strength

Cedrik Meier[1], Stephan Lüttjohann[1], Matthias Offer[1], Hartmut Wiggers[2], and Axel Lorke[1]

[1] Department of Physics and CeNIDE, University of Duisburg-Essen, Lotharstr. 1, 47057 Duisburg, Germany
cedrik.meier@uni-due.de
[2] Combustion & Gas Dynamics and CeNIDE, University of Duisburg-Essen, Lotharstr. 1, 47057 Duisburg, Germany

Abstract. In this review, recent results on optical spectroscopy on silicon nanoparticles are summarized. We will demonstrate the quantum size effect observed in the photoluminescence for nanoparticles with diameters below 10 nm. Moreover, the excitonic fine structure splitting caused by the exchange interaction is investigated using time-resolved and magnetic-field-dependent photoluminescence measurements. From these results, it is possible to estimate the rate of non-radiative recombinations in these nanoparticles, which allows to determine the oscillator strength and the quantum yield independently.

1 Introduction

Silicon has in the past been the most important material for modern microelectronics. In the last two decades, however, interest in nanostructured silicon has increased significantly. This development is on the one hand triggered by the ongoing miniaturization of silicon based integrated circuits for electronic applications, where critical dimensions $CD < 100\,\mathrm{nm}$ can be reached in large scale production processes. On the other hand, the finding of photoluminescence in porous silicon [1] has sparked the hope for silicon as a material also for optoelectronic applications. Until today, photoluminescence from nanostructured silicon could be demonstrated not only from porous silicon, but also silicon nanocrystals formed in an SiO_2 matrix by implantation of Si ions and subsequent annealing [2–4] and isolated silicon nanoparticles [5–8]. Recently, in silicon nanocrystal based devices amplified stimulated emission [9] and field-injection based electroluminescence [10] could be demonstrated.

As an important prerequesite for optoelectronic device applications, one needs to have detailed knowledge of the recombination dynamics. In the case of silicon nanoparticles, this is of special importance, as bulk silicon is an indirect semiconductor and only shows vanishing near band-edge photoluminescence. In this paper, we will review results obtained on the size-dependence of the photoluminescence, the recombination dynamics, which can be explained using the excitonic fine structure, and on the oscillator strength of the radiative recombinations in these systems.

2 Experimental details

All the results presented here have been obtained using Si nanoparticles fabricated in a low-pressure microwave plasma using silane (SiH_4) as a precursor gas. Details of the fabrication process can be found in [11]. The photoluminescence measurements have been performed under excitation from a 532 nm laser in continuous operation in a confocal photoluminescence setup using a Czerny-Turner monochromator and a liquid-nitrogen-cooled charge-coupled device. For the time-resolved photoluminescence measurements, an acousto-optical modulator in combination with an avalanche photodiode was used. The time resolution of this setup is in the range of 4 ns.

3 Results and discussion

In Fig. 1, the room temperature photoluminescence of a bulk silicon sample and silicon nanoparticles with $d \approx 4.5$ nm are shown. From these spectra, one can make the following observations: First, the photoluminescence emission from the bulk silicon sample is significantly sharper than the emission from the silicon nanoparticle sample. The full width of half maximum (FWHM) of the Si bulk sample is $\Delta E \approx 120$ meV, while the Si nanoparticles show a FWHM of about $\Delta E \approx 400$ meV. This spectral broadening is caused by the size distribution of the silicon nanoparticles, which is intrinsic to the fabrication procedure [11]. The second noteworthy fact is the spectral shift in the Si nanoparticle emission towards higher energies compared to the Si bulk sample. This is a consequence of the quantum size effect typically observed in nanocrystalline Si samples. The final observation is related to the photoluminescence intensity. Taking into account the spectral response of the different detectors and the influence of the other optical components (instrument function) and the acquisition time, one finds that the intensity of the photoluminescence emission of the Si nanoparticle sample is about 1500× higher than that of the Si bulk sample. The increase in photoluminescence intensity with respect to the bulk is due to the stronger localization of electrons and holes inside the nanoparticles rather than due to a transition from an indirect semiconductor to a direct semiconductor as was predicted theoretically for small particle diameters [12]. For particles with diameters larger than $d = 2.0$ nm, the Si nanoparticles remain indirect semiconductors, as could be verified by time-resolved photoluminescence and absorption measurements [7].

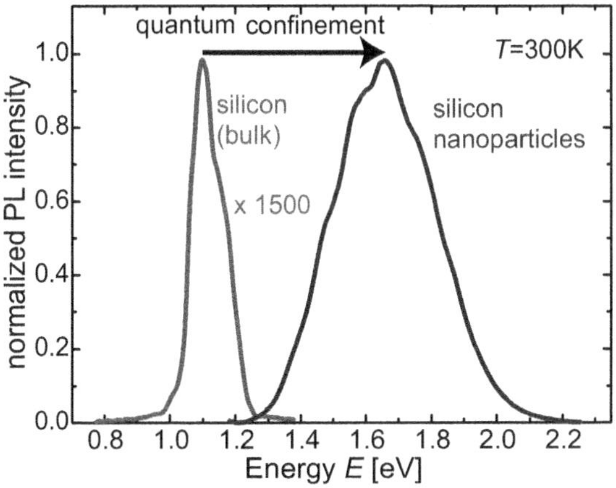

Fig. 1. Room temperature photoluminescence of bulk silicon and silicon nanoparticles.

As mentioned above, the reason for the blue-shift of the PL emission energy is the quantum size effect observed in the Si nanoparticles. The true origin of this is still under debate, see [13] and references therein. However, the effect of the particle size on the PL emission is quite significant. In Fig. 2 the PL from nanoparticle samples with different mean particle diameters is shown. The mean particle diameter is varied between $d = 4.1\,\mathrm{nm}$ and $d = 5.2\,\mathrm{nm}$. This comparably small change in particle size leads to significant shifts in the PL emission energy, from $E \approx 1.4\,\mathrm{eV}$ for the largest particles to about $E = 1.67\,\mathrm{eV}$ for the particles with the smallest diameter. The overall peak shape stays mostly unaffected, except for the sample with the smallest particles, where a stronger high-energy wing is observed. This could be due to the enhanced oscillator strength of the smaller particles, as discussed later in this review. In Fig. 3, the photoluminescence is plotted for three different temperatures. As expected from Varshni's law, one observes a blueshift of the PL emission peak energy with decreasing temperatures. At the same time, however, the intensity exhibits a non-monotonic behaviour: Cooling the sample from 300 to 80 K, the PL intensity increases first, reaches a maximum at $T \approx 80\,\mathrm{K}$ and then decreases again when the temperature is decreased further. This is different from what is observed in most other semiconductor quantum dots or nanocrystals made from direct semiconductors: in such systems, the exciton population increases as the thermal energy of the system is reduced, leading to an increase in the PL intensity with decreasing temperature [14–16]. A systematic study of the temperature dependence of the Si nanoparticle photoluminescence intensity is shown in Fig. 4. It can clearly be seen that the PL intensity maximum is around $T \approx 80\,\mathrm{K}$. At the same time, the PL does not quench for $T \rightarrow 0$. Similar results have been obtained for Si nanocrystals embedded in a SiO_2 matrix [3].

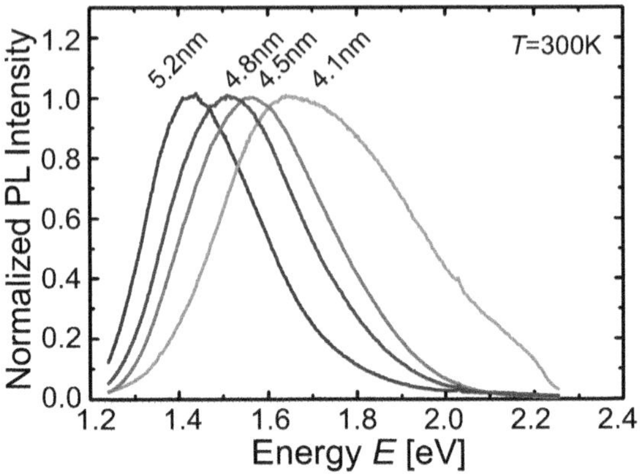

Fig. 2. Room temperature photoluminescence of silicon nanoparticles with different mean particle diameters.

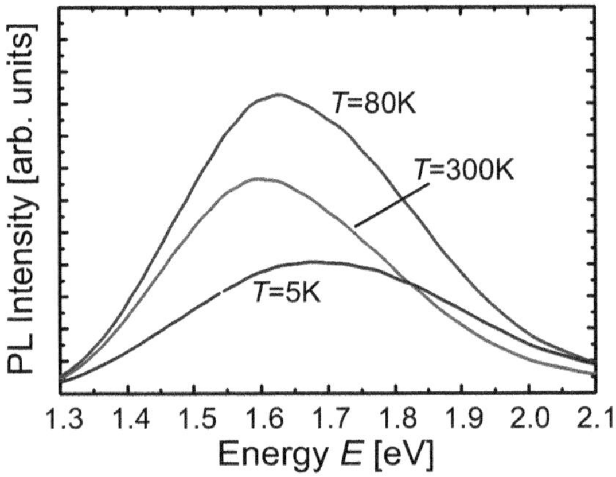

Fig. 3. Photoluminescence of silicon nanoparticles at room temperature, $T = 80\,\mathrm{K}$ and $T = 5\,\mathrm{K}$.

The reason for this behaviour is the excitonic fine structure in the silicon nanoparticles. The left part of Fig. 5 shows the schematic band structure $E(k)$ for the silicon nanoparticles. It should be noted that in this study, only particles with diameters larger than $d = 2.5\,\mathrm{nm}$ are studied, where one can still apply the Bloch equation based band structure model. For smaller clusters, the situation changes and a HOMO/LUMO approach has to be applied [12]. The quantum confinement in the nanoparticles leads to a lifting of the degeneracy of the heavy-hole and the light-hole band at the Γ-point, as the energy shift is inversely proportional to the effective mass. Therefore, for the optical transitions in the silicon nanoparticles one needs to take into account only the conduction band and the heavy-hole band. The conduction band is

s-like, corresponding to an angular momentum component of $J_z^{\mathrm{el}} = \pm 1/2$. The heavy-hole band is p-like and is characterized by an angular momentum component of $J_z^{\mathrm{hh}} = \pm 3/2$. For the formation of excitons, four possibilities exist to align the respective angular momentum components: Two parallel configurations and two antiparallel configuration, leading to a twofold degenerate exciton state with $J_{\mathrm{exc}} = 1$ and another twofold degenerate exciton state with $J_{\mathrm{exc}} = 2$. Because the excitons with $J_{\mathrm{exc}} = 2$ cannot transfer their angular momentum to a single photon with $J = 1$, these exciton states are called "dark excitons" and the other states "bright excitons". Due to the exchange interaction, the degeneracy between the bright and the dark states is lifted, so that the dark states are lower in energy with respect to the bright states by a splitting energy Δ as indicated in the right part of Fig. 5.

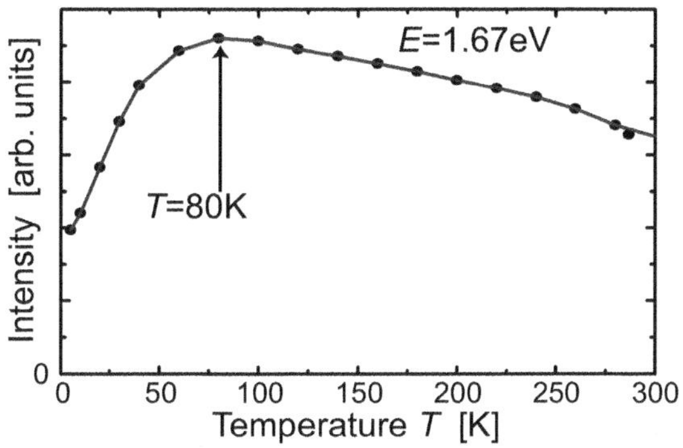

Fig. 4. Photoluminescence intensity of silicon nanoparticles as a function of the temperature.

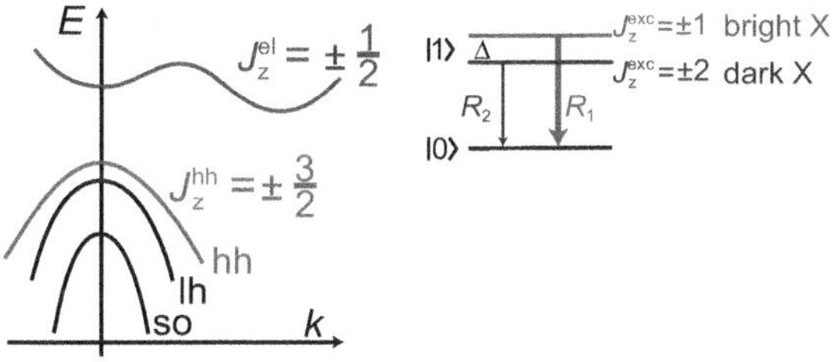

Fig. 5. (*Left*) Schematic band structure for silicon nanoparticles. The quantum confinement leads to a lifting of the degeneracy of the heavy and light hole bands at Γ. (*Right*) Excitonic fine structure.

To study the excitonic fine structure of the silicon nanoparticles in greater detail, one needs to look at the dynamics of the recombination. Therefore, we studied the photoluminescence decay as a function of emission and temperature. As the emission energy scales directly with the particle size, this is equivalent to studying the size dependence of the recombination dynamics. The temperature can be used to change the population of the fine structure of the split states, as will be discussed later. The results of these measurements

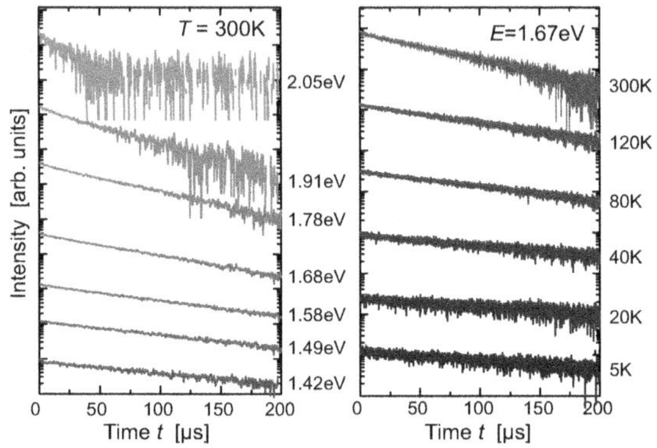

Fig. 6. (*Left*) Photoluminescence decay measured at different emission energies at room temperatures. (*Right*) PL decay at $E = 1.67\,\text{eV}$ for different temperatures.

are shown in Fig. 6. In the left part, the PL decay at $T = 300\,\text{K}$ is shown for different emission energies, corresponding to different particle sizes. As one can see, the PL decay shows a monoexponential behaviour with decay times τ ranging from 40 to 200 μs. This monoexponential decay is indicative of a high degree of crystallinity in the samples used, which suggests that the plasma-based fabrication leads to nanoparticles with better optical properties than those generated, e.g., by photoelectrochemical etching routes, where a stretched exponential decay with a disorder parameter β is reported [17]. The reason for the decrease of PL decay time with decreasing nanoparticle diameter is the increase in phonon-assisted recombinations with decreasing particle size due to the increased electron-phonon interaction in small particles [18]. At the same time, the contributions from the phonon-less transitions also increase due to the enhanced electron-hole wave function overlap [19].

While the PL intensity has a maximum around $T = 80\,\text{K}$, the PL decay times τ_{PL} exhibit a monotonic behaviour: As shown in the right part of Fig. 6 and for more temperatures in the upper part of Fig. 7, τ_{PL} increases with decreasing temperatures.

To isolate the temperature dependence of the radiative recombinations from the total PL decay, one can combine the information from the temperature dependence of the photoluminescence intensity $I(T)$ with the temperature

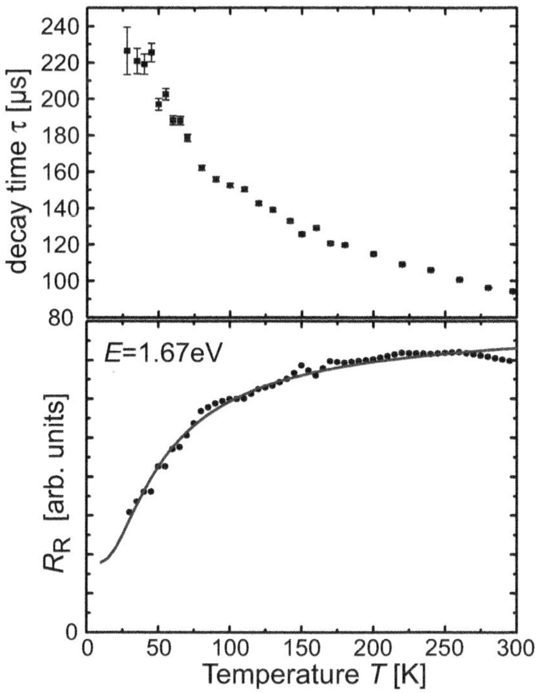

Fig. 7. (*Upper part*) Measured PL decay times. (*Lower part*) Temperature dependence of the radiative recombinations.

dependence of the PL decay rate $R_{\mathrm{PL}} = 1/\tau_{\mathrm{PL}}$. The PL decay rate consists of the radiative and the non-radiative recombination rate:

$$R_{\mathrm{PL}} = R_{\mathrm{R}} + R_{\mathrm{NR}}.$$

The intensity is proportional to the quantum efficiency η:

$$I \propto \eta = \frac{R_{\mathrm{R}}}{R_{\mathrm{R}} + R_{\mathrm{PL}}}$$

Therefore, the product of intensity and PL decay rate is proportional to the radiative recombination rate alone: $\rightarrow I(T) \cdot R_{\mathrm{PL}} \propto R_{\mathrm{R}}$. The result of this analysis is shown in the lower part of Fig. 7. The circles represent data points, while the solid line stems from the model described below. One can see, that the radiative recombination rate decreases as the temperature is reduced, contrary to what is observed to direct semiconductor quantum dot/nanocrystal systems such as InAs quantum dots or CdSe/CdS nanocrystals.

The fact that the radiative recombination rate *decreases* towards lower temperatures suggests that in this regime the dark states play a dominant role. The fact, that the dark state is lower in energy than the bright state (see

Fig. 5) is a hint that the excitons can thermalize into the lower (dark) state before recombining. In other, direct gap semiconductor quantum systems such a thermalization between the bright and the dark state is not possible as the exciton lifetimes are much shorter, typically in the range of few nanoseconds. In the case of silicon nanoparticles however, the lifetimes are much longer, suggesting that the excitons can scatter and a thermal equilibrium between the dark and the bright state is established. Therefore, we can use a thermal distribution of the states to calculate the total radiative recombination rate R_R as a function of the temperature using only the parameters R_1 and R_2 for the individual recombination rates from the bright and the dark state, respectively, and the exchange interaction energy Δ. The radiative recombination rate is then given by:

$$R_\mathrm{R} = \frac{2R_2 + 2R_1 \cdot \exp\left(-\frac{\Delta}{k_\mathrm{B}T}\right)}{2 + 2 \cdot \exp\left(-\frac{\Delta}{k_\mathrm{B}T}\right)}$$

However, due to the fact that the intensity is only proportional but not equal to the quantum efficiency, only the temperature dependence of the radiative recombination rate $R_\mathrm{R}(T)$ is known, but not the absolute values. Therefore, by fitting the above equation to the data points in the lower part in Fig. 7, we can only obtain the ratio of R_1/R_2 and not the individual values. Moreover, we can deduce the exchange interaction energy Δ from the above model. Fitting the data to the above equation yields an exchange energy of $\Delta = 5.8\,\mathrm{meV}$ and a ratio of bright state/dark state recombination rate of $R_1/R_2 = 8$. The first value is significantly higher than the value for bulk silicon, for which $\Delta = 140\,\mu\mathrm{eV}$ has been reported [20]. However, the splitting energy is mostly given by short-range exchange interaction, which is enhanced in confined systems [21]. The result for the ratio between the bright and dark state recombination rate is more surprising. As discussed before, one expects the dark exciton states to be optically inactive due to the total angular momentum conservation rule. In such a case, however, one should obtain $R_1/R_2 \to \infty$ as the recombination from the dark states should be $R_2 \approx 0$. Our experiments, however, strongly suggest that the dark state recombination plays an important role for the PL at low temperatures. Indeed, applying the above model, we can even estimate that for $T < 40\,\mathrm{K}$ the luminescence rate from the dark excitons is larger than the one from the bright states [8] due to the preferential occupation of the lower dark state over the energtically higher bright state. To check these findings, we analyzed the photoluminescence decay as a function of the emission energy for different temperatures. The results are plotted in the upper part of Fig. 8. While for temperatures $T > 40\,\mathrm{K}$ we find the expected decrease in PL decay time τ_PL with increasing emission energy (decreasing particle size), the results for $T < 40\,\mathrm{K}$ show an entirely different behaviour: In this regime, the measured PL decay times are nearly independent of particle size and temperature and have a nearly constant value of $\tau_\mathrm{PL} \approx 200\,\mu\mathrm{s}$. The reason for this is a non-radiative decay mechanism which limits all measurable decay times.

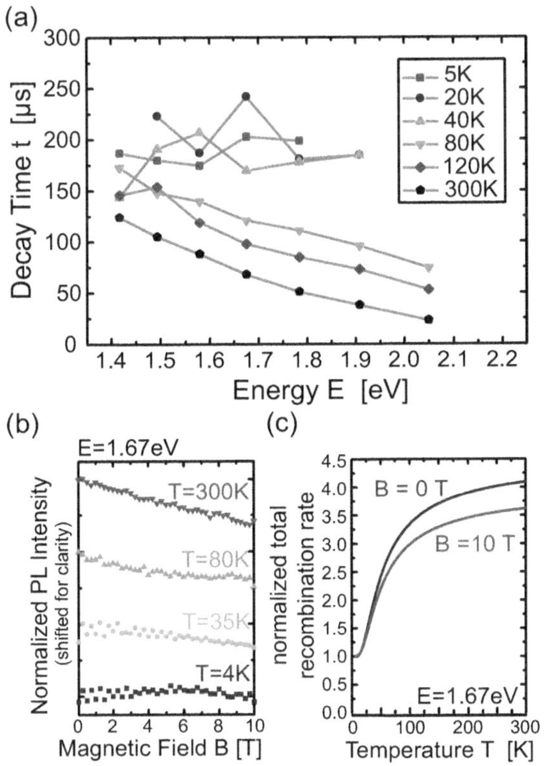

Fig. 8. (**a**) PL decay times for different temperatures. (**b**) PL intensity as a function of the magnetic field for different temperatures. (**c**) Calculated recombination from the dark and the bright state based on a thermal distribution between the states.

In principle, we would expect also for the dark excitons an increase in the recombination rate with decreasing particle size.

Another test for the above discussed model is to investigate the PL intensity at different temperatures as a function of the magnetic field. The magnetic field causes a mixing of the dark and the bright state. This should lead to a significant effect at higher temperatures, where both states are occupied, and a change in the recombination rate of the bright exciton should have a greater impact. At low temperatures, where the excitons are mostly in the dark state, one only expects a small effect on the intensity. The results of the corresponding measurements are shown in the lower part of Fig. 8. Indeed, one finds that for larger temperature the PL intensity decreases, when the magnetic field is increased. In the same manner, the intensity is nearly independent of the magnetic field at low temperature.

The above measurements gives an excellent estimate for the non-radiative recombination time τ_{NR} in this system, a quantity, which is always present in semiconductor quantum systems, but usually difficult to access experimentally. Using these results, we can now immediately calculate the *radiative* recombination times from the measured PL decay times τ_{PL}: $\tau_{\mathrm{R}}^{-1} = \tau_{\mathrm{PL}}^{-1} - \tau_{\mathrm{NR}}^{-1}$. From this, we can then determine the oscillator strength $f_{\mathrm{osc}}(\omega)$ that describes the strength of an optical transition:

$$f_{\mathrm{osc}}(\omega) = \frac{2\pi\varepsilon_0 mc^3}{e^2 n\omega^2}\frac{1}{\tau_{\mathrm{R}}}$$

In the above equation, n is the refractive index and $m = m_e + m_h$ is the exciton mass in the weak confinement regime, given by the sum of the individual masses $m_e = 0.19m_0$ and $m_h = 0.286m_0$ [22, 23].

The results of the above calculations are shown in Fig. 9 together with the radiative recombination times derived via the above route. The oscillator strengths found for the present silicon nanoparticles are in the range of $f \approx 10^{-5}$ for emission energies between 1.4 and 2.1 eV. The oscillator strength increases with increasing emission energy/decreasing particle size. This is due to the increased localization of the electron and hole wavefunctions in the nanocrystals, leading to larger dipole matrix elements and thus larger oscillator strength. These results are in excellent agreement with recent theoretical results obtained using the tight-binding method [12].

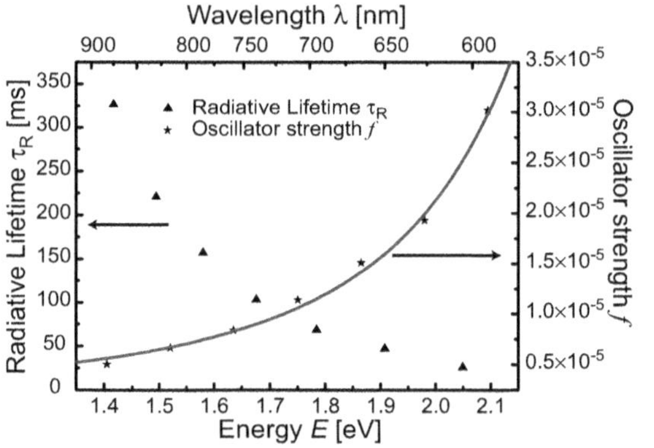

Fig. 9. PL decay and oscillator strength over the PL emission energy.

Using the values for both the radiative and non-radiative recombination rates, it is also possible to estimate the quantum efficiency for the optical transitions in these particles. The quantum efficiency is given by:

$$\frac{\tau_{\mathrm{R}}^{-1}}{\tau_{\mathrm{R}}^{-1} + \tau_{\mathrm{NR}}^{-1}}$$

Using this technique, one obtains quantum efficiencies between $\eta = 34\%$ and $\eta = 86\%$, where the quantum efficiency increases with decreasing particle sizes. It is interesting to point out that this method does not rely on either an Ulbricht sphere or the use of a calibrated standard sample, but just on the time-resolved PL measurements. Similar values for silicon nanoparticles have also been reported by other groups [6, 24].

4 Conclusion

In the present paper, we have reviewed some recent results on the photoluminescence properties of silicon nanoparticles. We find that the excitonic fine structure can be used to describe the experimentally found temperature dependence of the stationary and dynamic photoluminescence results. While at room temperature, the photoluminescence is clearly governed by recombinations from the energetically higher bright states, at about $T = 40\,\mathrm{K}$ the dark states start to dominate and govern the low temperature photoluminescence properties. The reason for this is the ability of the exciton system to thermalize due to the large exciton lifetimes in this system. By analyzing the PL decay results, we can also demonstrate that the behaviour of bright and dark states is clearly different. From these results we can deduce the non-radiative lifetime, which allows us finally to experimentally deduce the osciallator strength. The obtained values are in excellent agreement with theory.

5 Acknowledgments

This work was supported by the Deutsche Forschungsgemeinschaft (DFG) via grant "SFB 445-Nanoparticles from the gas phase" and "GRK 1240-Nanotronics".

References

1. L. Canham, Appl. Phys. Lett. **57**, 1046 (1990).
2. J. Linnros, N. Lalic, A. Galeckas, and V. Grivickas, J. Appl. Phys. **86**, 6128.
3. M. L. Brongersma, P. G. Kik, A. Polman, K. S. Min, and H. A. Atwater, Appl. Phys. Lett. **76** 351 (2000).
4. J. Heitmann, F. Müller, M. Zacharias, and U. Gösele, Adv. Mater. **17**, 795 (2005).
5. G. Ledoux, O. Guillois, D. Porterat, C. Reynaud, F. Huisken, B. Kohn, and A. Paillard, Phys. Rev. B **62**, 15942 (2000).
6. L. Mangolini, E. Thimsen, and U. Kortshagen, Nano Lett. **5**, 655 (2005).
7. C. Meier, A. Gondorf, S. Lüttjohann, and A. Lorke, J. Appl. Phys. **101**, 103112 (2007).
8. S. Lüttjohann, C. Meier, M. Offer, and H. Wiggers, Eurphys. Lett. **79**, 37002, (2007).

9. L. Pavesi, L. DalNegro, C. Mazzoleni, G. Franzo, and F. Priolo, Nature **408**, 440 (2000).
10. R. J. Walters, G. I. Bourianoff, and H. A. Atwater, Nat. Mater. **10**, 1038 (2005).
11. J. Knipping, H. Wiggers, B. Rellinghaus, P. Roth, D. Konjhodzic, and C. Meier, J. Nanosci. Nanotechnol. **4**, 1039 (2004).
12. F. Trani, G. Cantele, D. Ninno, and G. Iadonisi, Phys. Rev. B **72**, 075423 (2005).
13. S. Godefroo, M. Hayne, M. Jivanescu, A. Stesmans, M. Zacharias, O. I. Lebedev, G. van Tendeloo, and V. V. Moshchalkov, Nat. Nantechnol. **10**, 1038 (2008).
14. J. C. Kim, H. Rho, L. M. Smith, H. E. Jackson, S. Lee, M. Dobrowolska, and J. K. Furdyna, Appl. Phys. Lett. **75**, 214 (1999).
15. Y. G. Kim, Y. S. Joh, J. H. Song, K. S. Baek, S. K. Chang, and E. D. Sim, Appl. Phys. Lett. **83**, 2656 (2003).
16. E. C. Le Ru, J. Fack, and R. Murray, Phys. Rev. B **67**, 245318 (2003).
17. L. Pavesi and M. Ceschini, Phys. Rev. B **48**, 17625 (1993).
18. D. Kovalev, H. Heckler, M. Ben-Chorin, G. Polisski, M. Schwartzkopf, and F. Koch, Phys. Rev. Lett. **81**, 2803 (1998).
19. M. S. Hybertsen, Phys. Rev. Lett. **72**, 1514 (1994).
20. J. C. Merle, M. Capizzi, P. Fiorini, and A. Frova, Phys. Rev. B **17**, 4821 (1978).
21. D. H. Feng, Z. Z. Xu, T. Q. Jia, X. X. Li, and S. Q. Gong, Phys. Rev. B **68**, 035334 (2003).
22. J. B. Xia, Phys. Rev. B **40**, 8500 (1989).
23. A. D. Yoffe, Adv. Phys. **42**, 173 (1993).
24. D. Jurbergs, E. Rogojina, L. Mangolini, and U. Kortshagen, Appl. Phys. Lett. **88**, 233116 (2006).

Intrinsic Non-Exponential Decay of Time-Resolved Photoluminescence from Semiconductor Quantum Dots

Jan Wiersig, Christopher Gies, Norman Baer, and Frank Jahnke

Institute for Theoretical Physics, University of Bremen, Otto-Hahn Allee, 28334 Bremen, Germany
jwiersig@itp.uni-bremen.de

Abstract. A general introduction is presented to the recently observed intrinsic non-exponential and excitation intensity-dependent decay of time-resolved photoluminescence from semiconductor quantum dots. The commonly used two-level approximation fails in this situation since it relies on fully correlated carriers. In a semiconductor, however, the correlations are subject to scattering and dephasing processes. Hence, carriers are in general not fully correlated. It is shown that this effect leads to a non-exponential and excitation intensity-dependent decay of photoluminescence. The origin of the phenomenon is discussed in detail for a simplified situation. The full problem is studied numerically on the basis of a microscopic theory that includes Coulomb and carrier-photon correlation effects.

1 Introduction

The potential device applications of semiconductor quantum dots (QDs) are diverse, ranging from quantum information processing and non-classical light sources to lasers [1, 2]. With dimensions of only a few nanometers, semiconductor QDs are often described as "artificial atoms". However, the validity of this picture is limited. The Coulomb interaction of the carriers and the coupling to the environment, e.g., to phonons, leads to very specific properties and interaction mechanisms. These properties can be investigated, for example, in time-resolved photoluminescence (PL) measurements by optical excitation with short laser pulses [3]. Nowadays, this type of experiments is extensively used to study the modification of the spontaneous emission from QDs in optical microcavities due to the Purcell effect [2, 4, 5]. In order to extract a spontaneous emission lifetime from time-resolved measurements, often the exponential decay known from the two-level system is used.

Recently, a non-exponential PL decay from self-assembled (In,Ga)As/GaAs QDs embedded in GaAs-based micropillars has been observed experimentally [6]. The non-exponential decay is accompanied by a strong dependence on the excitation intensity. These two effects make it impossible to assign a single spontaneous emission lifetime to the QDs. A microscopic theory beyond the simple two-level system approximation has been used to demonstrate that both the non-exponential decay and its dependence on the excitation intensity are intrinsic semiconductor effects [6, 7].

In this paper, we give a general introduction to this phenomenon. We will show that the origin for the non-exponential and excitation intensity-dependent decay is the lack of full correlations between carriers in the QDs. The paper is organized as follows: In Sect. 2 we discuss the basics of time-resolved PL from semiconductor QDs. Section 3 introduces the concept of correlations. In Sect. 4 the cluster expansion truncation method is explained. Numerical results for the non-exponential PL decay are discussed in Sect. 5. The excitation-intensity dependence is the subject of Sect. 6.

2 Time-resolved photoluminescence

We consider self-assembled QDs, where the discrete states, corresponding to three-dimensional carrier confinement, are located energetically below a quasi-continuum of delocalized states, which corresponds to the two-dimensional motion of carriers in a wetting layer (WL). A sketch of the energy levels of conduction-band electrons and valence-band holes is shown in Fig. 1. To investigate PL properties, the system may be off-resonantly excited by an optical pulse, which creates carriers in the WL, from where they relax quickly into the localized QD states [8, 9]. At low temperatures and at low to moderate carrier densities, the carriers populate solely the QD states. Then the WL states are mainly important for carrier-scattering processes if the excitation involves the quasi-continuum. For the recombination dynamics due to carrier-photon interaction, the unpopulated WL states are of negligible importance and can therefore be ignored in the following discussion. In order to simplify the discussion, we restrict ourselves in this section to one single-particle QD state both for electrons and holes, and we consider only one spin polarization. Because of the fermionic character of the carriers, we have in total four many-particle states $|n_e, n_h\rangle$ with $n_e = 0, 1$ and $n_h = 0, 1$ as is illustrated in Fig. 2: $|0, 0\rangle$ is the vacuum state, $|1, 0\rangle$ is a single-electron state, $|0, 1\rangle$ is a single-hole state, and $|1, 1\rangle$ is an electron-hole-pair state. A general pure state can be written as superposition of the four basis states

$$|\psi\rangle = a_{00}|0,0\rangle + a_{10}|1,0\rangle + a_{01}|0,1\rangle + a_{11}|1,1\rangle \tag{1}$$

with the normalization condition

$$\sum_{i,j=0}^{1} |a_{ij}|^2 = 1 . \tag{2}$$

Note that a description with a density matrix would be more general, however for didactical reasons we use the wave function (1).

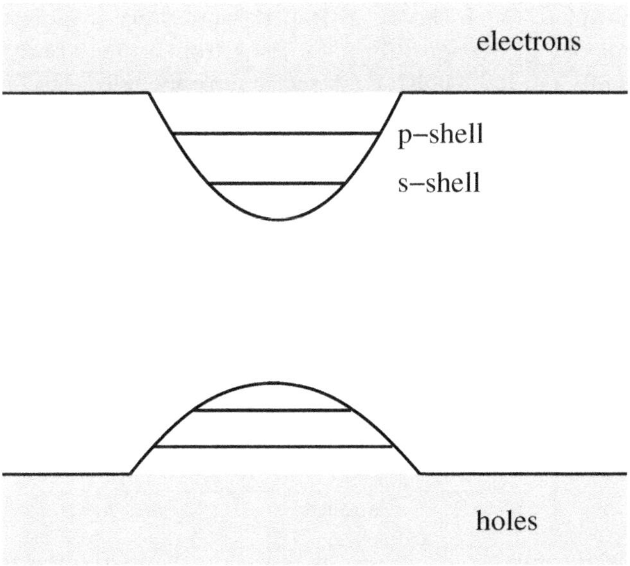

Fig. 1. Schematic representation of energy levels in a quantum dot with two shells for electrons and holes. The quasicontinuum of the wetting layer is shown as shaded areas.

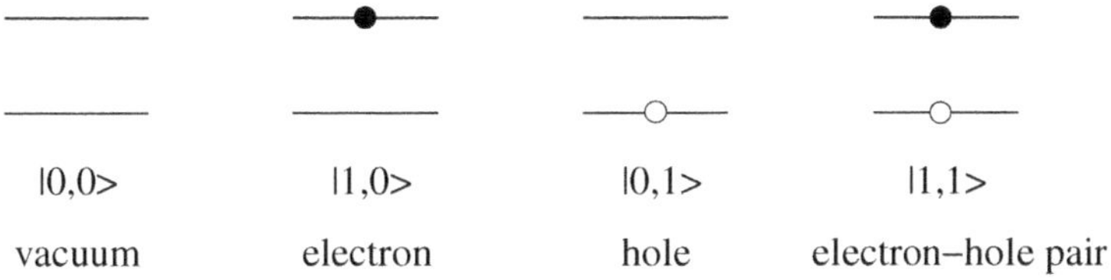

Fig. 2. Schematic level diagram of a quantum dot with one shell for electrons and holes only.

In the formalism of second quantization the states $|n_e, n_h\rangle$ are related by creation and annihilation operators. The fermionic operator e ($e^\dagger$) annihilates (creates) an electron. The corresponding operators for holes are h and $h^\dagger$. For instance, applying the creation operator $e^\dagger$ to the vacuum state $|0,0\rangle$ yields the one-electron state $|1,0\rangle$. In the same way, the creation operator $h^\dagger$ turns the vacuum into the one-hole state $|0,1\rangle$. The formalism of second quantization allows to describe many-particle effects in an elegant way. The experimentally relevant quantities can be expressed as expectation values of products of creation and annihilation operators. For example, the luminescence is determined by the dynamics of the photon number $\langle b^\dagger b\rangle$, where the bosonic operator b ($b^\dagger$) annihilates (creates) a photon in a given optical mode. For simplicity, in this section we consider only one optical mode. Other relevant quantities are the occupations of electrons $f^e = \langle e^\dagger e\rangle$ and holes $f^h = \langle h^\dagger h\rangle$.

From the microscopic theory in Ref. [7] equations of motion are derived for such kinds of expectation values. Starting from a many-body Hamiltonian where Coulomb and light-matter interaction are included, the so-called semiconductor luminescence equations (SLE) are obtained. To keep the discussion as simple as possible, we consider the special case of a single shell both for electrons and holes. As a general result, we find

$$\frac{d}{dt}\langle b^\dagger b\rangle = -\frac{d}{dt}f^{(e,h)}\Big|_{\text{spont}} \tag{3}$$

which shows that whenever a photon is created, an electron and a hole is destroyed due to spontaneous emission and vice versa. Ignoring Coulomb interaction and eliminating the polarization adiabatically, the results of Ref. [7] give for the change of electron and hole occupation

$$\frac{d}{dt}f^{(e,h)}\Big|_{\text{spont}} = -\frac{\langle e^\dagger e\ h^\dagger h\rangle}{\tau}\ , \tag{4}$$

where $1/\tau$ is the rate of spontaneous emission. Even though formula (4) is based on a considerable simplification in comparison to the general theory [7], it contains the basic physics behind the intrinsic non-exponential PL decay. To see this, let us express $\langle e^\dagger e\ h^\dagger h\rangle$ in terms of the four basis states illustrated in Fig. 2 (the photonic part of the states is here of no relevance and, therefore, not shown). We first note that $e|0, n_h\rangle = 0$ as one cannot remove an electron from a state that does not contain one. Equally, $h|n_e, 0\rangle = 0$. From this follows immediately $e^\dagger e\ h^\dagger h|n_e, n_h\rangle \neq 0$ only if $n_e = n_h = 1$. Using this relation we find

$$\begin{aligned}\langle e^\dagger e\ h^\dagger h\rangle &= \langle\psi|e^\dagger e\ h^\dagger h|\psi\rangle = \langle\psi|a_{11}|1,1\rangle \\ &= |a_{11}|^2 = P_{11}\ .\end{aligned} \tag{5}$$

This is a very intuitive relation, simply meaning that the PL decay described in Eq. (4) is proportional to the probability of finding an electron-hole pair, P_{11}. Only in state $|1,1\rangle$ the electron and the hole can recombine via emission of a photon.

For the following discussion it is essential to realize that the probability of observing an electron-hole pair is in general different from the probability of observing an electron (or a hole). This can be seen by rewriting the occupation of electrons as

$$\begin{aligned}f^e &= \langle e^\dagger e\rangle = \langle\psi|e^\dagger e|\psi\rangle = \langle\psi|\big(a_{10}|1,0\rangle + a_{11}|1,1\rangle\big) = |a_{10}|^2 + |a_{11}|^2 \\ &= P_{10} + P_{11}\ .\end{aligned} \tag{6}$$

Along the same lines one obtains for the holes

$$f^h = P_{01} + P_{11}\ . \tag{7}$$

With this, f^e can be interpreted as the probability of finding an electron in the system. Comparing Eq. (6) with (5) reveals that the probability of finding an electron-hole pair is smaller than the probability of finding an electron. The same holds for the holes. The probabilities are the same only in the special case $P_{10} = P_{01} = 0$. In this particular case we can write Eq. (4) as

$$\frac{d}{dt} f^{(e,h)}\Big|_{\text{spont}} = -\frac{f^{(e,h)}}{\tau} \ . \tag{8}$$

Assuming that no other mechanism contributes to the change of the population, it follows an exponential decay with rate $1/\tau$ for the population. As a consequence of Eq. (3) the PL also shows this exponential decay. The conditions $P_{10} = P_{01} = 0$ reduce the four-state system for the electron and hole to a *two-level system* for the electron-hole pair with basis states $|0,0\rangle$ and $|1,1\rangle$. This situation corresponds to *fully correlated carriers*: the absence (presence) of an electron implies the absence (presence) of a hole and vice versa, cf., Fig. 2.

In the opposite limiting case of *uncorrelated carriers* the two-particle quantity $\langle e^\dagger e\, h^\dagger h\rangle$ can be factorized into one-particle quantities

$$\langle e^\dagger e\, h^\dagger h\rangle_{\text{HF}} = \langle e^\dagger e\rangle\langle h^\dagger h\rangle = f^e f^h \ . \tag{9}$$

This is the Hartree-Fock (HF) factorization. The product $f^e f^h$ is the uncorrelated electron-hole population, also called electron-hole plasma contribution. Equations (5–7) show that one can interpret Eq. (9) as the factorization of the probability of finding an electron-hole pair into the product of the individual probabilities of finding an electron and a hole. Note that we have ignored polarization-like averages of the form $\langle e^\dagger h\rangle$ in Eq. (9) which vanish in the incoherent regime.

Replacing $\langle e^\dagger e\, h^\dagger h\rangle$ in Eq. (4) by its HF-factorization (9) gives

$$\frac{d}{dt} f^{(e,h)}\Big|_{\text{spont}} = -\frac{f^e f^h}{\tau} \ . \tag{10}$$

From Eq. (10) it is obvious that the decay of the population f^e is non-exponential, unless f^h is held constant by some mechanism, like background doping. Furthermore, the rate of decay depends on the carrier density and is higher for larger population.

In reality the carriers in a semiconductor QD are neither fully correlated nor uncorrelated. To quantify the deviation from the discussed limiting cases, one can calculate the degree of correlation within a cluster expansion that is applied to the hierarchy of equations of motion.

3 Correlations

The electron-hole correlation is defined as

$$\delta\langle e^\dagger e\, h^\dagger h\rangle = \langle e^\dagger e\, h^\dagger h\rangle - \langle e^\dagger e\, h^\dagger h\rangle_{\mathrm{HF}} = \langle e^\dagger e\, h^\dagger h\rangle - \langle e^\dagger e\rangle\langle h^\dagger h\rangle \ . \tag{11}$$

The relation to classical correlation functions can be seen more clearly in the following representation

$$\delta\langle e^\dagger e\, h^\dagger h\rangle = \langle (e^\dagger e - \langle e^\dagger e\rangle)\ (h^\dagger h - \langle h^\dagger h\rangle)\rangle \ . \tag{12}$$

While a variance like $\langle (e^\dagger e - \langle e^\dagger e\rangle)^2\rangle$ quantifies the fluctuations around the expectation value of a single quantity, the correlation function (12) is a covariance that quantifies how correlated the fluctuations of two different quantities are. It can have positive and negative contributions depending on the relative sign of the two brackets in Eq. (12). A positive correlation function here means that on average the fluctuations of electron and hole number around their respective expectation values have the same sign. In the case of a negative correlation function, the fluctuations have mostly opposite sign. We can identify the sources of positive and negative correlations in terms of the basis states $|n_e, n_h\rangle$ using the normalization condition (2)

$$\delta\langle e^\dagger e\, h^\dagger h\rangle = \sum_{n_e,n_h=0}^{1} P_{n_e,n_h}(n_e - f^e)(n_h - f^h) = P_{00}P_{11} - P_{10}P_{01} \ . \tag{13}$$

The contributions from the states $|0,0\rangle$ and $|1,1\rangle$ enter with positive sign: if an electron is absent (present) the hole is absent (present). In contrast, the contributions from the states $|1,0\rangle$ and $|0,1\rangle$ enter the correlation function with negative sign: if an electron is present (absent) the hole is absent (present). In other words, positive $\delta\langle e^\dagger e\, h^\dagger h\rangle$ implies that if we detect an electron then it is likely to find also a hole. Negative $\delta\langle e^\dagger e\, h^\dagger h\rangle$ implies that if we detect an electron then it is unlikely to find a hole. Vanishing correlation $\delta\langle e^\dagger e\, h^\dagger h\rangle$ means that the two events of detecting an electron and detecting a hole are uncorrelated. In this particular case, the HF factorization (9) is exact.

4 Cluster expansion method

The microscopic theory in Ref. [7] directly formulates equations of motion for correlations. Here a difficulty arises: the equation of motion of an average of N operators couples to $N+2$ operator averages due to the Coulomb and light-matter interactions. This hierarchy of equations must be truncated in an unambiguous way. One useful approach is the cluster expansion method [10]. As in Eq. (11), one schematically decomposes a four-operator average into

$$\langle a^\dagger a^\dagger a a\rangle = \langle a^\dagger a\rangle\langle a^\dagger a\rangle + \delta\langle a^\dagger a^\dagger a a\rangle \ , \tag{14}$$

denoting the fermionic operators e, h by a. The first term on the right hand side is called singlet contribution as it contains only single-particle quantities.

For a four-operator average, the singlet factorization corresponds to the HF approximation (9). The second term is the correlation, which is a doublet contribution. It describes genuine two-particle effects.

The equation of motion for the four-operator correlation couples to averages of six operators, which are schematically factorized according to

$$\langle a^\dagger a^\dagger a^\dagger aaa\rangle = \langle a^\dagger a\rangle\langle a^\dagger a\rangle\langle a^\dagger a\rangle + \langle a^\dagger a\rangle\,\delta\langle a^\dagger a^\dagger aa\rangle + \delta\langle a^\dagger a^\dagger a^\dagger aaa\rangle \qquad (15)$$

into singlet, singlet-doublet and triplet contributions. Note that all possible combinations of averages and correlations must be taken, meaning that each of the three terms on the right hand side may represent several terms of the same order. The triplet contribution, the last term in Eq. (15), contains only genuine three-particle effects. In the so-called singlet-doublet truncation scheme, used in Ref. [7], the triplet contribution is neglected. Hence, in this scheme all two-particle effects are consistently included.

5 Numerical results: non-exponential PL decay

We now present numerical results of the microscopic theory in Ref. [7] for flat QDs with rotational symmetry; for details of the model, see [8]. In contrast to the previous sections we consider QDs with two confined shells both for electrons and holes, see Fig. 1, which are denoted by s and p according to their in-plane symmetry. The s-shell is only spin-degenerate, while the p-shell has an additional angular-momentum two-fold degeneracy. To model the strong confinement in growth direction, an infinite potential well is used. Only the energetically lowest state due to the confinement in growth direction will be considered. We assume that PL takes place in the incoherent regime where the influence of a coherent polarization can be neglected. Examples of such a situation are incoherent carrier excitations or coherent excitation of higher states with rapid dephasing and carrier relaxation.

The material parameters are those of Ref. [11] for an InGaAs QD system. We consider a density of QDs on the WL of $3 \cdot 10^{10}\,\text{cm}^{-2}$, a gap energy of 1.52 eV, and a temperature of 30 K. We assume that the excitation involves only carriers with one spin polarization, e. g., due to excitation with circular polarized light. Since carrier generation and relaxation are much faster than the recombination, we assume a quasi-equilibrium distribution of carriers with given carrier density and temperature as the initial state for our calculation.

We define the time-dependent luminescence spectrum according to [12, 13] and consider the limit of high frequency resolution of a detector to obtain

$$I(\omega) = \frac{d}{dt} \sum_{\boldsymbol{q}} \langle b^\dagger_{\boldsymbol{q}} b_{\boldsymbol{q}}\rangle\Big|_{|\boldsymbol{q}|=\omega/c} . \qquad (16)$$

The label $\boldsymbol{q}$ contains the wave vector and the polarization vector of the optical modes in free space. The total photon number is calculated according to

$$I_{\rm tot} = \int {\rm d}\omega \; I(\omega) \; . \tag{17}$$

Figure 3 displays the evolution of the time-resolved PL for an initial carrier density of $1.5 \cdot 10^{10}\,\mathrm{cm}^{-2}$. The result of a calculation in the HF approximation, corresponding to uncorrelated carriers, is shown as dashed line. In this case, the decay shows a pronounced non-exponential behavior, which is in accordance with Eq. (10). For the solid line in Fig. 3 all correlations up to the singlet-doublet level have been included. In contrast to the HF-case we observe an exponential decay. From this finding together with Eq. (8) we conclude that carriers are strongly correlated on the singlet-doublet level.

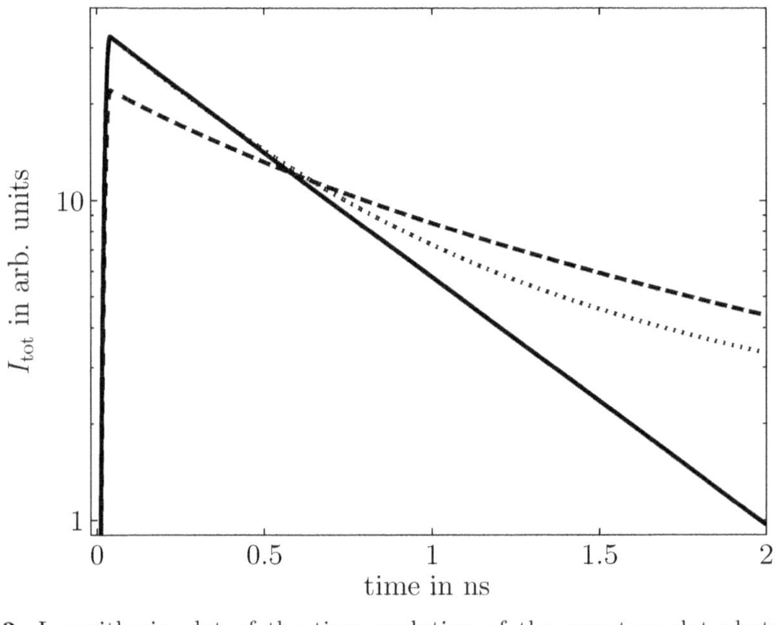

Fig. 3. Logarithmic plot of the time evolution of the quantum dot photoluminescence. The *dashed line* corresponds to the calculation in singlet (Hatree-Fock) factorization and the *solid line* to the singlet-doublet level. For the *dotted line* the described phonon contributions were added to the singlet-doublet calculation.

However, it turns out that the correlations are highly sensitive to even a weak dephasing of correlations. A physical mechanism providing such dephasing is phonon scattering. In the low-temperature regime, interaction of carriers with LA-phonons [14, 15] provide the dominant dephasing mechanism while at elevated temperatures the interaction of carriers with LO-phonons leads to very efficient dephasing [9]. Hoyer et al. have studied phonon scattering on a microscopic level for a quantum-well system. It is shown that dephasing of correlations is indeed provided, although this enters only via higher-order triplet terms [16, 17]. In the following, we account for the main features of phonon scattering, namely dephasing of correlations and relaxation of the carrier population towards thermal equilibrium at the lattice temperature, on a phenomenological level. Constant dephasing terms like $-i\gamma\langle e^\dagger e\, h^\dagger h\rangle$ are added to the equations of motion of the correlations, using a very weak constant dephasing rate of $\gamma = 0.001\,\text{meV}$. This constant dephasing may cause unphysical heating of the system [16, 17]. However, this effect is only weak for a small value of the dephasing and, additionally, the scattering counteracts the heating. We treat the scattering within relaxation-time approximation by introducing

$$\frac{d}{dt} f_\nu^{(e,h)}\Big|_{\text{relax}} = -\frac{f_\nu^{(e,h)} - F_\nu^{(e,h)}(T)}{\tau_{\text{relax}}^{e,h}}\,, \tag{18}$$

where $F_\nu^{(e,h)}(T)$ is a Fermi-Dirac distribution at temperature T. For the relaxation time $\tau_{\text{relax}}^{e,h}$ we take 1 ps for electrons and holes, see [9]. Figure 3 shows that the weak dephasing drastically reduces the correlations so that a non-exponential signature of the decay is regained on longer timescales. This result shows that the two-level approximation for electron-hole pairs is expected to fail for the description of photoluminescence from semiconductor QDs.

6 Numerical results: excitation-intensity dependence

In this section we discuss the dependence of the PL decay on the excitation density that determines the initial carrier density. Figure 4 shows the PL for various initial carrier densities for the three different levels of approximation: singlet level (top panel), singlet-doublet level without dephasing (middle) and with dephasing (bottom). For the singlet level we find a strong dependence on the initial carrier density. This is in agreement with Eq. (10). In the case of the singlet-doublet level without dephasing, Fig. 4 demonstrates that there is almost no dependence on the initial carrier density. This again shows that on this level the correlations are strong since for fully correlated carriers we expect to see no dependence on the initial carrier density, cf., Eq. (8). As we have already seen in the previous section, a very weak constant dephasing rate of $\gamma = 0.001\,\text{meV}$ strongly reduces the correlations. We therefore expect to obtain in this case a dependence on the initial carrier density. This is confirmed by the bottom panel of Fig. 4 which shows a pronounced dependence on the initial carrier density. This result again demonstrates that the non-exponential decay and its dependence on excitation intensity are directly related.

7 Conclusion

We have discussed the influence of carrier correlations on the time-resolved photoluminescence from semiconductor quantum dots. For fully correlated

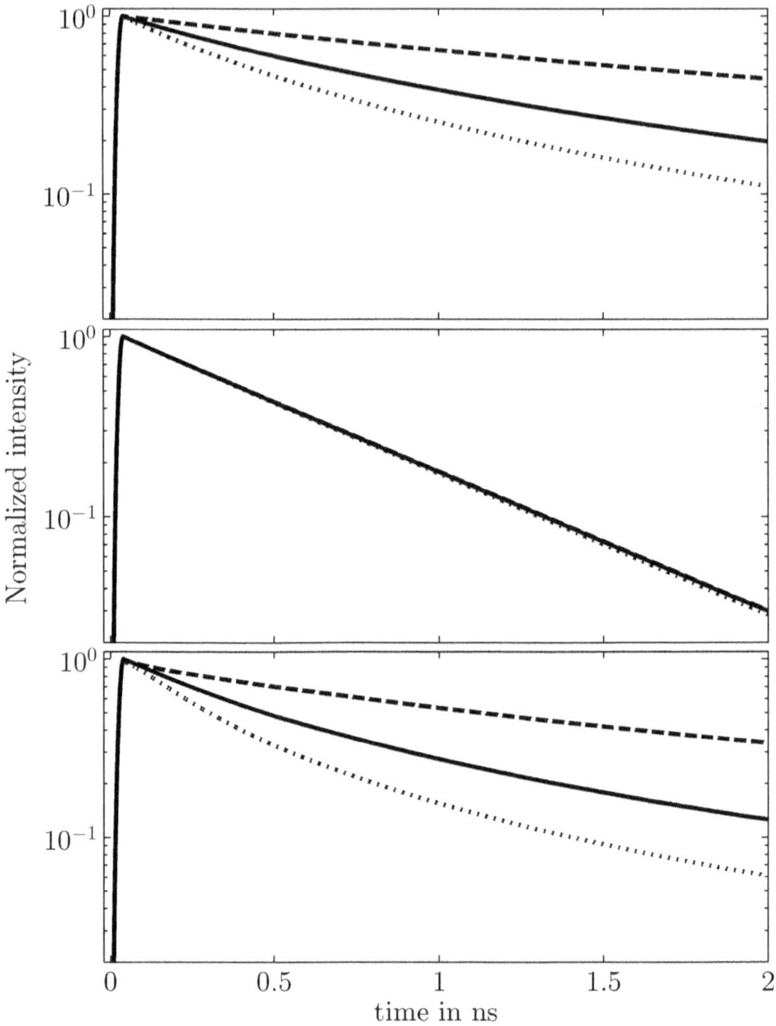

Fig. 4. Time evolution of the photoluminescence calculated on singlet level (*top panel*), on singlet-doublet level without dephasing (*middle*), and on singlet-doublet level with small dephasing (*bottom*) for various initial carrier densities: $0.6 \cdot 10^{10}\,\mathrm{cm}^{-2}$ (*dashed line*), $1.5 \cdot 10^{10}\,\mathrm{cm}^{-2}$ (*solid*), and $2.4 \cdot 10^{10}\,\mathrm{cm}^{-2}$ (*dotted*). For better comparison the *curves* are normalized.

carriers (two-level system approximation) an exponential decay is found. The case of uncorrelated carriers (Hartree-Fock approximation), however, exhibits significant deviations from an exponential decay and, moreover, a dependence on the excitation intensity. Numerical results show that correlations are strong in the quantum dot system when dephasing processes are ignored. Weak dephasing of carrier correlations, however, leads to a clear non-exponential and excitation-intensity-dependent decay of the photoluminescence signal.

8 Acknowledgments

We gratefully acknowledge financial support from the DFG research group "Quantum optics in semiconductor nanostructures" and a grant for CPU time at the Forschungszentrum Jülich (Germany).

References

1. P. Michler: *Single Quantum Dots: Fundamentals, Applications, and New Concepts*, Topics in Applied Physics (Springer, Berlin, 2003)
2. P. Lodahl, A. Floris van Driel, I. S. Nikolaev, A. Irman, K. Overgaag, D. Vanmaekelbergh, W. L. Vos: Controlling the dynamics of spontaneous emission from quantum dots by photonic crystals, Nature (London) **430**, 654 (2004)
3. D. Morris, N. Perret, S. Fafard: Carrier energy relaxation by means of Auger processes in InAs/GaAs self-assembled quantum dots, Appl. Phys. Lett. **75**, 3593 (1999)
4. I. L. Krestnikov, N. N. Ledentsov, A. Hoffmann, D. Bimberg, A. V. Sakharnov, W. V. Lundin, A. S. Tsatsul'nikov, A. F. Usikov, Z. I. Alferov, Y. G. Musikhin, D. Gerthsen: Quantum dot origin of luminescence in InGaN-GaN structures, Phys. Rev. B **66**, 155310 (2002)
5. R. A. Oliver, G. A. D. Briggs, M. J. Kappers, C. J. Humphreys, S. Yasin, J. H. Rice, J. D. Smith, R. A. Taylor: InGaN quantum dots grown by metalorganic vapor phase epitaxy employing a post-growth nitrogen anneal, Appl. Phys. Lett. **83**, 755 (2003)
6. M. Schwab, H. Kurtze, T. Auer, T. Berstermann, M. Bayer, J. Wiersig, N. Baer, C. Gies, F. Jahnke, J. Reithmaier, A. Forchel, M. Benyoucef, P. Michler: Radiative emission dynamics of quantum dots in a single cavity micropillar, Phys. Rev. B **74**, 045323 (2006)
7. N. Baer, C. Gies, J. Wiersig, F. Jahnke: Luminescence of a semiconductor quantum dot system, Eur. Phys. J. B **50**, 411–418 (2006)
8. T. R. Nielsen, P. Gartner, F. Jahnke: Many-body theory of carrier capture and relaxation in semiconductor quantum-dot lasers, Phys. Rev. B **69**, 235314 (2004)
9. J. Seebeck, T. R. Nielsen, P. Gartner, F. Jahnke: Polarons in semiconductor quantum-dots and their role in the quantum kinetics of carrier relaxation, Phys. Rev. B **71**, 125327 (2005)
10. J. Fricke: Transport equations including many-particle correlations for an arbitrary quantum system: A general formalism, Ann. Phys. **252**, 479–498 (1996)

11. N. Baer, P. Gartner, F. Jahnke: Coulomb effects in semiconductor quantum dots, Eur. Phys. J. B **42**, 231–237 (2004)
12. M. Kira, F. Jahnke, W. Hoyer, S. W. Koch: Quantum theory of spontaneous emission and coherent effects in semiconductor microstructures, Prog. Quant. Electr. **23**, 189 (1999)
13. J. H. Eberly, K. Wódkiewicz: The time-dependent physical spectrum of light, J. Opt. Soc. Am. **67**, 1252 (1977)
14. B. Krummheuer, V. M. Axt, T. Kuhn: Theory of pure dephasing and the resulting absorption lineshape in semiconductor quantum dots, Phys. Rev. B **65**, 195313 (2002)
15. E. A. Muljarov, R. Zimmermann: Dephasing in quantum dots: Quadratic coupling to acoustic phonons, Phys. Rev. Lett. **85**, 1516 (2000)
16. W. Hoyer, M. Kira, S. Koch: Influence of Coulomb and phonon interaction on the exciton formation dynamics in semiconductor heterostructures, Phys. Rev. B **67**, 155113 (2003)
17. W. Hoyer: *Quantentheorie zu Exzitonbildung und Photolumineszenz in Halbleitern*, Ph.D. thesis, University of Marburg, Germany (2002)

Electrical Spin Injection into Single InGaAs Quantum Dots

Michael Hetterich[1], Wolfgang Löffler[1], Pablo Aßhoff[1], Thorsten Passow[1], Dimitri Litvinov[2], Dagmar Gerthsen[2], and Heinz Kalt[1]

[1] Institut für Angewandte Physik and DFG Center for Functional Nanostructures (CFN), Universität Karlsruhe (TH), Wolfgang-Gaede-Straße 1, 76131 Karlsruhe, Germany
michael.hetterich@physik.uni-karlsruhe.de

[2] Laboratorium für Elektronenmikroskopie (LEM) and CFN, Universität Karlsruhe (TH), Engesserstraße 7, 76131 Karlsruhe, Germany

Abstract. In the context of a potential future quantum information processing we investigate the concurrent initialization of electronic spin states in InGaAs quantum dots (QDs) via electrical injection from ZnMn(S)Se spin aligners. Single dots can be read out optically through metallic apertures on top of our spin-injection light-emitting diodes (spin-LEDs). A reproducible spin polarization degree close to 100% is observed for a subset of the QD ensemble. However, the average polarization degree is lower and drops with increasing QD emission wavelength. Our measurements suggest that spin relaxation processes outside the QDs, related to the energetic position of the electron quasi-Fermi level, as well as defect-related spin scattering at the III–V/II–VI interface should be responsible for this effect, leading us to an improved device design. Finally, we present first time-resolved electroluminescence measurements of the polarization dynamics using ns-pulsed electrical excitation. The latter should not only enable us to gain a more detailed understanding of the spin and carrier relaxation processes in our devices. They are also the first step towards future time-resolved optical and electrical spin manipulation experiments.

1 Introduction

While in conventional electronics exclusively the manipulation of *charge* is utilized, the rapidly evolving field of semiconductor spintronics tries to take advantage of the electron's *spin* degree of freedom as well. In this context, the prospect of a possible future spin-based quantum information processing is particularly fascinating [1]. It is obvious that the electron spin can be used to code a classical bit by identifying the spin-up ($|\uparrow\rangle$) and spin-down ($|\downarrow\rangle$) state with a logical "1" and "0", respectively. Quantum mechanically, however, arbitrary coherent superpositions

$$|b\rangle = \alpha|\uparrow\rangle + \beta|\downarrow\rangle \quad (1)$$

can of course be formed, i.e., spin states can be used to code quantum bits (qubits), the basis of any quantum computer. The main advantage of this approach is the fact, that spin-qubits in semiconductors could be very robust. For instance, coherent electron spin transport over distances in the μm range has been observed [2, 3].

In order to realize a spin-based quantum information processing, suitable concepts for the initialization of spin states and their storage at well-defined sites are needed, as well as techniques to manipulate spins (qubit operations) and finally read out the result of the calculations performed. Some of the steps towards this aim have already been demonstrated: Quantum dots (QDs) – in particular the InGaAs/GaAs dots used in our experiments – could be identified as promising candidates for quantum information storage due to their long spin coherence times for electrons and even excitons [4–7]. (Excitons form for instance when holes are injected into the dots to optically read out the electron spin state, see Sect. 2). Furthermore, the creation of spin-polarized states (i.e., qubit initialization) in quantum dots or wells utilizing electrical spin injection as well as their optical readout could be successfully demonstrated by several groups, using either a diluted magnetic semiconductor or a ferromagnetic metal as spin aligner (see, e.g., [8–20], the references therein, and Sect. 2). A review is given in [21]. However, in most experiments carried out to date, large ensembles of spin-polarized electrons have been measured (although *optical* spin injection into single self-assembled CdSe/ZnSe quantum dots has recently been demonstrated [22]). On the other hand it is clear, that a future quantum information processing would not only require high initialization fidelities but also the ability to address single spin-qubits stored at individual localized sites.

In our own work summarized below we investigate *electrical* spin-injection into InGaAs quantum dots using a dilute magnetic ZnMn(S)Se spin aligner. Particularly, the concurrent *electrical* initialization of *several* spin-qubits in one device with polarization degrees close to 100% is demonstrated (Sect. 2, [9, 14, 15, 20]). Individual spin states in *single* quantum dots can be optically addressed and read out through metallic micro- or nano-apertures in a defined and reproducible way. The spin loss mechanisms in our devices and possible design optimizations are discussed in Sect. 3. Finally, we present first time-resolved electroluminescence (EL) measurements of the polarization dynamics using ns-pulsed electrical excitation in Sect. 4. The latter should not only enable us to gain a more detailed understanding of the spin and carrier relaxation processes in our devices. They are also the first step towards future time-resolved optical and electrical spin manipulation experiments.

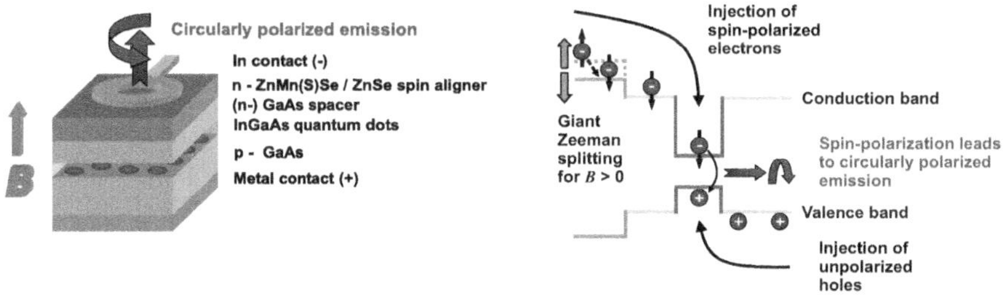

Fig. 1. *Left*: Basic design of the InGaAs quantum-dot spin-LEDs used in our investigations. *Right*: Corresponding flat band diagram.

2 Initialization and readout of spins in single dots

In this section we discuss how spin-qubits in a quantum dot ensemble can be initialized concurrently by electrical injection of spin-polarized electrons from a ZnMn(S)Se spin aligner and how the spin states in single quantum dots can then be addressed and read out optically.

The basic design of the spin-injection light-emitting diodes (spin-LEDs) used for our experiments is similar to [10] and shown in Fig. 1, together with the corresponding flat band diagram (see also Fig. 6 and [9] for further details). Electrons supplied through an In contact and an n-type ZnSe:Cl layer are injected into a dilute magnetic n-$Zn_{0.95}Mn_{0.05}S_ySe_{1-y}$:Cl spin aligner. In an externally applied magnetic field B and at low temperatures, the conduction and valence bands of the latter show a giant Zeeman splitting, resulting from the strong (s, p)–d exchange interaction of the charge carriers with the aligned spins in the half-filled d-shell of the Mn atoms [23, 24]. As a result, the electrons fed into the spin aligner quickly thermalize into the lower $-1/2$ spin state, thus leading to a spin-polarized current (qubit initialization). The spin-polarized carriers are then injected into InGaAs/GaAs quantum dots[3] (qubit storage). The g factor in the QDs is opposite to that in ZnMn(S)Se [9], which means the electrons are injected into the higher energy spin state (see Fig. 2). To optically read out the spin-qubits, unpolarized holes are fed into the dots from the bottom p-GaAs layer. Due to the strong strain-induced heavy-hole/light-hole splitting only the $\pm 3/2$ heavy-hole QD states are populated and lead to optical transitions. Electrons with spin polarization $-1/2$ ($+1/2$) can only recombine with $-3/2$ ($+3/2$) holes, emitting circularly polarized σ^+ (σ^-) photons in Faraday geometry (Fig. 2). Therefore, the circular polarization degree

$$CPD = \frac{I_{\sigma^+} - I_{\sigma^-}}{I_{\sigma^+} + I_{\sigma^-}} \qquad (I_{\sigma^\pm} : \text{ luminescence intensities}) \tag{2}$$

[3] Alternatively, GaInNAs [9] or InGaAs [13] quantum wells can be used, e.g., to realize long-wavelength light sources with circularly polarized emission.

of the emitted light directly indicates the type and degree of electron spin polarization in the QDs.

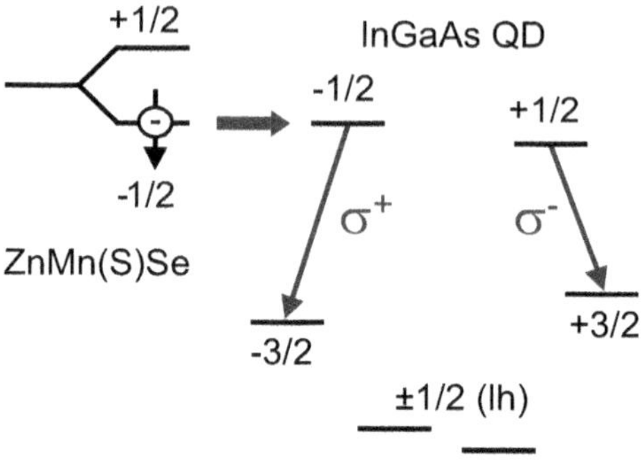

Fig. 2. Zeeman levels in the ZnMn(S)Se spin aligner and the InGaAs QDs, respectively. Spin-down electrons generated by the spin aligner are injected into the *upper* electronic/excitonic spin state of the dots. The possible σ^+ and σ^- polarized optical transitions in the QDs are indicated by arrows.

The concept of electrical spin injection has the advantage, that many qubits in different QDs can be initialized simultaneously. This would be difficult to achieve with all-optical techniques (e.g., resonant excitation of the dots with circularly polarized light), because the involved electronic transition energies vary from dot to dot. On the other hand it is clear, that within the context of quantum information processing a defined optical access to single QDs is required as well, in particular to read out individual qubits. In order to address single dots in our spin-LEDs with EL, we use metallic micro- or nano-apertures on top of our structures as well as high spatial-resolution spectroscopy [9, 14]. Figure 3 shows micro-electroluminescence (μ-EL) spectra of such a single QD spin-LED for different applied magnetic fields. At $B = 0\,\mathrm{T}$, a sharp emission peak (line width resolution-limited) from a single dot with no circular polarization is observed. For non-vanishing magnetic fields, the Zeeman splitting of the QD transition can be observed. When the magnetic field is increased, the electrons injected into the dot become more and more spin-polarized. As a result, the σ^+ transition, corresponding to the optically active exciton state for the injected spin-down electrons, grows, while the spin-up-related σ^- peak drops strongly. Finally, at about $B = 7\,\mathrm{T}$, the σ^- emission nearly disappears, indicating that the electrons in the QD are highly spin-polarized.

From these observations we can conclude that initialization of electronic spin-qubits in QDs utilizing electrical spin injection is indeed possible with high fidelity. Furthermore, the robustness of the spin states involved becomes evident, because spin polarization is conserved during the whole process of electrical injection, transport, storage in QDs, and final optical readout. In particular, no significant spin relaxation takes place in the dots, although the electrons are injected into the upper Zeeman level (see Figs. 2 and 3). Finally, our results also show that individual spin states in single quantum dots can be optically addressed and read out selectively, an important prerequisite for quantum information processing.

As already mentioned, an advantage of the electrical spin injection concept is the fact that initialization of many qubits at the same time should be possible. Indeed we have recently demonstrated the simultaneous initialization of 5 different spin-qubits in a single device with close to 100% polarization degree [14]. However, as shown in Fig. 4, the achieved CPD varies strongly from dot to dot, even within the same device and for similar emission wavelengths, although the obtained value is always reproducible over time for a given dot. This suggests that local effects must be responsible for the loss of spin polarization. Even more interestingly, the average polarization increases for higher energy QDs, a trend also found in ensemble measurements [9, 16]. Possible mechanisms behind these observations are discussed in the following section.

3 Spin loss mechanisms and device optimization

Despite the fact, that efficient spin injection into InGaAs QDs is obviously possible, Fig. 4 clearly shows that the obtained polarization degree for individual dots in the ensemble varies strongly and can be very low. At first it seems tempting to ascribe this variation to structural differences (morphology, composition) between the individual QDs that might influence the spin relaxation

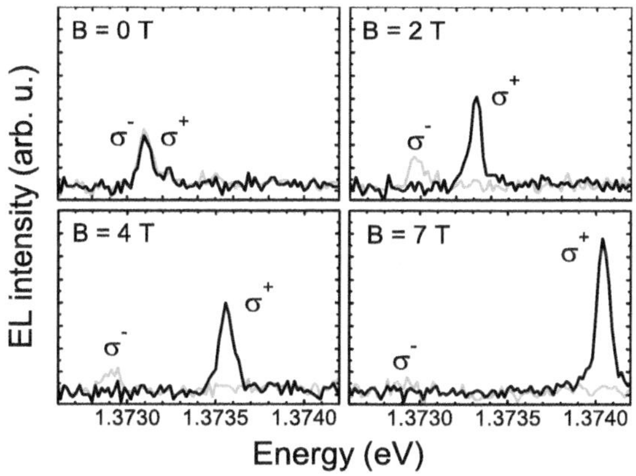

Fig. 3. Magnetic-field dependent μ-EL spectra of a single quantum-dot spin-LED ($I = 4\,\mathrm{mA}$, $U = 2.1\,\mathrm{V}$, $T = 5\,\mathrm{K}$). For each field value, the σ^+ and σ^- component of the QD emission have been measured.

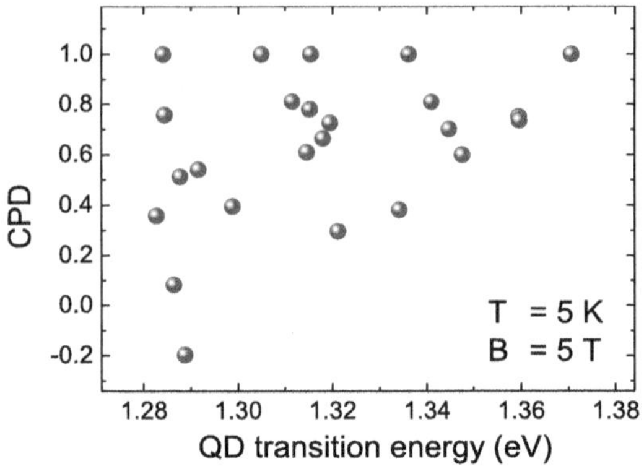

Fig. 4. *CPD* of the EL for different QDs in the same spin-LED.

time.[4] On the other hand, all-optical measurements suggest no significant spin relaxation during exciton lifetime in QDs [4], i.e., spin polarization should be essentially preserved in the dot, even if a hole is injected to read out the spin state. This expected robustness is also confirmed by our own investigations, as discussed in the previous section (emission from the energetically higher spin state). Indeed, we have strong experimental evidence that spin polarization must already be lost *before* the electrons are captured in the dots, i.e., when they are still mobile or only weakly localized (see [9] and discussion below).

An obvious source for spin relaxation outside the QDs are defect-related scattering processes, in particular due to dislocations at the III–V/II–VI interface [12]. Indeed, we have optimized the Mn concentration in our $Zn_{1-x}Mn_xSe$ spin aligners in order to obtain a low density of stacking faults/misfit dislocations and other defects while preserving a sufficiently large Zeeman splitting (see [9, 25–28]). However, despite the good quality achieved [9, 13] misfit dislocations resulting from the lattice mismatch between $Zn_{1-x}Mn_xSe$ and GaAs cannot be completely avoided. Therefore, we have recently started to develop *lattice-matched* $Zn_{1-x}Mn_xS_ySe_{1-y}$ spin aligners. First results of these investigations are shown in Fig. 5, where quantum-dot *ensemble* measurements of two spin-LEDs are compared. The devices are identical apart from the spin-aligner material used (750 nm $Zn_{0.95}Mn_{0.05}Se$ and $Zn_{0.95}Mn_{0.05}S_{0.1}Se_{0.9}$, respectively). As can clearly be seen, the utilization of a ZnMnSSe spin-aligner layer approximately lattice-matched to GaAs improves the achieved *CPD* significantly. Based on these preliminary results, the development of lattice-matched spin aligners with higher Mn concentrations seems feasible as well.

[4] However, the *CPD* varies smoothly between the QD and wetting layer region of the spectrum, suggesting that the dimensionality of the injection target is not important [9]. This is consistent with the model presented below, that spin relaxation actually takes place outside the dots.

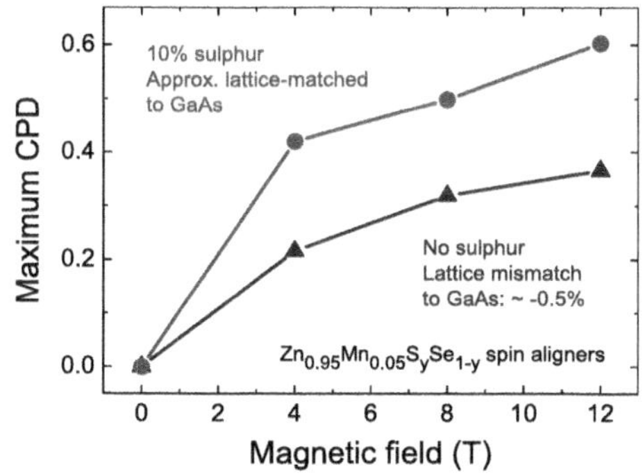

Fig. 5. Magnetic-field dependent *CPD* of the QD ensemble EL ($T = 5\,\mathrm{K}$) for two spin-LEDs. The devices are identical apart from the spin-aligner material used. The *CPD* has been measured at its spectral maximum, i.e., at the short-wavelength end of the QD ensemble emission.

The latter would provide a larger Zeeman splitting and thus enable operation of the devices at lower magnetic fields and/or higher temperatures.

As already mentioned at the end of Sect. 2, the achieved *average CPD* depends strongly on QD transition energy (see Figs. 4, 7(b), and 8(a) as well as [9, 14, 15, 17]). We believe that this effect is related to some peculiarities in the band structure of our devices. The latter is shown for typical operating conditions in Fig. 6. As can be seen, the band bending and the position of the electronic quasi-Fermi level $E_F(\mathrm{e})$ give rise to the formation of a two-dimensional electron gas (2DEG) at the III–V/II–VI interface. Furthermore, the injected spin-polarized electrons have to tunnel through a potential barrier in order to reach the wetting layer and the quantum dots. Since the average effective barrier is larger for low-energy dots, tunnelling will be slower, thus giving more opportunity for spin relaxation, i.e., the *CPD* should drop for these quantum dots, as observed experimentally. In reality, the problem is

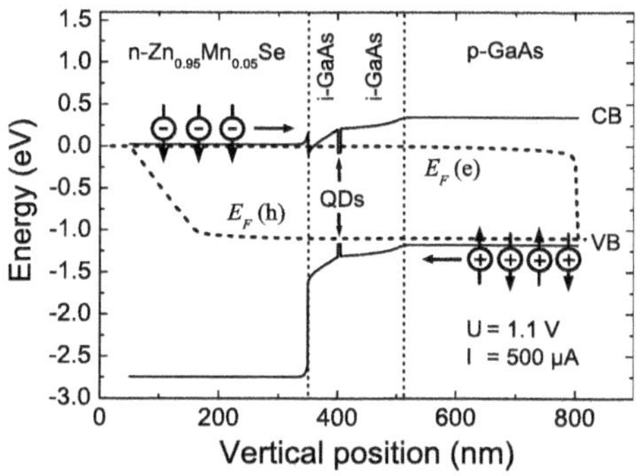

Fig. 6. Spin-LED band structure for typical operating conditions.

somewhat more complex. Both the effective tunnel barrier and $E_F(\mathrm{e})$ show *local* fluctuations, e.g., due to variations in the properties of the spin aligner (Mn concentration, defects) as well as the QDs (density, morphology, and composition, all modifying the density of states) but also due to the influence of nearby impurities or defects. As a result, the *CPD* achieved should depend on the local environment of the individual dot considered. In particular, QDs with high spin polarization would be expected to be found in the whole spectral emission range, although the probability to obtain efficient injection should drop with decreasing transition energy, which is indeed observed experimentally (Fig. 4).

If the outlined model is correct, spin relaxation should essentially occur in the GaAs spacer layer between the spin aligner and the QDs (see Figs. 1 and 6) as a result of only slow electron tunnelling into the dots. In order to test this hypothesis we have measured the magnetic-field dependent spectrally integrated *CPD* for spin-LEDs with different spacer thicknesses [9]. The results are shown in Fig. 7(a). As can be seen, even a moderate increase of spacer thickness from 25 to 75 nm decreases the achieved *CPD* from 30% to only 10%. This pronounced dependence gives clear evidence that spin relaxation within the GaAs spacer is very important in our devices. Indeed, the at first sight surprisingly strong drop can easily be understood as a consequence of the larger tunnel barrier.

Further evidence for the correctness of our model comes from temperature-dependent investigations of the quantum-dot ensemble EL [9, 15, 17]. As an example, Fig. 7(b) shows the spectrally resolved *CPD* for a spin-LED held at temperatures between 10 and 70 K. As can be seen, the high-energy polarization degree drops when the device is heated up. This results from the fact, that the Zeeman splitting in the spin aligner decreases with temperature and the upper electron spin level is thermally populated, thus leading to a lower initial spin polarization of the injected carriers [9]. However, more importantly the slope of the energy-dependent *CPD* strongly decreases when going from $T = 10$–$30\,\mathrm{K}$, suggesting spin relaxation to become less efficient. In a certain

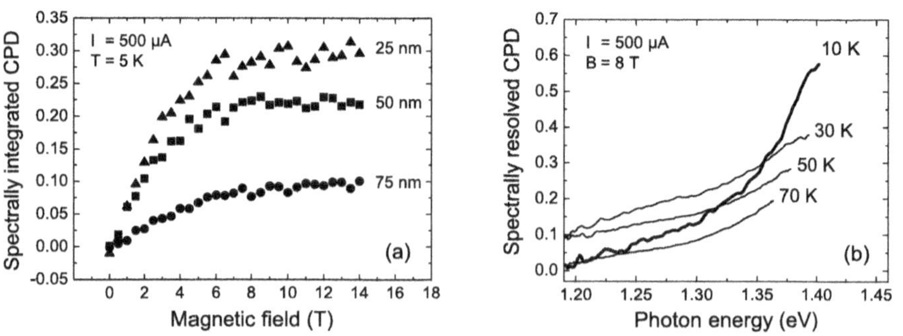

Fig. 7. *CPD* dependence on spacer thickness (**a**) and temperature (**b**) for QD ensembles.

spectral region this actually implies an initial CPD rise with temperature. We attribute this effect to phonon-assisted tunnelling and motional narrowing processes, typical for, e.g., the D'yakonov-Perel' mechanism. Therefore, mobile states (in the GaAs spacer but also the wetting layer) should indeed play a major role in the spin relaxation process, consistent with the outlined model.

From the discussion above it is obvious that the strategy to improve the CPD of our devices must be to shift the electron Fermi level to lower energies in order to prevent the formation of a 2DEG at the III–V/II–VI interface and the injection of poorly polarized electrons into the QDs. One possibility to achieve this is to rise the effective QD density of states via an increased dot density. In fact, preliminary experiments seem to indicate an improved spin injection efficiency when this approach is used. Alternatively, the spin aligner doping can be reduced to lower the electron Fermi level and potentially avoid the formation of a 2DEG altogether. Figure 8(a) compares the spectrally resolved CPD of two spin-LEDs with identical structure but different spin aligner doping levels ($1 \times 10^{17}\,\text{cm}^{-3}$ and $1 \times 10^{18}\,\text{cm}^{-3}$, respectively). As can

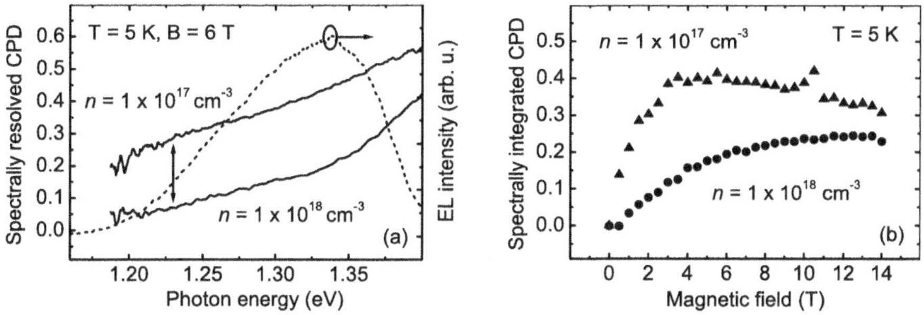

Fig. 8. Comparison of two spin-LEDs with identical structure but different spin aligner doping levels. (**a**) EL and spectrally resolved CPD; (**b**) magnetic-field dependence of spectrally integrated CPD (both QD ensemble measurements).

be seen, the predicted improvement in spin polarization is indeed observed, in particular for low QD transition energies [16]. An additional advantage of the lower Fermi level in the spin aligner is the fact, that a smaller Zeeman splitting in the latter suffices to prevent population of the upper conduction band state, i.e., to enable efficient spin polarization. As a result, high CPD values can already be achieved for relatively weak magnetic fields, as shown in Fig. 8(b), or for increased temperatures. (The decreasing CPD for very high fields in Fig. 8(b) is attributed to spin relaxation in the Zeeman-split conduction band of the III–V structure.)

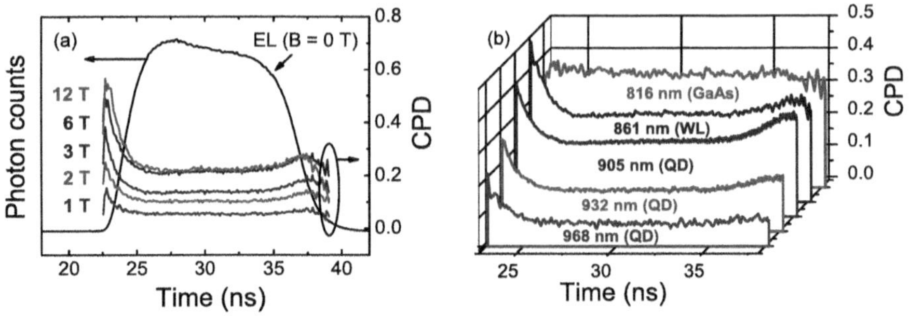

Fig. 9. (**a**) Temporal behavior of the emission intensity and magnetic-field dependent *CPD* (measured at the spectral maximum) for a spin-LED under pulsed ($\sim$ 20 ns) electrical excitation. (**b**) Time-resolved *CPD* (B = 6 T), measured at different emission wavelengths of the same device. The latter correspond to the spectral range of GaAs, the wetting layer (WL), and the QDs, respectively.

4 Time-resolved measurements

In order to gain a more detailed understanding of the spin and carrier relaxation processes in our devices but also as a first step towards future optical and electrical spin manipulation experiments we have recently started to perform time-resolved EL studies of the polarization dynamics using pulsed ($\sim$ 20 ns) electrical excitation and time-correlated single photon counting. First preliminary results of these measurements are shown in Fig. 9. Surprisingly, the determined *CPD* is not constant but shows a pronounced maximum at the rising edge of the pulse (to a lesser degree also at the falling edge), while the magnetic-field dependent polarization degrees found in the plateau region of the traces essentially agree with those obtained in continuous current measurements (Fig. 9(a)). The observed feature only appears for confined states (QDs, wetting layer), but not for the GaAs emission (Fig. 9(b)), thus excluding experimental artifacts like, e.g., delay effects. It cannot be explained by the initially lower current through the device either, because the *CPD* peak at the end of the pulse is much less pronounced for equally low currents. Moreover, the current dependence of the polarization degree is too small to fully account for the increased *CPD*. Currently we assume that spin relaxation processes in the wetting layer due to many-carrier effects could be reduced at the rising edge of the pulse, since initially the carrier density in the device is still low. However, further investigations are required in order to better understand the observed phenomena.

5 Conclusions

We demonstrated the concurrent initialization of electronic spin states in InGaAs QDs with near 100% fidelity based on electrical injection from a

ZnMn(S)Se spin aligner. Individual QD states could be read out through metallic apertures. Based on our results suggesting that spin relaxation mainly occurs outside the QDs, we realized optimized devices with improved electron quasi-Fermi level and lattice-matched spin aligners. Finally, we presented first time-resolved EL measurements of the polarization dynamics using ns-pulsed electrical excitation. The latter should not only enable us to gain a more detailed understanding of the spin and carrier relaxation processes in our devices. They are also the first step towards future time-resolved optical and electrical spin manipulation experiments.

6 Acknowledgements

The authors gratefully acknowledge contributions by J. Müller, H. Flügge, N. Höpcke, C. Mauser, J. Fallert, H. Burger, B. Daniel, J. Lupaca-Schomber, J. Hetterich, S. Li, B. Ramadout, D. Z. Hu, and D. M. Schaadt at the University of Karlsruhe. This work has been performed within project A2 of the DFG Research Center for Functional Nanostructures (CFN). It has been further supported by a grant from the Ministry of Science, Research and the Arts of Baden-Württemberg (Az.: 7713.14-300).

References

1. D. Loss, D. P. DiVincenzo: Phys. Rev. A **57**, 120 (1998)
2. J. M. Kikkawa, D. D. Awschalom: Nature (London) **397**, 139 (1999)
3. D. Hägele, M. Oestreich, W. W. Rühle, N. Nestle, K. Eberl: Appl. Phys. Lett. **73**, 1580 (1998)
4. M. Paillard, X. Marie, P. Renucci, T. Amand, A. Jbeli, J. M. Gérard: Phys. Rev. Lett. **86**, 1634 (2001)
5. M. Scheibner, G. Bacher, S. Weber, A. Forchel, Th. Passow, D. Hommel: Phys. Rev. B **67**, 153302 (2003)
6. T. Flissikowski, A. Hundt, M. Lowisch, M. Rabe, F. Henneberger: Phys. Rev. Lett. **86**, 3172 (2001)
7. E. Tsitsishvili, R. v. Baltz, H. Kalt: Phys. Rev. B **67**, 205330 (2003)
8. S. Fathpour, M. Holub, S. Chakrabarti, P. Bhattacharya: Electron. Lett. **40**, 694 (2004)
9. M. Hetterich, W. Löffler, J. Fallert, N. Höpcke, H. Burger, T. Passow, S. Li, B. Daniel, B. Ramadout, J. Lupaca-Schomber, J. Hetterich, D. Litvinov, D. Gerthsen, C. Klingshirn, H. Kalt: Phys. Stat. Sol. B **243**, 3812 (2006)
10. R. Fiederling, M. Keim, G. Reuscher, W. Ossau, G. Schmidt, A. Waag, L. W. Molenkamp: Nature (London) **402**, 787 (1999)
11. Y. Ohno, D. K. Young, B. Beschofen, F. Matsukura, H. Ohno, D. D. Awschalom: Nature (London) **402**, 790 (1999)
12. R. M. Stroud, A. T. Hanbicki, Y. D. Park, G. Kioseoglou, A. G. Petukhov, B. T. Jonker, G. Itskos, A. Petrou: Phys. Rev. Lett. **89**, 166602 (2002)

13. W. Löffler, D. Tröndle, J. Fallert, H. Kalt, D. Litvinov, D. Gerthsen, J. Lupaca-Schomber, T. Passow, B. Daniel, J. Kvietkova, M. Grün, C. Klingshirn, M. Hetterich: Appl. Phys. Lett. **88**, 062105 (2006)
14. W. Löffler, M. Hetterich, C. Mauser, S. Li, T. Passow, H. Kalt: Appl. Phys. Lett. **90**, 232105 (2007)
15. M. Hetterich, W. Löffler, J. Fallert, T. Passow, B. Daniel, J. Lupaca-Schomber, J. Hetterich, S. Li, C. Klingshirn, H. Kalt: Electrical spin injection into InGaAs quantum dot ensembles and single quantum dots, Proc. 28th Int. Conf. on the Physics of Semiconductors, Vienna, Austria, 2006, AIP Conf. Proc. **893**, 1285 (2007)
16. W. Löffler, N. Höpcke, C. Mauser, J. Fallert, T. Passow, B. Daniel, S. Li, D. Litvinov, D. Gerthsen, H. Kalt, M. Hetterich: Spin and carrier relaxation dynamics in InAs/GaAs quantum-dot spin-LEDs, Int. Conf. on Nanoscience and Technology (ICN+T) 2006, Basel, Switzerland, J. Phys.: Conf. Series **61**, 745 (2007)
17. W. Löffler, D. Tröndle, J. Fallert, E. Tsitsishvili, H. Kalt, D. Litvinov, D. Gerthsen, J. Lupaca-Schomber, T. Passow, B. Daniel, J. Kvietkova, M. Hetterich: Electrical spin injection into InGaAs quantum dots, 8th Int. Workshop on Nonlinear Optics and Excitation Kinetics in Semiconductors (NOEKS 8), Münster, Germany, 2006, Phys. Stat. Sol. C **3**, 2406 (2006)
18. W. Löffler, D. Tröndle, H. Kalt, D. Litvinov, D. Gerthsen, J. Lupaca-Schomber, T. Passow, B. Daniel, J. Kvietkova, M. Hetterich: Electrical spin injection from ZnMnSe into InGaAs-based quantum structures, Proc. 12th Int. Conf. on Modulated Semiconductor Structures (MSS-12), Albuquerque, 2005, Physica E **32**, 434 (2006)
19. T. Passow, S. Li, D. Litvinov, W. Löffler, J. Fallert, B. Daniel, J. Lupaca-Schomber, J. Kvietková, D. Gerthsen, H. Kalt, M. Hetterich: Investigation of InAs quantum dot growth for electrical spin injection devices, 4th Int. Conf. on Quantum Dots (QD2006), Chamonix-Mont Blanc, France, Phys. Stat. Sol. C **3**, 3943 (2006)
20. W. Löffler, M. Hetterich, C. Mauser, S. Li, J. Leuthold, H. Kalt: Spin-polarized excitonic emission from quantum dots after electrical injection, 10th Conf. on the Optics of Excitons in Confined Systems (OECS 10), Patti, Messina, Italy, 2007, Phys. Stat. Sol. B, DOI: 10.1002/pssb.200777612
21. G. Schmidt: J. Phys. D **38**, R107 (2005)
22. J. Seufert, G. Bacher, H. Schömig, A. Forchel, L. Hansen, G. Schmidt, L. W. Molenkamp: Phys. Rev. B **69**, 035311 (2004)
23. J. K. Furdyna: J. Appl. Phys. **64**, R29 (1988)
24. O. Goede, W. Heimbrodt: Phys. Stat. Sol. B **146**, 11 (1988)
25. D. Litvinov, D. Gerthsen, B. Daniel, C. Klingshirn, M. Hetterich: J. Appl. Phys. **100**, 023523 (2006)
26. M. Hetterich, B. Daniel, C. Klingshirn, P. Pfundstein, D. Litvinov, D. Gerthsen, K. Eichhorn, D. Spemann: Lattice parameter and elastic constants of cubic $Zn_{1-x}Mn_xSe$ epilayers grown by molecular-beam epitaxy, Proc. 11th Int. Conf. on II–VI compounds, Niagara Falls, New York, USA, 2003, Phys. Stat. Sol. C **1**, 649 (2004)
27. J. Kvietkova, B. Daniel, M. Hetterich, M. Schubert, D. Spemann, D. Litvinov, D. Gerthsen: Phys. Rev. B **70**, 045316 (2004)
28. B. Daniel, K. C. Agarwal, J. Lupaca-Schomber, C. Klingshirn, M. Hetterich: Appl. Phys. Lett. **87**, 212103 (2005)

Part III

Spin and Magnetism

Spintronic and Electro-Mechanical Effects in Single-Molecule Transistors

Maarten R. Wegewijs[1,2], Felix Reckermann[1,2], Martin Leijnse[2], and Herbert Schoeller[2]

[1] Institut für Festkörper-Forschung, Forschungszentrum Jülich, Leo-Brandt-Strasse, 52425 Jülich, Germany
[2] Institut für Theoretische Physik A, RWTH Aachen, Huyskenweg, 52056 Aachen, Germany

Abstract. We investigate electron transport through a mixed-valence molecular dimer, where an excess electron is delocalized over equivalent monomers, which can be locally distorted. In this system the *Born-Oppenheimer approximation breaks down*, resulting in quantum entanglement of the mechanical and electronic motion. We show that this breakdown results in distinct features in the transport spectrum that can be measured in recently developed three-terminal junctions with mechanical control. Additionally, for monomers with fixed localized spins, we show that the interplay of spin and vibrational motion allows the molecular spin parameters to be *detected in situ*, without an applied magnetic field. Conversely, the spin state of the entire molecule can be *controlled* via the non-equilibrium quantized molecular vibrations due to a novel *vibration-induced spin-blockade*.

1 Introduction

Nano-electromechanical devices (NEMS) electrically detect and control mechanical motion with great precision [1] and can be constructed in various nanostructures, including macromolecules such as suspended carbon nanotubes [2, 3]. Nowadays even nanometer sized molecules are within reach of experimental investigation. Successful three-terminal transport measurements [4] have been reported, detecting the quantized vibrational [5, 6], spin [7] and magnetic [8, 9] excitations of a single molecule. Quantum limited operation of NEMS is thus a starting point, rather than a goal, in the single-molecule regime. More challenging is achieving control over such devices. Recently, electrical three-terminal devices have been demonstrated with additional mechanical control [10, 11]. Here the size of the nanogap in which the molecule is embedded can be adjusted with sub-Ångstrom precision, thereby changing the capacitive and resistive coupling of the molecule to the source and drain electrodes, as well as the intrinsic molecular properties. The interesting question arises how single-molecule quantum states involving electronic and mechanical degrees of freedom may be detected and controlled using transport. This has been addressed in several theoretical studies, e.g. [12–15]. Fundamental to nearly all of these works is the adiabatic Born-Oppenheimer (BO) approximation, where one separates the timescales of the fast electronic motion from the slow dynamics of the nuclei. The transport is then governed by the Franck-Condon (FC) principle, where the tunneling of an electron onto the molecule simultaneously induces a change of the electron number $N \rightarrow N+1$ and the vibrational state $\chi \rightarrow \chi'$. The amplitude for this process factorizes and is proportional to the overlap integral of the mechanical wave-functions $\langle \chi | \chi' \rangle$ which in general is non-zero and strongly depends on both of the vibrational states. In contrast to spin-related tunneling, there are thus no strict selection rules and new interesting non-equilibrium transport mechanisms and characteristics arise. By proper choice and design of the vibrational properties of molecular transistors (number of modes, adiabatic potentials, etc.), one may realize electro-mechanical sensing devices.

A novel aspect of *molecule-based NEMS* is that they may display strong *vibronic* effects (distinct from vibrational) due to the non-trivial coupled quantum dynamics of the electronic and nuclear degrees of freedom, see [16] for a review. Here the Born-Oppenheimer separation of the time-scales of the nuclear and electron motion breaks down. The molecule is only adequately described by so-called vibronic states, in which the quantum entanglement renders the distinction between electronic and nuclear motion meaningless. The most prominent and well studied vibronic effect is the *dynamical* Jahn-Teller (JT) effect, which occurs in molecules where the electronic ground state is degenerate due to a high spatial symmetry of the static nuclear framework of the (non-linear) molecule. In such systems, there always exist a static symmetry-breaking coordinate which would lower the total molecular energy and lift the degeneracy. However, in a single molecule, the distortions are dynamical and the electronic degeneracy is transformed into a vibronic degeneracy [16] i.e. of the quantum-mechanical *molecular* eigenstates. Recently, the selection rules for tunneling encoded in these molecular eigenstates by the high symmetry were predicted to block electron transport through a JT active molecule [17]. An important issue now arises: how can one distinguish such vibronic blockade from spin- [18], magnetic [19] or Franck-Condon [14, 20] blockade effects? More generally, the BO-approximation can break down already when electronic levels only come close in energy on the scale of the vibrational frequency, referred to as the dynamical *pseudo-Jahn-Teller* (pJT) effect. This occurs in many molecular systems [16] and the question is which transport effects allow this to be detected. Here we address these issues for a generic model of a *mixed-valence molecular dimer*. It describes the intramolecular delocalization of an excess electron, a fundamental issue in physical chemistry (e.g. for the classification of mixed-valence compounds by the Robin-Day scheme) which has been studied extensively, see [21] and the references therein. We predict characteristic non-equilibrium transport signatures of the pJT dynamics of such a single-molecule dimer utilized as a transistor. In particular, in the experimental setup of [10, 11] one may identify the breakdown of the Born-Oppenheimer principle by non-linear conductance peaks with a sharp dependence of their *position, magnitude and sign* on the electro-mechanical parameters of the molecule.

A second main issue naturally arises, namely the role of localized spins which are often present in (p)JT-active molecules. Mixed-valence molecules exhibit an interplay of vibrational and spin degrees of freedom which makes them interesting candidates for single-molecule devices with coupled electro-mechanical and spintronic functionality. Importantly, this is not based on weak spin-orbit effects, but rather on strong direct, kinetic and double-exchange mechanisms. Here we demonstrate that in a single-molecule transistor configuration, the non-linear transport current can both detect and control the molecular spin solely due to the vibrational motion. This is possible due to a spin-dependence of the pseudo-Jahn-Teller effect and a novel *vibration-induced spin-blockade* of transport.

2 Mixed-valence dimer transistor

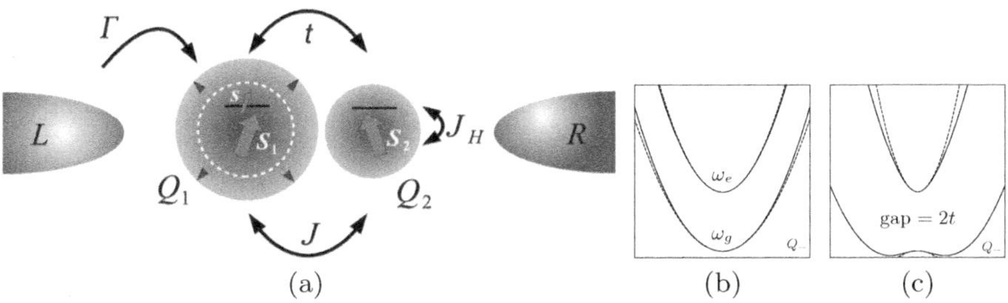

Fig. 1. (*Color online*) (**a**) Sketch: dimer molecule trapped in a nanogap between two voltage biased electrodes, $V_b = \mu_L - \mu_R$, and capacitively coupled to a back-gate (not shown) at voltage V_g, which shifts the effective molecular energy levels. The monomers (*blue*) can vibrate along the local totally symmetric breathing mode and are connected by a mechanically stiff bridge (not shown). (**b**–**c**) Adiabatic potentials W_g and W_e for the symmetry-breaking molecular distortion Q_- (*full lines*) and harmonic expansions around $Q_- = 0$ (*dashed lines*) for (**b**) weak ($\lambda = 0.7, t = 2.15\omega$) and (**c**) intermediate strength of the pJT effect ($\lambda = 1.93$, $t = 2.15\omega$).

We first formulate a basic non-equilibrium transport model for a mixed-valence molecule placed in a gap between two conducting electrodes as sketched in Fig. 1. We start from a model capturing the generic features of a mixed-valence molecule [21], such as, for example, a $Ru^{2+,3+}$ dimer with pyridine organic ligands. To describe the transport we need to extend this model to account for at least two charge states of the molecule. The generic properties are encoded in Hamiltonians H^N for each electron number N and give rise to transport effects which strikingly deviate from simpler models studied so far. These models are discussed in some detail in Sects. 2.1 and 2.2. The resulting molecular Hamiltonian H_{MV} is then incorporated in a transport Hamiltonian $H = H_{\mathrm{MV}} + H_{\mathrm{T}} + H_{\mathrm{res}}$. Here $H_{\mathrm{MV}} = \sum_{N=0,1} X^N H^N X^N + E_{\mathrm{add}}(V_g)N$, where X^N projects onto states with electron number N and the addition energy $E_{\mathrm{add}}(V_g)$ is controlled linearly with the voltage V_g on the gate electrode. The voltage-biased (V_b) electrodes $r = L, R$, kept at temperature T and chemical potential $\mu_r = \mu \pm V_b/2$, are described by

$$H_{\mathrm{res}} = \sum_{r=L,R} \sum_{k,\sigma} (\epsilon_k + \mu_r) c^{\dagger}_{rk\sigma} c_{rk\sigma}. \tag{1}$$

They couple to the molecule by the tunneling Hamiltonian

$$H_{\mathrm{T}} = \sum_{r=L,R} \sum_{i=1,2} \sum_{k,\sigma} T^i_r d^{\dagger}_{i\sigma} c_{rk\sigma} + h.c., \tag{2}$$

where $c^{\dagger}_{rk\sigma}$ creates an electron with spin projection σ in state k of electrode r, $d^{\dagger}_{i\sigma}$ creates an electron in state $|i\sigma\rangle$ and T^i_r is the amplitude for tunneling onto monomer i. We assume a sequential arrangement as sketched in Fig. 1, i.e. $T^1_L = T^2_R = \sqrt{\Gamma/(2\pi\rho)}$ and $T^2_L = T^1_R = 0$, where Γ denotes the tunneling rate and ρ the density of states.

2.1 Pseudo-Jahn-Teller effect

We consider the simplest model of a mixed-valence dimer molecule consisting of two identical monomers, labelled by $i = 1, 2$, which can vibrate along their totally symmetric ("breathing") mode Q_i about the potential minimum at $Q_i = 0$ with frequency ω. Each monomer accommodates one electronic orbital state, $|i\sigma\rangle$, for an excess electron with spin projection σ that can tunnel between the two monomers with amplitude t via a mechanically stiff bridging ligand. The electron significantly distorts the monomer it occupies along coordinate Q_i due to a change of the bond-lengths (c.f. Fig. 1(a)). The resulting shift of the potential minimum is $\sqrt{2}\lambda$, expressed in units of the zero-point motion energy of the vibration of the undistorted monomers. Thus λ is the dimensionless electron–vibration coupling. The local distortion of the monomer is dragged along by the electron as it becomes delocalized over the dimer. This results in coherent electro-mechanical motion and the breakdown of the Born-Oppenheimer separation. The central quantity controlling the character of the molecular states is the delocalization energy t relative to the coupling to the localized distortion $\lambda\omega$. We assume charging effects (Coulomb blockade) to be strong enough that only two molecular charge states participate in transport processes, which we label by the number of excess electrons on the molecule $N = 0, 1$. The Hamiltonians H^N for $N = 0, 1$ excess electrons on the molecule reduce to the vibrational parts i.e. we set $H^N = H^N_{\mathrm{vib}}$,

$$H^0_{\text{vib}} = \sum_{i=\pm} \tfrac{1}{2}\omega \left(P_i^2 + Q_i^2\right) \tag{3}$$

$$H^1_{\text{vib}} = H^0_{\text{vib}} - \lambda\omega Q_+ + \lambda\omega Q_-(\hat{n}_1 - \hat{n}_2) + t\sum_\sigma (d^\dagger_{1\sigma} d_{2\sigma} + h.c.) \tag{4}$$

It is written in the molecular vibrational coordinates $Q_\pm = (Q_2 \pm Q_1)/\sqrt{2}$ and $d^\dagger_{i\sigma}$ creates an electron in state $|i\sigma\rangle$ with occupation operator $\hat{n}_i = \sum_\sigma d^\dagger_{i\sigma} d_{i\sigma}$. The symmetric coordinate, Q_+, corresponds to the *molecular* breathing mode where the monomers vibrate in phase, i.e. the molecule as a whole changes its size. It couples to the total excess charge, N, of the molecule, resulting in a shift of the potential surface along Q_+ by an amount λ (linear term in Eq. (4)). The Franck-Condon (FC) transport effects resulting from this type of coupling have been calculated [13–15, 20] and found experimentally [5, 6, 22]. In contrast, the anti-symmetric mode, Q_-, corresponds to the monomers vibrating with opposite phase. This molecular shape distortion couples to the internal charge imbalance $\hat{n}_1 - \hat{n}_2$ if an excess electron is present. Due to the intra-molecular tunneling, t, the Hamiltonian (4) mixes electronic and vibrational states of the mode Q_- prohibiting a factorization of the molecular wave-function into a Q_--vibrational and an electronic part. Despite this breakdown of the BO-approximation, it is instructive to consider the adiabatic potentials for the Q_- vibrations, obtained by neglecting the nuclear kinetic energy operator ($P_- \to 0$) in Eq. (4) and to find the electronic eigenstates as function of Q_-, while neglecting Q_+. The resulting ground (g) and excited (e) adiabatic potentials $W_{g,e}(Q_-) = \frac{1}{2}\omega Q_-^2 \mp \sqrt{(\lambda\omega Q_-)^2 + t^2}$, are plotted in Fig. 1(b–c).

2.2 Spin-dependent Pseudo-Jahn-Teller effect

If each monomer consists of a transition metal ion there is a finite localized spin at each center. In this case, the non-trivial dynamics of the symmetry-breaking distortion of the ligand shells, Q_-, depends on the relative orientation of the local ionic spins and hence on the total molecular spin. This coupling arises due to a dominant local direct exchange interaction with the ionic spins (Hund's rule), which forces the excess electron spin to align with the ionic spin of the monomer it occupies. This makes the kinetic energy gained by this electron depend on the relative orientation of the ionic spins. The resulting *double-exchange* mechanism [23] competes with other types of exchange interaction. However, peculiar to this spin-coupling mechanism is that it makes the vibronic pJT mixing, discussed above, depend on the total spin of the molecule. The molecular Hamiltonian for $N = 0, 1$ excess electrons reads

$$H^N = H^N_{\text{vib}} - J\boldsymbol{S}_1 \cdot \boldsymbol{S}_2 - J(\boldsymbol{s}_1 \cdot \boldsymbol{S}_2 + \boldsymbol{S}_1 \cdot \boldsymbol{s}_2) - \sum_{i=1,2} J_H \boldsymbol{S}_i \cdot \boldsymbol{s}_i. \tag{5}$$

The intra-ionic Hund interaction, J_H, couples the spin of the excess electron, $\boldsymbol{s}_i = \frac{1}{2}\sum_{\sigma,\sigma'} d^\dagger_{i\sigma} \boldsymbol{\sigma}_{\sigma\sigma'} d_{i\sigma'}$, to the spin of the transition-metal ions, $\boldsymbol{S}_i$, where

$\boldsymbol{\sigma}$ denotes the vector of Pauli-matrices and $\boldsymbol{s}_i = 0$ if no electron is present on ion i. The intra-molecular tunneling, t, is incorporated in (5) through (4). Importantly, we assume J_H to be much larger than the other energy scales, including the transport voltage, and to be ferromagnetic in sign (less than half filled ionic shell): $J_H \gg |J|, \omega, t > 0$. The excitations where $\boldsymbol{S}_i$ and $\boldsymbol{s}_i$ are anti-parallel can therefore be neglected. Together with the intra-molecular tunneling t the intra-ionic Hund interaction J_H results in a double-exchange splitting $2t_S$ of eigenstates with $N = 1$ and total molecular spin S [23]:

$$\frac{t_S}{t} = \frac{S + \frac{1}{2}}{2S_1 + 1} \leq 1, \tag{6}$$

where $S_1 = S_2$ denotes the spin-length of the equivalent ions. This represents an effective, spin-dependent, tunneling strength. In the semi-classical limit of large ionic spins [23], $S_1 \gg 1$, this reduces to $t_S/t = S/2S_1 = \cos(\theta/2)$, where θ is the angle between the two classical ionic spins. Due to the strong intra-ionic coupling the kinetic energy which can be gained by the excess electron is maximal for parallel ionic spins and suppressed with the electron spin-eigenfunction component $\cos(\theta/2)$ quantized in the direction of the *local* ionic spin. Equation (5) also incorporates the intra-molecular coupling J of the spins of different ions. In the simple case of $S_1 = S_2 = 1/2$, studied from hereon, the Hamiltonian for the charged molecule consists of an $S = 3/2$ and an $S = 1/2$ diagonal block, with $2S + 1$ sub-blocks on the diagonal, given by

$$H_S^1 = H_{\mathrm{vib}}^1|_{t=t_S} - \tfrac{1}{2} J S(S+1) + \mathrm{const.} \tag{7}$$

This makes explicit the interesting property of mixed-valence molecules, that the strength of the pJT effect depends on the total molecular spin S [21] through the effective delocalization energy of the electron, t_S, which competes with the local distortion energy $\lambda\omega$.

3 Transport spectroscopy

We focus on the regime of voltages and temperatures where single-electron tunneling dominates the transport as is the case in many experiments [5–9, 22]. The transport rates can then be evaluated in lowest non-vanishing order in H_T (i.e. Fermi's Golden Rule). Using a master equation we calculate the non-equilibrium stationary state occupations of the molecule for each charge (N) and spin multiplet (S), keeping track of the symmetric vibrational (Q_+) quantum number as well as the quantum number for the entangled state of the electron and the anti-symmetric vibrations (Q_-). In Sect. 3.1 we first discuss the breakdown of the BO approximation for spin-less monomers ($S_1 = S_2 = 0$) and neglecting small effects due to the excess electron-spin (i.e. replacing $|i\sigma\rangle \rightarrow |i\rangle$ for $i = 1, 2$). The effects of finite ionic and electron spins are then considered separately in Sect. 3.2.

3.1 Spinless monomers

In order to appreciate the breakdown of the BO separation, we first discuss a case where it has approximate validity. This is the case when the distortion of the monomers is sufficiently weak compared to the tunneling between the monomers. In Fig. 2(a) we show the differential conductance as function of the applied voltages. Many excitations appear, involving the Q_+ and/or the

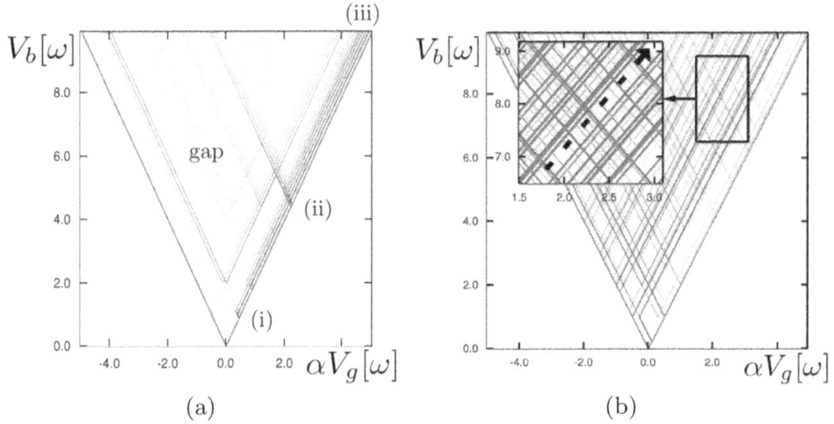

Fig. 2. (*Color online*) dI/dV_b ($\Gamma = 2.5 \cdot 10^{-5}\omega$, $T = 4 \cdot 10^{-3}\omega$) for (**a**) weak pJT mixing ($\lambda = 0.7$, $t = 2.15\omega$) (**b**) moderate pJT mixing ($\lambda = 1.93$, $t = 2.15\omega$). Inset: high contrast, *dashed black line* marking the $V_b - V_g$ trace taken in Fig. 3(a). For convenience the gate voltage is defined such that $V_g = 0$ corresponds to the charge degeneracy point and $\alpha = dE_{\mathrm{add}}/dV_g$. Due to symmetric biasing, energy scales appear at twice the separation on the voltage axis.

Q_- mode, which are separated in bias voltages by multiples of 2ω (due to symmetric biasing). The first of these excitations starts out at the marker (i) in Fig. 2(a). However, in contrast to usual FC transport spectra [13–15], the Q_- excitations are suppressed within the noticeable gap of $4t$, and enhanced conductance peaks (starting out from marker (ii)) delimit the upper boundary of the gap. One can thus directly estimate the strength of the delocalization of the excess electron. Furthermore, the excitations spaced in voltage V_b by 2ω also have a detailed substructure of a dense series of conductance peaks, for instance along the right edge of the transport region, terminating at the marker (iii). The latter correspond to tunnel processes between Q_- excitations of the $N = 0$ and $N = 1$ ground potential with the same vibrational quantum number $m_- = 0, 1, 2, \ldots$. For $\lambda^2\omega \ll t$, the potential $W_g(Q_-)$ is approximately harmonic, but with reduced frequency $\omega_g/\omega \approx 1 - \omega\lambda^2/t$ due to the pJT interaction, see Fig. 1(b). Therefore the resonances corresponding to subsequent values of m_- occur at slightly different positions [15, 24]. The equidistant energy spacings correspond to $\omega - \omega_g \approx (\lambda\omega)^2/t \ll \omega$. Similarly,

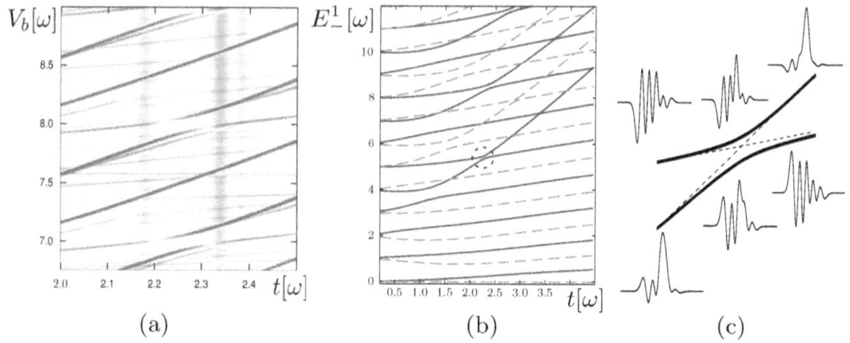

Fig. 3. (*Color online*) Signature of the pJT effect: anti-crossings as the intramolecular delocalization t is varied due to a mechanical change of the nanogap size. (**a**) Evolution with t of the $\mathrm{d}I/\mathrm{d}V_b$ trace along the line in the V_b, V_g plane marked in the inset of Fig. 2(b). (**b**) Evolution with t of energies of the vibronic states for $N = 1$ ($\lambda = 1.93$), with the harmonic Q_+-vibration energies subtracted. The *green* (*dashed*)/red (*solid*) line indicates positive/negative parity. The anti-crossing in the transport in (**a**) corresponds to the marked anti-crossing around $t \approx 2.3\omega$ and $E^1_- \approx 5\omega$. (**c**) Evolution with t of the vibrational parts $\chi^1_{m_-}(Q_-)$ of the vibronic states $|m_-,\rangle = |\chi^1_{m_-}\rangle|1\rangle + |\chi^2_{m_-}\rangle|2\rangle$, due to the symmetry of the dimer, $\chi^1_{m_-}(Q_-) = \pi\chi^2_{m_-}(-Q_-)$ where $\pi = \pm$ is the parity w.r.t. exchange of the nuclei.

above the gap near marker (ii), a dense series of conductance peaks with negative V_g dependence indicates that the excited adiabatic potential has a higher frequency $\omega_e/\omega = 1 + \omega\lambda^2/t$.

For larger values of λ, the adiabatic potentials become increasingly anharmonic due to the pJT interaction resulting in markedly non-equidistant spacing of the dense series of conductance peaks. However, the most dramatic effect of the pJT interaction is the complete breakdown of the BO separation when excited states of the two adiabatic potentials W_e and W_g come close in energy. This occurs at high bias voltage $V_b \sim 4t$, i.e. above the gap. At a first glance, the transport spectrum in this case, shown in Fig. 2(b), seems inextricably complex. However, if one is able to vary the size of the nanogap in gated break-junction as described in [10, 11], clear signatures of the pJT effect are revealed. We assume that only the intramolecular hopping, t, will vary as the gap size is changed and the molecule is strained, leaving the mechanical parameters ω and λ of the monomers unaltered. (We checked that similar qualitative results follow if this assumption is relaxed). In Fig. 3(a) we show the corresponding evolution of the $\mathrm{d}I/\mathrm{d}V_b$ trace taken along the dashed black line in the inset of Fig. 2(b), as t is varied. If the BO separation were valid, the $\mathrm{d}I/\mathrm{d}V_b$ resonances with a weak t-dependence could be assigned to highly excited vibrations in the lower adiabatic electronic state, W_g, whereas those with a strong linear t-dependence would correspond to the lowest excitations in the excited adiabatic electronic state, W_e. However, this distinction is completely lost at the anti-crossings visible in Fig. 3(a). The anti-crossing of conductance lines at $t \approx 2.3\omega$ maps out the corresponding anti-crossing in the evolution of the vibronic energy spectrum with t which is shown in Fig. 3(b). Strikingly, the *sign* of the conductance of the two anti-crossing transport resonances is different. This directly relates to the large difference in the kinetic energy of the nuclear motion of the two anti-crossing adiabatic electronic states. In Fig. 3(c) we show vibrational components of the vibronic wave-functions of the involved states. Far from the anti-crossing their spatial variation is clearly different, resulting in different transport rates and therefore non-equilibrium occupations and conductance. At the anti-crossing the strong pJT mixing causes the components of *both* vibronic wave-functions to rapidly vary. As a result the conductance peaks disappear in a narrow range of t values in the anti-crossing region. Note that all other resonances, which follow from the BO approximation and the FC-principle, smoothly depend on t, making the pJT effect clearly stand out. Strikingly, the transport anti-crossings seen in Fig. 3(a) are *replicas* of one and the same anti-crossing marked in Fig. 3(b), due to the simultaneous excitation of the Q_+ mode. Thus, interestingly, the *pJT-inactive mode* proliferates the violation of the adiabatic BO separation in the transport. The many other anti-crossings in Fig. 3(b) result in pJT resonances at different voltages (not shown).

3.2 Monomers with spin

We now turn to the effects of finite ionic spins, $S_1 = S_2 = 1/2$. The differential conductance map is shown in Fig. 4(a), using a set of parameters representative for the generic behavior of the model for weak electron–vibration coupling and ferromagnetic intra-molecular coupling: $J = 2.9\omega$, $t = 1.5\omega$, $\lambda = 0.5$. The transport spectrum displays a number of sharp, well-separated resonance lines, which are "dressed" by many more lines with small separations (discussed above). Two pronounced pairs of excitations appear, which derive from the $S = 1/2$ and $S = 3/2$ spin-multiplets and are split by $3J/2$ due to the intra-molecular spin-coupling (c.f. (7)). These multiplets are further split by approximately $2t_S$, see also Fig. 4(c). This double-exchange gap is larger for the high spin $S = 3/2$ state, $t_{3/2}/t_{1/2} = 2$ and this ratio allows the spin to be determined using Eq. (6). A central result of this work is that an independent check of this assignment of the spin is provided by the vibronic lines "dressing" the spin-excitations. Their occurrence indicates a significant pJT mixing in the $N = 1$ charge state which changes the frequency and additionally induces an-harmonicity in the effective adiabatic potentials. If the pJT effect is weak (as for $S = 3/2$ due to large $t_{3/2}$), these potentials are approximately harmonic but with a charge-dependent frequency. The spacing between these lines (due to transitions between excited vibrational and vibronic states) is then even and equals the small frequency *difference*. For a stronger pJT effect ($S = 1/2$) the potential in the $N = 1$ charge state becomes anharmonic and the lines are unevenly spaced. Clearly, in Fig. 4(a) the "dressing" of the lower

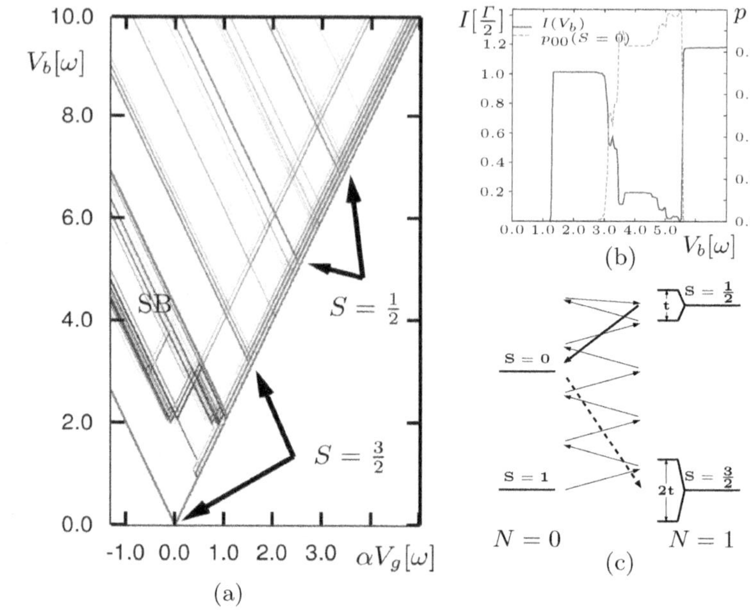

Fig. 4. (*Color online*) (**a**) Differential conductance (dI/dV_b) for V_g vs. V_b ($J = 2.9\omega$, $t = 1.5\omega$, $\lambda = 0.5$, $\Gamma = 2.5{\cdot}10^{-5}\omega$, $\gamma_0^{\mathrm{v}} = 3.6{\cdot}10^{-3}\Gamma$, $\gamma_0^{\mathrm{s}} = 3.6{\cdot}10^{-9}\Gamma$, $\alpha = dE_{\mathrm{add}}/dV_g$, $T = 4 \cdot 10^{-3}\omega$) (*red*: $dI/dV_b > 0$, *blue*: $dI/dV_b < 0$). The double-exchange coupling leads to a spin-dependent gap size of the vibronic spectrum (see *arrows*). The spectrum of $S = 3/2$ is harmonic (signalled by equidistant resonance lines of small energy separation), while the one of $S = 1/2$ is anharmonic (non-equidistant lines). In the low bias region (marked as "SB") the current is strongly suppressed due to the spin-blockade. We used a typical energy-dependent density of states in the relaxation rate $\gamma^{\mathrm{s(v)}}(E) = \gamma_0^{\mathrm{s(v)}} \cdot (E/\omega)^2$ between states with equal ($\gamma_0^{\mathrm{v}} \ll \Gamma$) and different spin ($\gamma_0^{\mathrm{s}} \ll \gamma_0^{\mathrm{v}}$). (**b**) Current vs. V_b (*red solid*) at $V_g = -0.6\omega$ and probability of $S = 0$ multiplet with no vibrational/vibronic quanta excited (*green dashed*). (**c**) Spin-blockade mechanism: sketch of the $N = 0, 1$ energy spectrum. The molecule is "pumped" by a sequence of tunneling events which change the charge and excite vibrational and/or vibronic quanta. Vibrational states are omitted for clarity.

pair of lines is more evenly spaced than the upper set of lines, confirming the assignment of high spin state at low energy. This spin identification by vibrations works very well for $t \gtrsim \lambda^2\omega$, i.e. when the pJT effect is weak to moderate. Finally, we note that the energy average of the total S multiplets split by double exchange follows $J\,S$, providing a third, independent check of the spin value assignment. Thus the intra-molecular ferromagnetic coupling is revealed by the transport spectrum at zero magnetic field, by double-exchange and vibronic effects.

We have not yet discussed the most pronounced structure in the $\mathrm{d}I/\mathrm{d}V_b$ stability plot appearing in the low energy gap in Fig. 4(a) as a sequence of excitations with negative differential conductance (NDC) (blue lines) where the current, plotted in Fig. 4(b), is stepwise reduced. Simultaneously, the occupation of the molecular state with zero-spin and no vibrational excitations grows, reaching close to 100%. Within this region the current is strongly suppressed due to a pronounced population inversion that stabilizes the charge to $N = 0$ and the spin to $S = 0$ (c.f. Fig. 4(b)). This *vibration-induced spin-blockade* provides another indication for the spin properties of the mixed-valence molecule and additionally allows the spin to be controlled electrically. The effect is readily understood by considering the non-equilibrium vibrations induced by the electric current. First we note that in the low-bias region where the spin-blockade occurs, the direct transition from $S = 1 \rightarrow S = 1/2$ states is energetically not yet possible and the transition $S = 0 \leftrightarrow S = 3/2$ is generally forbidden by spin-selection rules. Now consider an electron which has just enough energy to excite a vibrational, Q_+, (or vibronic, Q_-) quantum, when entering/leaving the molecule ($N = 0 \leftrightarrow 1$). If the molecule does not immediately relax between subsequent tunneling events due to the coupling to the environment it can accumulate additional quanta in subsequent tunneling processes as sketched in Fig. 4(c). Eventually, the molecule is vibrationally excited enough that a tunnel process can bring it to a low spin $S = 1/2$ state at high energy: using all the accumulated vibrational energy a low energy electron is assisted to highly excite the electronic spin-system (despite the insufficient voltage). Finally, the molecule can relax to the $N = 0$, $S = 0$ state by a single tunnel process in which the excess energy is dissipated into the electrodes. Here the molecule is trapped and the current is suppressed: neither the $S = 3/2$ states (due to the spin selection rule for tunneling $\Delta S = \pm 1/2$) nor the $S = 1/2$ state are accessible (due to the insufficient bias voltage). Figure 4(b) shows that the spin-blockade is lifted at higher bias $V_b \approx (J + t) - 2V_g$ where the direct process back to $S = 1/2$ becomes possible, thereby confirming the above mechanism. Clearly, the vibration induced spin-blockade is expected to break down when escape processes from $S = 0$ become dominant already at low bias voltages. However, such processes have to change the spin by *at least* 1 quantum and are therefore parametrically weak, since they must relate to spin-orbit coupling or higher order tunnel processes. For instance, in Fig. 4(a–b) we have accounted phenomenologically for relaxation between states with equal spin and, with a much smaller rate, between states with different spin. The finite latter rate leads to the small remnant current in the region of spin-blockade.

Thus the single-electron transport current "pumps" the vibrational system and thereby drives a pronounced population inversion among the spin-states. This allows the molecule to be switched to a state with a well-defined charge and spin by adjusting the applied voltages. For this the temperature and the tunnel rate have to be sufficiently small to prevent escape from the $S = 0$ state by thermal or quantum fluctuations ($\Gamma \ll T \ll \omega$). Crucial is that the intra-molecular vibrations are weakly damped (as expected), allowing the $S = 0$ state to be efficiently reached. Finally, we note that the vibration-induced spin-blockade is generic and also occurs for anti-ferromagnetic intra-molecular coupling $J < 0$ provided that $t > |J|$. In this case it leads to a stabilized *excess charge* $N = 1$ and *high spin* $S = 3/2$ and analogous pronounced transport effects.

4 Acknowledgement

We acknowledge K. Flensberg, P. Kögerler, and H. Luecken for discussions and the financial support from DFG SPP-1243, the NanoSci-ERA, the Helmholtz Foundation, and the FZ-Jülich (IFMIT).

References

1. K. L. Ekinci, M. L. Roukes: Rev. Sci. Instrum. **76**, 061101 (2005)
2. V. Sazonova, et al.: Nature **431**, 284 (2004)
3. S. Sapmaz, et al.: Phys. Rev. Lett. **96**, 026801 (2006)
4. H. Park, et al.: Appl. Phys. Lett. **75**, 301 (1999)
5. H. Park, et al.: Nature **407**, 57 (2000)
6. A. N. Pasupathy, et. al.: Nano Lett. **5**, 203 (2005)
7. E. A. Osorio, et al.: Nano Lett. **7**, 3336 (2007)
8. H. B. Heersche, et al.: Phys. Rev. Lett. **96**, 206801 (2006)
9. M.-H. Jo, et al.: Nano Lett. **6**, 2014 (2006)
10. A. R. Champagne, A. N. Pasupathy, D. C. Ralph: Nano Lett. **5**, 305 (2005)
11. J. J. Parks, et al.: Phys. Rev. Lett. **99** (2007)
12. S. Braig, K. Flensberg: Phys. Rev. B **68**, 205324 (2003)
13. A. Mitra, I. Aleiner, A. J. Millis: Phys. Rev. B **69**, 245302 (2004)
14. J. Koch, F. von Oppen: Phys. Rev. Lett. **94**, 206804 (2005)
15. M. R. Wegewijs, K. C. Nowack: New J. Phys. **7**, 239 (2005)
16. I. B. Bersuker, V. Z. Polinger: *Vibronic Interactions in Molecules and Crystals* (Springer, 1989)
17. M. G. Schultz, T. S. Nunner, F. von Oppen: Phys. Rev. B **77**, 075323 (2008)
18. C. Romeike, M. R. Wegewijs, H. Schoeller: Phys. Rev. B **75**, 064404 (2007)
19. C. Romeike, M. R. Wegewijs, H. Schoeller: Phys. Rev. Lett. **96**, 196805 (2006)
20. S. Braig, K. Flensberg: Phys. Rev. B **68**, 205324 (2003)
21. I. Bersuker, S. Borshch: *Vibronic Interactions in Polynuclear Mixed-Valence Clusters*, vol. 81, Advances in Chemical Physics (Wiley, 1992) Chap. 6, p. 703
22. E. A. Osorio, et al.: Adv. Mater. **19**, 281 (2007)
23. P. W. Anderson, H. Hasegawa: Phys. Rev. **100**, 675 (1955)
24. J. Koch, F. von Oppen: Phys. Rev. B **72**, 113308 (2005)

Transport in 2DEGs and Graphene: Electron Spin vs. Sublattice Spin

Maxim Trushin and John Schliemann

Institute for Theoretical Physics, University of Regensburg, Universitaetsstrasse 31, 93053 Regensburg, Germany
maxim.trushin@physik.uni-regensburg.de

Abstract. We propose a quasiclassical model which describes kinetics of classical particles having an additional quantum degree of freedom. The later might be an electron spin in two dimensional electron gases, or sublattice index in graphene. As an application, we focus on current-induced spin accumulation in a 2DEG with spin-orbit coupling of both the Rashba and the Dresselhaus type. This phenomenon sometimes also referred to as the kinetic magnetoelectric or inverse spin-galvanic effect and shows for the system under study significant anisotropies. The approach developed here is also applied to the description of carrier transport in graphene where low energy excitations have, with respect to the sublattice degree of freedom, a similar chiral structure as the usual 2DEG with Rashba spin-orbit interaction.

1 Introduction

Boltzmann kinetic equation is a powerful tool for the investigation of electron transport in semiconductors. There are, however, obvious restrictions which are inhereted in the Boltzmann approach due to its quasiclassical origin. Indeed, we could not apply Boltzmann equation in its conventional form for the description of a quantum mechanical degree of freedom. The electron spin in presence of spin-orbit coupling is one of such a quantum mechanical quantity: an electron can not only be in one of two possible spin eigen states of the free Hamiltonian H_0 but in an arbitrary superposition of them. The latter is not possible in classical physics where a given particle always has a definite position in the phase space. Another example of such a quantum mechanical degree of freedom in solid state physics might be the sublattice index for carriers in graphene. Indeed, the eigen states of the free effective Hamiltonian for carriers in graphene are chiral, i.e. momentum and sublattice degree of freedom are coupled with each other. The latter makes it difficult to distiguish which sublattice a given particle belong to. To describe the kinetics of chiral particles properly we should consider the sublattice degree of freedom as a truely quantum number. The latter allows us to derive the conductivity formula which reproduces many peculiar experimental features including the conductivity minimum.

The paper is organized as follows. First, we derive the kinetic equation with an additional quantum degree of freedom which can be recognized as an electron spin or sublattice index. It is important that the particle momentum is still assumed to be a classical variable, i.e. our kinetic equation remains classical in this sense. Second, we apply our approach to investigate the current-induced spin accumulation (or magneto-electric effect) in two-dimensional electron gas (2DEG) in presence of both Rashba and Dresselhaus spin-orbit interactions. The magneto-electric effect in a 2DEG, sometimes also referred to as the inverse spin-galvanic effect, has attracted particular interest from an both experimental [1–5] and theoretical [6–10] point of view. This effect amounts in spin accumulation as a response to an applied in-plane electric field and is therefore a possible key ingredient towards all-electrical spin control in semiconductor structures, a major goals in today's spintronics research. Third, we derive an analytical formula for the electrical conductivity in graphene and compare this with the experimental data [11–15]. In particular, we focus on the minimum conductivity phenomena — a non-vanishing electrical conductivity even at zero carrier concentration. At the end we conclude with a few remarks.

2 Derivation of the kinetic equation

The change of the density matrix $\hat{\rho}$ in presence of the electric field $\mathbf{E}$ and scattering potential U can be described by means of the von Neumann equation

$$\frac{e\mathbf{E}}{\hbar}\frac{\partial\hat{\rho}}{\partial\mathbf{k}} + \frac{i}{\hbar}[\hat{H}_0, \hat{\rho}] = -\frac{i}{\hbar}[U, \hat{\rho}]. \tag{1}$$

The Hamiltonian $\hat{H}_0$ describes a free particle which is characterized by the momentum $\mathbf{k}$ and a quantum number κ. The eigen values and states of $\hat{H}_0$ are $E_{\mathbf{k}\kappa}$ and $|\mathbf{k}\kappa\rangle$ respectively. Let us now rewrite Eq. (1) in the basis of the eigen states $|\mathbf{k}\kappa\rangle$, i.e. for the left hand side of Eq. (1) we have

$$\frac{e\mathbf{E}}{\hbar}\frac{\partial\hat{\rho}}{\partial\mathbf{k}} + \frac{i}{\hbar}[\hat{H}_0, \hat{\rho}] \to \frac{e\mathbf{E}}{\hbar}\frac{\partial}{\partial\mathbf{k}}\langle\mathbf{k}'\kappa'|\hat{\rho}|\mathbf{k}\kappa\rangle + \frac{i}{\hbar}\rho_{\mathbf{k}\kappa\mathbf{k}'\kappa'}\left(E_{\mathbf{k}\kappa} - E_{\mathbf{k}'\kappa'}\right). \tag{2}$$

To transform the right hand side of Eq. (1) into the collision integral we substitute the density matrix by the following expression describing its time evolution after each scattering event

$$\hat{\rho} \to \hat{\rho}(t=0) + \frac{i}{\hbar}\int\limits_0^\infty dt \mathrm{e}^{-\frac{i}{\hbar}\hat{H}_0 t}[\hat{\rho}(t=0), U]\mathrm{e}^{\frac{i}{\hbar}\hat{H}_0 t}. \tag{3}$$

The integrals over t can be taken introducing a small parameter λ

$$\int_0^\infty dt e^{\frac{i}{\hbar}(E_{\mathbf{k}'\kappa'}-E_{\mathbf{k}\kappa})t-\lambda t} = P\frac{i\hbar}{E_{\mathbf{k}'\kappa'}-E_{\mathbf{k}\kappa}} + \pi\hbar\delta(E_{\mathbf{k}'\kappa'}-E_{\mathbf{k}\kappa}). \tag{4}$$

Here, P denotes "principal value". To simplify the subsequent formulas we retain only leading δ-functional terms in Eq. (4). The collision integral can be then written as

$$\begin{aligned}
&-\frac{i}{\hbar}[U,\hat{\rho}] \rightarrow \\
&\sum_{\kappa_1\kappa_1'}\int\frac{d\mathbf{k}_1}{(2\pi)^2}\int\frac{d\mathbf{k}_1'}{(2\pi)^2}\frac{\pi}{\hbar}\left[u_{\mathbf{k}\kappa\mathbf{k}_1'\kappa_1'}\rho_{\mathbf{k}_1'\kappa_1'\mathbf{k}_1\kappa_1}u_{\mathbf{k}_1\kappa_1\mathbf{k}'\kappa'}\delta(E_{\mathbf{k}'\kappa'}-E_{\mathbf{k}_1\kappa_1})\right.\\
&-u_{\mathbf{k}\kappa\mathbf{k}_1'\kappa_1'}u_{\mathbf{k}_1'\kappa_1'\mathbf{k}_1\kappa_1}\rho_{\mathbf{k}_1\kappa_1\mathbf{k}'\kappa'}\delta(E_{\mathbf{k}_1\kappa_1}-E_{\mathbf{k}_1'\kappa_1'})\\
&-\rho_{\mathbf{k}\kappa\mathbf{k}_1\kappa_1}u_{\mathbf{k}_1\kappa_1\mathbf{k}_1'\kappa_1'}u_{\mathbf{k}_1'\kappa_1'\mathbf{k}'\kappa'}\delta(E_{\mathbf{k}_1'\kappa_1'}-E_{\mathbf{k}_1\kappa_1})\\
&\left.+u_{\mathbf{k}\kappa\mathbf{k}_1\kappa_1}\rho_{\mathbf{k}_1\kappa_1\mathbf{k}_1'\kappa_1'}u_{\mathbf{k}_1'\kappa_1'\mathbf{k}'\kappa'}\delta(E_{\mathbf{k}_1\kappa_1}-E_{\mathbf{k}\kappa})\right].
\end{aligned} \tag{5}$$

Now we assume that $\mathbf{k}$ is the classical variable, whereas κ is the quantum number. Thus, we take the trace on $\mathbf{k}$ (i.e. $\rho_{\mathbf{k}\kappa\mathbf{k}'\kappa'} \rightarrow f_{\kappa\kappa'}(\mathbf{k})\delta_{\mathbf{k}\mathbf{k}'}$) and get the final formula for the colission integral

$$\begin{aligned}
I[f]_{\kappa\kappa'} &= \sum_{\kappa_1\kappa_1'}\int\frac{d\mathbf{k}_1'}{(2\pi)^2}\left[K^{\kappa\kappa'}_{\kappa_1'\kappa_1}(\mathbf{k}_1',\mathbf{k})f_{\kappa_1'\kappa_1}(\mathbf{k}_1')\delta(E_{\mathbf{k}\kappa'}-E_{\mathbf{k}_1'\kappa_1})\right.\\
&-K^{\kappa\kappa_1}_{\kappa_1'\kappa_1'}(\mathbf{k}_1',\mathbf{k})f_{\kappa_1\kappa'}(\mathbf{k})\delta(E_{\mathbf{k}\kappa_1}-E_{\mathbf{k}_1'\kappa_1'})\\
&-K^{\kappa_1\kappa'}_{\kappa_1'\kappa_1'}(\mathbf{k}_1',\mathbf{k})f_{\kappa\kappa_1}(\mathbf{k})\delta(E_{\mathbf{k}_1'\kappa_1'}-E_{\mathbf{k}\kappa_1})\\
&\left.+K^{\kappa\kappa'}_{\kappa_1\kappa_1'}(\mathbf{k}_1',\mathbf{k})f_{\kappa_1\kappa_1'}(\mathbf{k}_1')\delta(E_{\mathbf{k}_1'\kappa_1}-E_{\mathbf{k}\kappa})\right],
\end{aligned} \tag{6}$$

where $K^{\kappa\kappa'}_{\kappa_1\kappa_1'}(\mathbf{k}_1',\mathbf{k}) = \frac{\pi}{\hbar}u_{\mathbf{k}\kappa\mathbf{k}_1'\kappa_1}u_{\mathbf{k}_1'\kappa_1'\mathbf{k}\kappa'}$.

The left hand side Eq. (2) is transformed accordingly. At the end of the day we obtain a kinetic equation similar to the Boltzmann one, but with a quantum degree of freedom incorporated.

3 2DEG with spin-orbit coupling

In order to describe the spin dynamics in 2DEG with spin-orbit interactions we consider the Hamiltonian as a sum of the kinetic energy and two spin-orbit coupling terms: Rashba [16] and Dresselhaus [17]. Then, the Hamiltonian takes the form

$$H = \frac{\hbar^2(k_x^2+k_y^2)}{2m} + \alpha(\sigma_x k_y - \sigma_y k_x) + \beta(\sigma_x k_x - \sigma_y k_y). \tag{7}$$

Here, $\sigma_{x,y}$ are the Pauli matrices, $k_{x,y}$ are the electron wave vectors, and m is the effective electron mass. Introducing the angle γ_k so, that

$$\tan \gamma_k = \frac{\alpha k_x + \beta k_y}{\beta k_x + \alpha k_y}, \tag{8}$$

we obtain the following spinors as the eigen functions of the Hamiltonian H

$$\Psi_\pm(x, y) = \frac{1}{\sqrt{2}} \mathrm{e}^{ik_x x + ik_y y} \begin{pmatrix} 1 \\ \pm \mathrm{e}^{-i\gamma_k} \end{pmatrix}. \tag{9}$$

The energy spectrum has the form

$$E_\pm = \frac{\hbar^2 k^2}{2m} \pm \sqrt{(\alpha k_x + \beta k_y)^2 + (\beta k_x + \alpha k_y)^2}, \tag{10}$$

where $k = \sqrt{k_x^2 + k_y^2}$. The velocity matrix in the helicity basis (9) is not diagonal, and its elements read

$$v_x^{11(22)} = \hbar k_x/m \pm (\beta \cos \gamma_k + \alpha \sin \gamma_k)/\hbar, \tag{11}$$
$$v_x^{12(21)} = \pm i(\beta \sin \gamma_k - \alpha \cos \gamma_k)/\hbar; \tag{12}$$
$$v_y^{11(22)} = \hbar k_y/m \pm (\alpha \cos \gamma_k + \beta \sin \gamma_k)/\hbar, \tag{13}$$
$$v_y^{12(21)} = \pm i(\alpha \sin \gamma_k - \beta \cos \gamma_k)/\hbar. \tag{14}$$

For the diagonal elements of the velocity matrix we also use simplified notations $\mathbf{v}^{\mathbf{11(22)}} \equiv \mathbf{v}_\pm$.

In the following, it will be convenient to use polar coordinates $k_x = k \cos \theta$, $k_y = k \sin \theta$. Then, the spectrum read $E_\pm = \hbar^2 k^2/(2m) \pm \mid k \mid \kappa_\theta$, where

$$\kappa_\theta = \sqrt{\alpha^2 + \beta^2 + 4\alpha\beta \sin \theta \cos \theta}$$

is the generalized spin-orbit interaction constant for a given direction of motion. The wave vectors for a given energy read

$$k_\pm = \mp \frac{m}{\hbar^2} \kappa_\theta + \sqrt{\left(\frac{m}{\hbar^2}\right)^2 \kappa_\theta^2 + \frac{2mE}{\hbar^2}}, \tag{15}$$

and the expression for γ_k takes the form

$$\tan \gamma_k = \frac{\alpha \cos \theta + \beta \sin \theta}{\beta \cos \theta + \alpha \sin \theta}.$$

We rewrite the kinetic equation (1) in the helicity basis where the Hamiltonian is diagonal, and the equation takes the form

$$\frac{e\mathbf{E}}{\hbar} \frac{\partial}{\partial \mathbf{k}} \begin{pmatrix} f_{11} & f_{12} \\ f_{21} & f_{22} \end{pmatrix} + \frac{ie\mathbf{E}}{2\hbar} \frac{\partial \gamma_k}{\partial \mathbf{k}} \begin{pmatrix} f_{21} - f_{12} & f_{22} - f_{11} \\ f_{11} - f_{22} & f_{12} - f_{21} \end{pmatrix}$$
$$+ \frac{i}{\hbar} \begin{pmatrix} 0 & f_{12}(E_+ - E_-) \\ f_{21}(E_- - E_+) & 0 \end{pmatrix} = I[\hat{f}(\mathbf{k})], \tag{16}$$

where

$$\frac{\partial \gamma_k}{\partial \mathbf{k}} = \frac{\alpha^2 - \beta^2}{k\kappa_\theta^2} \left(\mathbf{e}_x \sin\theta - \mathbf{e}_y \cos\theta \right). \tag{17}$$

The collision term is given by Eq. (6) with $K^{\kappa\kappa'}_{\kappa_1\kappa_1'}(\mathbf{k}_1', \mathbf{k})$ derived for δ-potential scattering. Thus, each $K^{\kappa\kappa'}_{\kappa_1\kappa_1'}(\mathbf{k}_1', \mathbf{k})$ contains either $\sin(\gamma_k - \gamma_{k_1'})$ or $1 \pm \cos(\gamma_k - \gamma_{k_1'})$ depending on the relation $\kappa\kappa'\kappa_1\kappa_1'$.

In order to simplify Eq. (16), we assume $k\kappa_\theta \ll T$, where T is the temperature. This assumption will be discussed later in details. Now, we expand the Fermi distribution function $f^0(E_{sk})$ in terms of $k\kappa_\theta/T$ so, that Eq. (16) in the linear responce regime takes the form

$$\begin{pmatrix} e\mathbf{E}\mathbf{v}^{\mathbf{11}} \frac{\partial f^0(E_{+k})}{\partial E_{+k}} & \frac{e\mathbf{E}}{2}\mathbf{v}^{\mathbf{12}} \left[\frac{\partial f^0(E_{+k})}{\partial E_{+k}} + \frac{\partial f^0(E_{-k})}{\partial E_{-k}} \right] \\ \frac{e\mathbf{E}}{2}\mathbf{v}^{\mathbf{21}} \left[\frac{\partial f^0(E_{+k})}{\partial E_{+k}} + \frac{\partial f^0(E_{-k})}{\partial E_{-k}} \right] & e\mathbf{E}\mathbf{v}^{\mathbf{22}} \frac{\partial f^0(E_{-k})}{\partial E_{-k}} \end{pmatrix}$$
$$+ \frac{i}{\hbar} \begin{pmatrix} 0 & f_{12}(E_+ - E_-) \\ f_{21}(E_- - E_+) & 0 \end{pmatrix} = I[\hat{f}(\mathbf{k})]. \tag{18}$$

Though Eq. (18) is still quite cumbersome, it is easy to check that $f_{12} = f_{21} = 0$ and $f_{11(22)}$ given by

$$f_{\kappa\kappa} = f^0_{\kappa\kappa} + (-e\mathbf{E})\mathbf{k}\frac{\hbar\tau}{m} \left[-\frac{\partial f^0(E_{\kappa k})}{\partial E_{\kappa k}} \right]. \tag{19}$$

represent the solution. Thus, the off-diagonal elements of the distribution function do not play a role at high temperatures. This can be explained in what follows. Note, that the off-diagonal elements of the distribution function have essentially quantum mechanical origin since they correspond to the off-diagonal elements of the density matrix. Classically, an electron can be in only one state of two, and, therefore, off-diagonal terms vanish here. In contrast, in the spin-coherent kinetic equation the off-diagonal terms can be essential. However, in the real samples the quantum effects are negligible at room temperatures because of the temperature smearing. Indeed, the spin-orbit coupling constant κ_θ is of the order of $10^{-11}\,\mathrm{eV}\cdot\mathrm{m}$ for typical InAs samples. To be specific, let us take n-type InAs quantum well containing the 2DEG used for photocurrent measurements at room temperature [18]. The parameters are as follows: $\alpha/\beta = 2.15$, mobility is about $2 \cdot 10^4\,\mathrm{cm}^2/(\mathrm{V}\cdot\mathrm{s})$ and free carrier density is $1.3 \cdot 10^{12}\,\mathrm{cm}^{-2}$. The latter allows us to estimate characteristic Fermi wave vector $k_F = \sqrt{2\pi n_e} \simeq 3 \cdot 10^6\,\mathrm{cm}^{-1}$. Thus, the spin-orbit splitting energy $k_F\kappa_\theta$ is about 3 meV, that is much smaller than $T_{\mathrm{room}} = 25\,\mathrm{meV}$, and our solution is suitable for description of a large variety of experiments.

To study the spin accumulation we calculate the net spin density, whose x, y-components read

$$\langle S_{x,y} \rangle = \sum_s \int \frac{d^2k}{(2\pi)^2} S_{x,y}(k,s) f_s(\mathbf{k}), \tag{20}$$

where $S_x(k,s) = \frac{s}{2}\cos\gamma_k$, $S_y = -\frac{s}{2}\sin\gamma_k$ are the spin expectation values. (We do not consider S_z-component since it is zero.)

The integral over $\mathbf{k}$ can be taken easily making the substitution $\varepsilon = E(s,k)$ and assuming that $-\partial f^0(\varepsilon)/\partial\varepsilon = \delta(E_F - \varepsilon)$. This assumption is reasonable with respect to the system studied in Ref. [18] from which one can deduce a Fermi energy of the order of 100 meV, which is clearly larger than room temperature. Since our solution is valid for temperatures much higher than the spin-orbit splitting energy $k_F\kappa_\theta$, the inequality describing the applicability of our results obtained below reads

$$k_F\kappa_\theta \ll T \ll E_F.$$

The rest integrals over the polar angle can be taken analytically. After some algebra we have

$$\langle \mathbf{S} \rangle = \frac{em\tau}{2\pi\hbar^3}\begin{pmatrix} \beta & \alpha \\ -\alpha & -\beta \end{pmatrix}\mathbf{E}. \tag{21}$$

Then, the magnitude of the spin accumulation $\langle S \rangle = \sqrt{\langle S_x\rangle^2 + \langle S_y\rangle^2}$ is given by

$$\langle S \rangle = \frac{eEm\tau}{2\pi\hbar^3}\sqrt{\alpha^2 + \beta^2 + 2\alpha\beta\sin(2\widehat{\mathbf{E}\mathbf{e}_x})}. \tag{22}$$

It is interesting to note, that $\langle S \rangle$ depends on the direction of the electric field, i.e. the spin accumulation is anisotropic. To our knowledge, this interesting feature was not noticed in the literature so far. If $\beta = 0$ then $\langle S \rangle = eEm\tau\alpha/(2\pi\hbar^3)$ that is in agreement with Refs. [6, 8]. If the electric field is applied along the x-direction (the case studied in Ref. [9]), then the spin density is $\langle S \rangle = eEm\tau\sqrt{\alpha^2+\beta^2}/(2\pi\hbar^3)$. This result contradicts to Ref. [9], where the spin density has some strange kinks as a function of $m\alpha^2/\hbar^2$ and $m\beta^2/\hbar^2$. However, one can see from Eq. (22) that the kinks in the dependencies of $\langle S \rangle$ on α (or β) could not take place.

Relying on the effect found the following novel spintronic device can be proposed. Let us attach two pairs of contacts to the 2DEG so that the first pair provides the electric current along the crystallographic axis corresponding to the minimal spin accumulation (i.e. $[1\bar{1}0]$-axis for [001]-grown InAs samples [18]), and the second one is connected along the perpendicular axis. Then, the spin accumulation depends strongly on which contacts the transport voltage is applied, and its anisotropic contribution can be extracted easily using optical methods [2–5] or just measuring the magnetization. To give an example, applying the electric field of 20 V/m (which corresponds to the current density of 1 mA/cm) we obtain the magnetization difference of the order of $10^6\mu_B$ per cm^2, where μ_B is the Bohr magneton. This is comparable to the Pauli magnetization at the magnetic fields of a few gauss. It is also interesting to note the small characteristic switching time $\tau \simeq 10^{-13}$ s of the device proposed. Therefore, besides the fundamental importance of such an experiment, our four terminal device could find some applications as a high-speed spin switch.

At the end of the discussion, let us turn to the charge current whose density can be found as

$$\mathbf{j} = -e \sum_s \int \frac{d^2k}{(2\pi)^2} \mathbf{v_s} f_s(\mathbf{k}). \tag{23}$$

The integral over $\mathbf{k}$ in Eq. (23) can be taken in the same manner as in Eq. (20), and after some algebra the conductivity tensor σ takes the form

$$\sigma_{yy(xx)} = \frac{e^2\tau}{\pi\hbar^2} \left(\frac{m(\alpha^2+\beta^2)}{\hbar^2} + E_F \right), \tag{24}$$

and, most surprisingly, $\sigma_{xy(yx)} = 0$. It is convenient to express $\sigma_{yy(xx)}$ via the electron concentration (25) in a 2DEG with spin-orbit interactions

$$n_e = \frac{mE_F}{\pi\hbar^2} + \left(\frac{m}{\hbar^2}\right)^2 \frac{\alpha^2+\beta^2}{\pi}. \tag{25}$$

Then, the conductivity takes much simpler form, namely $\sigma_{yy(xx)} = e^2 n_e \tau/m$. This is the Drude formula, i.e. the conductivity is just a number (not a tensor), though the distribution function (19) is anisotropic. This result is in contrast with the findings of Ref. [19], where a subtle approximation generated off-diagonal elements in the conductivity tensor. This artifact is absent in our truly exact solution: The electrical conductivity is isotropic for any relation between α and β. The latter simplifies essentially the relation between the spin and charge current densities, which takes the form

$$\langle \mathbf{S} \rangle = \frac{m^2}{2\pi\hbar^3 e n_e} \begin{pmatrix} \beta & \alpha \\ -\alpha & -\beta \end{pmatrix} \mathbf{j}. \tag{26}$$

We find it useful to write down Eq. (26) in the basis $\frac{1}{\sqrt{2}}(1;1)$, $\frac{1}{\sqrt{2}}(1;-1)$. Then the relation between $\langle \mathbf{S} \rangle$ and $\mathbf{j}$ takes the simpler form

$$\langle \mathbf{S} \rangle = \frac{m^2}{2\pi\hbar^3 e n_e} \begin{pmatrix} 0 & \beta-\alpha \\ \beta+\alpha & 0 \end{pmatrix} \mathbf{j}. \tag{27}$$

From this equation one can see easily, that the spin accumulation is strongly anisotropic if the constants α and β are close to each other. The reason of such an anisotropy is the angular dependence of the dispersion law (10). Indeed, the spin-orbit splitting is different for different direction of the momentum. Therefore, the spin precession frequency depends essentially on the direction of the electron motion. This leads to the anisotropic spin relaxation times (see e.g. Ref. [20]), and, thus, the anisotropy of the spin accumulation occurs. In particular, the electric current of arbitrary strength applied along the $[1\bar{1}0]$ crystallographic axis does not lead to any spin accumulation at $\alpha = \beta$. The latter is due to the vanishing spin splitting along the $[1\bar{1}0]$-axis (see e.g. Ref. [21]).

Equations (26) and (27) are the main result of this section which can be applied directly to experimental studies of current-induced spin accumulation in [001]-grown InAs samples. The theory developed here is, of course, applicable to arbitrary oriented heterostructures after a few minor changes regarding the spin-orbit coupling terms in the Hamiltonian.

4 Transport in graphene

The carriers in the π-system of graphene near half filling can be described by the Dirac Hamiltonian

$$H = \hbar v_0(\sigma_x k_x + \sigma_y k_y), \tag{28}$$

where $v_0 \approx 10^6 \mathrm{ms}^{-1}$ is the effective "speed of light", $\sigma_{x,y}$ are the Pauli matrices, and $\mathbf{k}$ is the two-component particle momentum. The eigenstates of Eq. (28) have the form

$$\Psi_{\mathbf{k}\pm}(x,y) = \frac{1}{\sqrt{2}} \mathrm{e}^{ik_x x + ik_y y} \begin{pmatrix} 1 \\ \pm \mathrm{e}^{i\theta} \end{pmatrix}, \tag{29}$$

where $\tan\theta = k_y/k_x$, and the energy spectrum reads $E_{k\pm} = \pm\hbar v_0 k$. The velocity matrix in the basis Eq. (29) is

$$\frac{\mathbf{v}}{v_0} = \mathbf{e}_x \begin{pmatrix} \cos\theta & -i\sin\theta \\ i\sin\theta & -\cos\theta \end{pmatrix} + \mathbf{e}_y \begin{pmatrix} \sin\theta & i\cos\theta \\ -i\cos\theta & -\sin\theta \end{pmatrix}. \tag{30}$$

Let us consider the Boltzmann equation for charge carriers in the presence of impurity scatterers described by the effective potential

$$V(\mathbf{r}) = \frac{qeZ}{r} \mathrm{e}^{-r/R} \tag{31}$$

where eZ, q are the impurity atom and carrier electrostatic charge respectively, and R is the screening radius. The potential (31) differs from its conventional short-range δ-function approximation [22] by the additional fitting parameter R. As we shall see below, the minimum conductivity value is governed by *both* the impurity concentration and screening radius whereas the carrier mobility turns out to be R-independent. It seems to be necessary to introduce such a parameter in order to explain the experimental picture [15] where two samples made out of the same graphene flake (i.e. having equal mobility) demonstrate essentially different minimum conductivity values. This difference is attributed to the screening parameter R in our model.

As it is derived in the Sect. 2, the kinetic equation in the linear responce regime explicitly reads

$$I[f] = \frac{i}{\hbar}\begin{pmatrix} 0 & f_{12}\left(E_{k+} - E_{k-}\right) \\ f_{21}\left(E_{k-} - E_{k+}\right) & 0 \end{pmatrix}$$

$$+q\mathbf{E}\begin{pmatrix} -\mathbf{v}_{11}\left[-\frac{\partial f^0(E_{k+})}{\partial E_{k+}}\right] & \frac{\mathbf{v}_{12}}{2E_{k+}}\left(f^0_{E_{k-}} - f^0_{E_{k+}}\right) \\ \frac{\mathbf{v}_{21}}{2E_k}\left(f^0_{E_{k+}} - f^0_{E_{k-}}\right) & -\mathbf{v}_{22}\left[-\frac{\partial f^0(E_{k-})}{\partial E_{k-}}\right] \end{pmatrix}, \tag{32}$$

where the collision term given by Eq. (6) with $K^{\kappa\kappa_1}_{\kappa'\kappa'_1}$ being

$$K^{\kappa\kappa_1}_{\kappa'\kappa'_1} = (\pi R^2 V_0^2/\hbar) \times \frac{1 + \kappa\kappa'\kappa_1\kappa'_1 + \kappa\kappa' \mathrm{e}^{i(\theta'-\theta)} + \kappa_1\kappa'_1 \mathrm{e}^{-i(\theta'-\theta)}}{1 + R^2\left[k^2 + k'^2 - 2kk'\cos(\theta'-\theta)\right]}. \tag{33}$$

Here $V_0 = 2\pi qeZ\sqrt{N}$ is an effective scattering potential with N being the impurity concentration,

In contrast to the previous section the high temperature approximation is not possible here since the "pseudospin splitting" $E_{k+} - E_{k-}$ is of the order of Fermi energy and, therefore, might be much larger than the temperature. However, by somewhat more tedious calculations one can again construct an analytical solution to Eq. (32) with nonequilibrium terms given by

$$f^1_{11} = q\mathbf{E}\mathbf{v}_{11}\tau(k)\left\{\left(1 + \frac{1}{2\alpha}\right)\left[-\frac{\partial f^0(E_{k+})}{\partial E_{k+}}\right] + \frac{1}{2\alpha}\left[-\frac{\partial f^0(E_{k-})}{\partial E_{k-}}\right] + \frac{1}{2\alpha E_{k+}}\left(f^0_{E_{k+}} - f^0_{E_{k-}}\right)\right\} \tag{34}$$

$$f^1_{12} = \frac{q\mathbf{E}\mathbf{v}_{12}\tau(k)\left(\frac{1}{2} + \frac{1}{2\alpha}\right)}{1 + 2iE_{k+}\tau(k)/\hbar}\left\{\frac{1}{E_{k+}}\left(f^0_{E_{k+}} - f^0_{E_{k-}}\right) + \left[-\frac{\partial f^0(E_{k+})}{\partial E_{k+}} - \frac{\partial f^0(E_{k-})}{\partial E_{k-}}\right]\right\}, \tag{35}$$

and f^1_{22}, f^1_{21} can be obtained from Eqs. (34–35) just exchanging the indices belong to E_k and $\mathbf{v}$ accordingly. Here we have introduced the novel electron-hole incoherence parameter $\alpha = 4E^2_{k+}\tau^2(k)/\hbar^2$. The momentum realxation time $\tau(k)$ is given by

$$\tau(k) = \frac{1}{k}\frac{\hbar^2 v_0}{4R^2V_0^2}\frac{2R^4k^4}{1 + 2R^2k^2 - \sqrt{1 + 4R^2k^2}} \tag{36}$$

$$\approx \frac{1}{k}\frac{\hbar^2 v_0}{4R^2V_0^2}, \quad Rk \ll 1. \tag{37}$$

In the limit case of weak scattering ($\alpha \gg 1$) the diagonal elements f^1_{11} and f^1_{22} are $f^1_{\kappa\kappa} = q\mathbf{E}\mathbf{v}_{\kappa\kappa}\tau(k)\left[-\partial f^0(E_{k\kappa})/\partial E_{k\kappa}\right]$, whereas the off-diagonal elements f^1_{12} and f^1_{21} are real and do not contribute to the current which reads

$$\mathbf{j} = 4q \int \frac{d^2k}{(2\pi)^2} \left(\mathbf{v_{11}} f_{11} + \mathbf{v_{22}} f_{22} + 2\Im \mathbf{v_{12}} \Im f_{12} \right), \tag{38}$$

where the factor 4 is due to the fourfold degeneracy.

Let us first concentrate on the electron-hole incoherent case when $\alpha \gg 1$. Utilising Eq. (37) we find from Eq. (38) for the conductivity

$$\sigma \equiv \sigma_0 = \frac{q^2 \hbar v_0^2}{4\pi R^2 V_0^2} \equiv \frac{q^2 \sqrt{\alpha}}{h}, \tag{39}$$

which, in particular, does neither depend on temperature nor on the carrier concentration. Thus, the most striking features of the experimental findings are reproduced: (i) the conductivity is not zero even at zero carrier concentration, (ii) this minimum conductivity does not depend on temperature. Note, that the above result is *not universal*, i.e. the minimum conductivity at zero Fermi energy can change from sample to sample in accordance with recent experimental reports [14, 15]. Moreover $\sqrt{\alpha} > 4$ (i.e. $1/\alpha < 0.0625$) for the vast majority of samples [14, 15].

Let us take into account higher-order terms in $\tau(Rk)$. Then instead of Eq. (39) we have at zero temperature

$$\sigma = \frac{q^2}{\pi \hbar^2} \tau(k_F) E_F, \tag{40}$$

$$\approx \sigma_0 + q^2 \frac{\hbar v_0^2}{2V_0^2} n, \quad 2\pi R^2 n \ll 1 \tag{41}$$

where the carrier (electron) concentration is given by $n = k_F^2/\pi$ with $k_F = E_F/\hbar v_0$ being the Fermi wave vector. Thus, the low-temperature conductivity at low doping increases linearly with carrier concentration, in accordance with the experiments. Deviations from linear dependency can be described taking into account k-dependence of the relaxation time given by Eq. (36). We emphasise, that in contrast with [15] we deal with two fitting parameters R and V_0 which can be deduced using Eq. (41) and the experimental data [11, 14, 15]. Indeed, from Eq. (41) we can define the electron mobility as $\mu = q\hbar v_0^2/(2V_0^2)$ which *does not depend on the screening radius R*. The effective scattering potential then reads $V_0 = v_0\sqrt{q\hbar/(2\mu)}$, and for the most common samples with μ ranged from $\sim 10^3$ to $2 \cdot 10^4\,\mathrm{cm}^2/(\mathrm{Vs})$ we have V_0 covering the range from 0.1 meV to 0.06 eV. The screening radius can be estimated from Eq. (39) assuming that σ_0 is of the order of $4q^2/h$. Then we have R of the order of $\hbar v_0/V_0$ covering the range from $10^{-3}\,\mathrm{cm}$ (high mobility samples) to $10^{-6}\,\mathrm{cm}$ (low mobility samples). Now one can see that the linear approximation (41) in terms of $2\pi R^2 n$ at $n \sim 10^{12}\,\mathrm{cm}^{-2}$ holds only in relatively low mobility samples, in accordance with the experimental report [14, 15].

To consider the low mobility samples properly, we should take into account the terms proportional to $1/\alpha$ in the solution (34–35). Note, above all, that the quasiclassical approach is doubtful at $\alpha < 1$ since $E_F\tau(k_F) < \hbar$, and the carrier mean free path becomes comparable with its de Broglie wavelength. Nevertheless, an asymptotic dependence of the solution close to $\alpha \sim 1$ can give us a clue to what happens in this regime. The direct integration of Eq. (38) with $\tau(k)$ given by Eq. (37) leads to the logarithmic divergence due to the terms proportional to $f^0_{E_{k+}} - f^0_{E_{k-}}$ in Eqs. (34–35). It is therefore essential to integrate over $\mathbf{k}$ *before* the subsequent approximations relying on a small R; we would obtain the logarithmic divergence in the conductivity integral otherwise [23, 24]. The correct formula for conductivity reads

$$\sigma = \sigma_0 + q^2 \frac{\hbar v_0^2}{2V_0^2} n + q^2 \frac{R^2 V_0^2}{\pi \hbar^3 v_0^2} \ln(\pi R^2 n), \tag{42}$$

Interrestingly, the conductivity minimum is uncertain, but since R is small the conductivity at $n \to 0$ is indeed indistinguishable from its residual value σ_0. This case corresponds to the chemically undoped graphene samples studied in Ref. [15] where the conductivity demonstrates rather sharp dip with nonuniversal minimum value strongly dependent on scattering parameters. Moreover, according to the very recent report [25], the conductivity of suspended graphene after annealing indeed demonstrates very sharp dip at zero gate voltages.

5 Conculsion remarks

We have proposed a quasiclassical model which describes kinetics of particles having an additional quantum degree of freedom. We have solved the kinetic equation analytically for 2DEGs with arbitrary large spin-orbit interactions of both Rashba and Dresselhaus type. This solution is suitable for samples at sufficiently high temperatures common to experiments. Using this solution, we discovered the anisotropy of the current-induced spin accumulation, though the conductivity remains isotropic. Our analytical study is expected to be a reliable starting point for further investigations of spin dependent electron transport. We have also solved the kinetic equation for carriers in a single graphene sheet including the off-diagonal elements of the distribution function in the helicity basis. The analytical solution allows us to investigate the influence of the electron-hole coherence on the minimum conductivity phenomena as well as to discover the limitations of previous studies based on Boltzmann equation. We have introduced a special parameter α distinguishing the electron-hole coherent and incoherent regimes. Our approach successfully describes the linear dependence of the conductivity above its minimum as well as the sharp dip at negligible carrier concentrartions.

References

1. S. D. Ganichev, E. L. Ivchenko, V. V. Bel'kov, S. A. Tarasenko, M. Sollinger, D. Weiss, W. Wegscheider, W. Prettl: Spin-galvanic effect, Nature **417**, 153 (2002)
2. S. D. Ganichev, S. N. Danilov, P. Schneider, V. V. Bel'kov, L. E. Golub, W. Wegscheider, D. Weiss, W. Prettl: Electric current-induced spin orientation in quantum well structures, J. Magn. Magn. Mater. **300**, 127 (2006)
3. A. Y. Silov, P. A. Blajnov, J. H. Wolter, R. Hey, K. H. Ploog, N. S. Averkiev: Current-induced spin polarization at a single heterojunction, Appl. Phys. Lett. **85**, 5929 (2004)
4. C. L. Yang, H. T. He, L. Ding, L. J. Cui, Y. P. Zeng, J. N. Wang, W. K. Ge: Spectral dependence of spin photocurrent and current-induced spin polarization in an InGaAs/InAlAs two-dimensional electron gas, Phys. Rev. Lett. **96**, 186605 (2006)
5. N. P. Stern, S. Ghosh, G. Xiang, M. Zhu, N. Samarth, D. D. Awschalom: Current-induced polarization and the spin hall effect at room temperature, Phys. Rev. Lett. **97**, 126603 (2006)
6. V. M. Edelstein: Spin polarization of conduction electrons induced by electric current in two-dimensional asymmetric electron systems, Solid State Comm. **73**, 233 (1990)
7. A. G. Aronov, Y. B. Lyanda-Geller, G. E. Pikus: The polarization of electrons by an electric current, JETP **100**, 973 (1991)
8. J.-I. Inoue, G. E. W. Bauer, L. W. Molenkamp: Diffuse transport and spin accumulation in rashba two-dimensional electron gas, Phys. Rev. B **67**, 33104 (2003)
9. Z. Huang, L. Hu: Controllable kinetic magnetoelectric effect in two-dimensional electron gases with both rashba and dresselhaus spin-orbit couplings, Phys. Rev. B **73**, 113312 (2006)
10. Y. Jiang, L. Hu: Kinetic magnetoelectric effect in a two-dimensional semiconductor strip due to boundary-confinement-induced spin-orbit coupling, Phys. Rev. B **74**, 75302 (2006)
11. K. S. Novoselov, A. K. Geim, S. V. Morozov, D. Jiang, M. I. Katsnelson, I. V. Grigorieva, S. V. Dubonos, A. A. Firsov: Two-dimensional gas of massless dirac fermions in graphene, Nature **438**, 197 (2005)
12. Y. Zhang, Y.-W. Tan, H. L. Stormer, P. Kim: Experimental observation of the quantum hall effect and berry's phase in graphene, Nature **438**, 201 (2005)
13. S. Cho, M. S. Fuhrer: Charge transport and inhomogeneity near the charge neutrality point in Graphene (unpublished, arXiv:0705.3239)
14. J. H. Chen, C. Jang, S. Adam, M. S. Fuhrer, E. D. Williams, M. Ishigami: Charged-impurity scattering in graphene, Nature Physics **4**, 377–381 (2008)
15. Y.-W. Tan, Y. Zhang, K. Bolotin, Y. Zhao, S. Adam, E. H. Hwang, S. D. Sarma, H. L. Stormer, P. Kim: Measurement of scattering rate and minimum conductivity in graphene, Phys. Rev. Lett. **99**, 246803 (2007)
16. Y. Bychkov, E. Rashba: Properties of 2D electron gas with lifted spectral degeneracy, JEPT Lett. **39**, 78 (1984)
17. G. Dresselhaus: Spin-orbit coupling effects in zinc blende structures, Phys. Rev. **100**, 580 (1955)

18. S. D. Ganichev, V. V. Bel'kov, L. E. Golub, E. L. Ivchenko, P. Schneider, S. Giglberger, J. Eroms, J. D. Boeck, G. Borghs, W. Wegscheider, D. Weiss, W. Prettl: Experimental separation of rashba and dresselhaus spin splittings in semiconductor quantum wells, Phys. Rev. Lett. **92**, 256601 (2004)
19. J. Schliemann, D. Loss: Anisotropic transport in a two-dimensional electron gas in the presence of spin-orbit coupling, Phys. Rev. B **68**, 165311 (2003)
20. N. S. Averkiev, L. E. Golub: Giant spin relaxation anisotropy in zinc-blende heterostructures, Phys. Rev. B **60**, 15582 (1999)
21. J. Schliemann, J. C. Egues, D. Loss: Nonballistic spin-field-effect transistor, Phys. Rev. Lett. **90**, 146801 (2003)
22. K. Nomura, A. H. MacDonald: Quantum transport of massless dirac fermions, Phys. Rev. Lett. **98**, 76602 (2007)
23. M. Auslender, M. I. Katsnelson: Generalized kinetic equations for charge carriers in graphene, Phys. Rev. B **76**, 235425 (2007)
24. M. Trushin, J. Schliemann: Minimum electrical and thermal conductivity of graphene: a quasiclassical approach, Phys. Rev. Lett. **99**, 216602 (2007)
25. K. I. Bolotin, K. J. Sikes, Z. Jiang, M. Klima, G. Fudenberg, J. Hone, P. Kim, H. L. Stormer Ultrahigh electron mobility in suspended graphene, Solid State Communications **146**, 351–355 (2008)

Spin Dynamics in High-Mobility Two-Dimensional Electron Systems

Tobias Korn, Dominik Stich, Robert Schulz, Dieter Schuh, Werner Wegscheider, and Christian Schüller

Institut für Experimentelle und Angewandte Physik, Universität Regensburg, Universitätsstrasse 31, 93040, Regensburg, Germany
tobias.korn@physik.uni-regensburg.de

Abstract. Understanding the spin dynamics in semiconductor heterostructures is highly important for future semiconductor spintronic devices. In high-mobility two-dimensional electron systems (2DES), the spin lifetime strongly depends on the initial degree of spin polarization due to the electron–electron interaction. The Hartree-Fock (HF) term of the Coulomb interaction acts like an effective out-of-plane magnetic field and thus reduces the spin-flip rate. By time-resolved Faraday rotation (TRFR) techniques, we demonstrate that the spin lifetime is increased by an order of magnitude as the initial spin polarization degree is raised from the low-polarization limit to several percent. We perform control experiments to decouple the excitation density in the sample from the spin polarization degree and investigate the interplay of the internal HF field and an external perpendicular magnetic field. The lifetime of spins oriented in the plane of a [001]-grown 2DES is strongly anisotropic if the Rashba and Dresselhaus spin-orbit fields are of the same order of magnitude. This anisotropy, which stems from the interference of the Rashba and the Dresselhaus spin-orbit fields, is highly density-dependent: as the electron density is increased, the kubic Dresselhaus term becomes dominant und reduces the anisotropy.

1 Introduction

In recent years, semiconductor spintronics [1–3] research has found increased interest, in part due to new materials like ferromagnetic semiconductors [4]. Among the key requirements for semiconductor spintronic devices is an understanding of the spin dephasing mechanisms in semiconductors. For GaAs, many experimental studies have focused on slightly n-doped bulk material, where extremely long spin lifetimes ($\geq 100\,\mathrm{ns}$) were observed [5, 6] for doping levels close to the metal-insulator transition. Even though this material has low mobility and poor conductivity properties, it was used in a number of spin injection [7] and transport [8, 9] experiments.

Relatively few studies have been performed on high-mobility two-dimensional electron systems (2DES): Brand et al. investigated the weak scattering regime and the temperature dependence of the D'Yakonov-Perel (DP) mechanism in a high-mobility 2DES [10], while Leyland et al. [11] experimentally showed the importance of electron–electron collisions for spin dephasing. Several groups performed experiments on [110]-grown quantum wells and 2DES, in which the DP mechanism can be suppressed for spins aligned along the growth direction [12], while it remains active for other spin orientations [13, 14]. In these systems, gate control of the spin lifetime was demonstrated by Karimov et al. [15]. Theoretical studies investigating the electron–electron interaction in 2DES were performed using perturbation theory by Glazov and Ivchenko [16], while Wu et al. developed a powerful microscopic many-body approach [17] to study 2DES far from thermal equilibrium, considering all relevant scattering mechanisms, including electron-hole and electron-phonon scattering [18–21].

2 Theory

2.1 Optical orientation of electrons in a 2D electron system

In GaAs, the conduction band has *s*-like character, while the valence bands have *p*-like character. The light-hole (LH $J_z = \pm\frac{1}{2}$) and the heavy-hole (HH $J_z = \pm\frac{3}{2}$) valence bands are degenerate at k=0 for bulk GaAs. Optical excitation above the bandgap can generate electron-hole pairs through photon absorption. Due to angular momentum conservation, a circularly-polarized photon ($s = 1$) may excite both, electrons with spin up (from a light-hole valence band state with $J_z = -\frac{1}{2}$) and spin down (from a heavy hole valence band state with $J_z = -\frac{3}{2}$). Due to the different probabilities of these transitions, a finite spin polarization may be created in bulk GaAs [22]. In a quantum well, the $\boldsymbol{k} = 0$ degeneracy of the light and heavy hole valence bands is lifted due to confinement. By resonantly exciting only the HH or LH transition in a quantum well (QW) with circularly-polarized light, almost 100 percent spin polarization may be generated [23]. If a 2DES is created within the QW by (modulation) doping, the electrons occupy conduction band states up to the Fermi energy. Optical absorption is only possible into states above the Fermi energy, and does not occur at $\boldsymbol{k} = 0$. For $\boldsymbol{k} \neq 0$, the valence band states in a QW consist of an admixture of heavy-hole and light-hole states and J_z is no longer a good quantum number [24]. Therefore, excitation with circularly polarized light will yield a mixture of spin-up and spin-down electrons in the conduction band, depending on the excitation wavelength. Additionally, the 2D electron system is typically unpolarized in thermal equilibrium. In order to create a large spin polarization in a 2D electron system by optical excitation, the excitation density therefore has to be on the order of the 2DES density if we assume that the 2D system returns to thermal equilibrium in between excitation pulses. The initial spin polarization degree P, created by a short optical pulse, may be calculated using the following formula:

$$P = \frac{n_{\text{ph}}}{n_{\text{e}} + n_{\text{ph}}^{\text{tot}}} \,. \tag{1}$$

Here, $n_{ph} = \xi \cdot n_{ph}^{tot}$ is the spin-polarized fraction ξ of the optically created electron density, $n_{\text{ph}}^{\text{tot}}$ is the total electron density that is optically created, and n_{e} is the background electron density of the 2DES.

2.2 Rashba and Dresselhaus spin-orbit fields in a [001] quantum well

In crystal structures which lack inversion symmetry, like GaAs, the spin-orbit interaction may be described by an intrinsic, $\boldsymbol{k}$-dependent magnetic field $\boldsymbol{B}_i(\boldsymbol{k})$, which causes a precession of the electron spin. Typically, a Larmor precession frequency vector corresponding to the electron precession about this internal field is defined as $\boldsymbol{\Omega}(\boldsymbol{k}) = \frac{g}{\hbar} \cdot \mu_B \boldsymbol{B}_i(k)$. In bulk structures, $\boldsymbol{\Omega}(\boldsymbol{k})$ is cubic in $\boldsymbol{k}$ and has the following form [25]:

$$\boldsymbol{\Omega}(\boldsymbol{k})_{BIA} = \frac{\gamma}{\hbar} \cdot [k_x(k_y^2 - k_z^2), k_y(k_z^2 - k_x^2), k_z(k_x^2 - k_y^2)] \,. \tag{2}$$

This term stems from the inversion asymmetry of the crystal lattice and is therefore often called bulk inversion asymmetry (BIA) term or Dresselhaus term. In a quantum well grown along the z-direction, the momentum along the growth direction is quantized due to confinement. In first approximation, the expectation value is $\langle k_z^2 \rangle = (\pi/d)^2$, where d is the quantum well thickness. It follows that:

$$\boldsymbol{\Omega}(\boldsymbol{k})_{BIA(2D)} = \frac{\gamma}{\hbar} \cdot [k_x(k_y^2 - \langle k_z^2 \rangle), k_y(\langle k_z^2 \rangle - k_x^2), 0] \,. \tag{3}$$

Typically in a 2D electron system, the in-plane momentum $(k_{\|})^2$ is smaller than $\langle k_z^2 \rangle$. Therefore, terms kubic in the in-plane momentum are often neglected, resulting in the following approximation:

$$\boldsymbol{\Omega}(\boldsymbol{k})_{\text{Dressel}} = \frac{\beta}{\hbar} \cdot [-k_x, k_y, 0] \,. \tag{4}$$

This is linear in the in-plane momentum $k_{\|}$ and therefore typically called linear Dresselhaus term. Its symmetry is shown in Fig. 1(a). Additionally, a lack of structure inversion symmetry along the quantum well growth direction causes a second intrinsic effective magnetic field, the so-called Rashba field [26].

$$\boldsymbol{\Omega}(\boldsymbol{k})_{\text{Rashba}} = \frac{\alpha}{\hbar} \cdot [k_y, -k_x, 0] \,. \tag{5}$$

Structure inversion asymmetry (SIA) along the growth direction may be induced by different barrier materials on either side of the quantum well, single-sided or asymmetric modulation doping resulting in an effective electric field due to ionized donors, or the application of an external electric field by a gate voltage. The symmetry of the Rashba field is shown in Fig. 1(b).

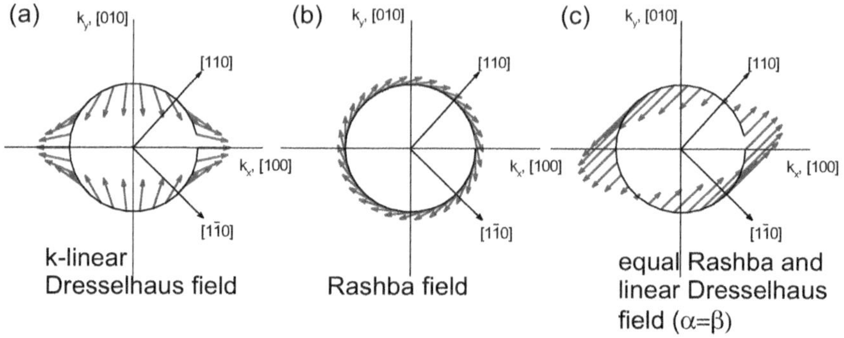

Fig. 1. (**a**) Symmetry of the linear Dresselhaus field in a 2D quantum well. The *red arrows* indicate the effective field direction and amplitude depending on the electron k vector in the xy plane. (**b**)Symmetry of the Rashba field in a 2D quantum well. (**c**) Vector sum of linear Dresselhaus and Rashba field (combined effective field, CEF) for identical amplitudes ($\alpha = \beta$).

2.3 Spin dephasing/relaxation

The main mechanism for spin dephasing in GaAs bulk and quantum well structures at low temperatures is the so-called D'Yakonov-Perel mechanism [27]. It is caused by the k-dependent spin-orbit fields which cause a precession of the electron spin. In an ensemble of electrons, the k values are distributed according to Fermi statistics, causing different precession frequencies and directions for the electron spins, which leads to dephasing. Two regimes can be distinguished according to the relationship between the average precession frequency $\bar{\Omega}$ and the momentum relaxation time τ_P:

1. weak scattering regime ($\bar{\Omega} \cdot \tau_P > 1$): here, the electron spins may process more than one full cycle about the spin-orbit field before they are scattered and their k value changes. A collective precession of electron spins may be observed in this regime.
2. strong scattering regime ($\bar{\Omega} \cdot \tau_P < 1$): here, the electrons are scattered so frequently that the spin-orbit field acts like a rapid fluctuation. In this regime, the spin relaxation time T_2^* is inversely proportional to the momentum relaxation time: $\frac{1}{T_2^*} \propto \tau_P$.

2.4 Magneto-Anisotropy

In samples where both the Rashba and the Dresselhaus terms are present and of the same magnitude, the effective spin-orbit field has to be calculated as the vector sum of the two terms:

$$\boldsymbol{\Omega}(\boldsymbol{k})_{CEF} = \frac{1}{\hbar} \cdot [(\alpha k_y - \beta k_x), (\beta k_y - \alpha k_x), 0] \; . \tag{6}$$

Due to the different symmetry of the two contributions, the resulting combined effective field (CEF) may show a well-defined preferential direction if $\alpha = \beta$:

$$\boldsymbol{\Omega}(\boldsymbol{k})_{CEF} = \frac{\beta}{\hbar} \cdot [(k_y - k_x), (k_y - k_x), 0] \equiv \frac{\beta}{\hbar}(k_y - k_x) \cdot [1, 1, 0] \,. \quad (7)$$

In this case, the CEF points along the in-plane [110] direction for any value of $\boldsymbol{k}$, as Fig. 1(c) shows. (For $\alpha = -\beta$ this preferential direction becomes $[1\bar{1}0]$). This means that electron spins that point along the [110] direction experience no torque and can therefore not dephase due to the D'Yakonov–Perel mechanism, which is effectively blocked for this spin orientation, while electron spins pointing along $[1\bar{1}0]$ or along the growth direction will experience a torque and start dephasing. This spin dephasing anisotropy was first pointed out by Averkiev and Golub [28], and recently observed experimentally [29, 30].

3 Sample structure and preparation

Our sample was grown by molecular beam epitaxy on a [001]-oriented semi-insulating GaAs substrate. The active region is a 20 nm-wide, one-sided modulation-doped GaAs-$Al_{0.3}Ga_{0.7}As$ single QW. The electron density and mobility at $T = 4.2\,K$ are $n_e = 2.1 \times 10^{11}\,cm^{-2}$ and $\mu_e = 1.6 \times 10^{6}\,cm^2/Vs$, respectively. These values were determined by transport measurements on an unthinned sample. For measurements in transmission geometry, the sample was glued onto a sapphire substrate with an optical adhesive, and the substrate and buffer layers were removed by selective etching.

4 Measurement techniques

4.1 Time-resolved Kerr/Faraday rotation

For both, the time-resolved Faraday rotation (TRFR) and the time-resolved Kerr rotation (TRKR) measurements, two laser beams from a mode-locked Ti:Sapphire laser, which is operated at 80 MHz repetition rate, were used. The laser pulses had a temporal length of about 600 fs each, resulting in a spectral width of about 3–4 meV, which allowed for near-resonant excitation. The laser wavelength was tuned to excite electrons from the valence band to states slightly above the Fermi energy of the host electrons in the conduction band. Both laser beams were focused to a spot of approximately 60 μm diameter on the sample surface. The pump pulses were circularly polarized by an achromatic $\frac{\lambda}{4}$ plate in order to create spin-oriented electrons in the conduction band, with spins aligned perpendicular to the QW plane. The weaker probe pulses were linearly polarized. The polarization rotation of the transmitted/reflected probe beam was analyzed by an optical bridge detector. In order to separate the time evolution of the spin polarization from the photocarrier dynamics, all measurements were performed using both helicities for the circularly-polarized pump beam [31]. The TRFR measurements were performed in a split-coil magnet cryostat with a ^{3}He insert, allowing for sample temperatures between 1.5 and 4.5 K. The TRKR measurements were performed in a continuous-flow He cold finger cryostat. In this cryostat, non-thinned samples from the same wafer were used. Unless otherwise stated, the experiments were carried out at a nominal sample temperature of T=4.5 K.

5 Experimental results

5.1 Variation of spin polarization degree

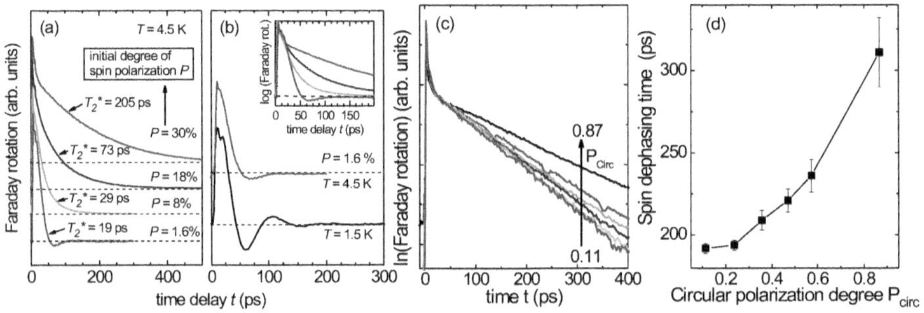

Fig. 2. (**a**) Normalized TRFR traces for different pump beam fluence and therefore different initial spin polarization. (**b**) Comparison of TRFR traces for low initial spin polarization at different temperatures. The inset shows the data from (**a**) with a log scale. (**c**) TRKR traces for varying circular polarization degree of the pump laser beam. The curves have been normalized for easier comparison. Note the log scale. (**d**) Spin relaxation time as a function of circular polarization degree of the pump beam, extracted from the TRKR traces. ((**a,b**) reprinted with permission from [32]. Copyright (2007) by the American Physical Society).

Typically, in time-resolved Faraday/Kerr rotation measurements on semiconductor heterostructures, the excitation density is kept very low to avoid heating the sample. Here, we present experiments in which the excitation density was considerable, resulting in a significant initial spin polarization. Figure 2(a) shows a series of TRFR measurements performed without external magnetic field. In this measurement, the excitation density, and therefore the initial spin polarization, was increased by increasing the laser pump fluence. The curves typically show a biexponential decay of the spin polarization, except for the lowest curve, where a strongly damped oscillation is visible. The oscillatory behavior will be discussed in the following subsection. We associate the fast decay with the spin dephasing of the photoexcited holes, which typically lose their spin orientation within a few picoseconds, and the slower decay with the spin dephasing time of the electrons, T_2^*. It is clearly seen how T_2^* is increased by more than an order of magnitude as the initial spin polarization is increased from the low-polarization limit to about 30%. This increase is due to the Hartree-Fock (HF) term of the Coulomb interaction (as predicted by Weng and Wu [18]), which acts as an out-of-plane, $\boldsymbol{k}$-dependent effective magnetic field. This effective field lifts the degeneracy of the spin-up and spin-down states, causing a spin-flip to require a change in energy. This significantly reduces the spin-flip rate and increases the spin dephasing time. In microscopic calculations including the HF term of the Coulomb interaction, the measurements shown in Fig. 2(a) could be reproduced with great accuracy.

In order to clearly demonstrate that the observed effect is due to the increase in spin polarization, and not caused by either a change in electron density or sample heating, we performed control experiments in which the initial spin polarization was varied while the excitation density was kept constant [33]. In order to achieve this, the circular polarization degree of the pump laser pulse was varied from almost 100% circular polarization to almost linear polarization, by gradual rotation of the $\frac{\lambda}{4}$ plate in the pump beam. By this means, the fraction of spin-polarized electrons created by the pump beam could be tuned. Figure 2(c) shows a series of TRKR measurements performed in this way. It is clearly visible how an increase of the circular polarization degree of the pump beam, and therefore an increase of the initial spin polarization, leads to reduced spin dephasing and therefore a shallower slope in the logarithmic plot of the TRKR trace. The TRKR traces have been normalized for easier comparison. In Fig. 2(d), the spin dephasing times extracted from the measurement series are plotted as a function of the circular polarization degree.

5.2 Coherent zero-field oscillation

In Fig. 2(b), two TRFR traces taken without external magnetic field at low excitation density are shown. The upper (red) curve, showing a strongly damped oscillation, represents the same data as the lowest (red) curve in Fig. 2(a). The lower curve in Fig. 2(b) was taken using the same excitation density, but at a lower sample temperature of 1.5 K. Here, the damping is significantly reduced. Both traces are representative of the weak scattering regime of the DP mechanism: an ensemble of electrons with $\boldsymbol{k}$ slightly above the Fermi wave vector is generated by the pump pulse. While the orientation of the effective spin orbit field for these electrons depends on the electron $\boldsymbol{k}$ vector, it generally lies in the sample plane, leading to spin precession. The z component of the electron spins oscillates, and the damped oscillation we observe, is the coherent superposition of these oscillations. We note that this effect was first reported by Brand et al. [10]. Two effects contribute to the damping of this coherent oscillation:

1. The amplitude of the effective spin-orbit field depends on $\boldsymbol{k}$ if both the Rashba and the Dresselhaus terms are present (see for example Fig. 1(c)), leading to a (*reversible*) dephasing of the electron spins due to different precession frequencies.
2. Initially, the electron spins are aligned along the growth direction by the laser pump pulse, therefore the effective spin-orbit field is perpendicular to the electron spin for the whole ensemble. Precession tilts the electron spins into the sample plane, and momentum scattering changes $\boldsymbol{k}$ and $\boldsymbol{\Omega}(\boldsymbol{k})$, leading to different angles between the electron spins and the effective spin-orbit fields. This causes (*irreversible*) dephasing due to different spin precession frequencies. Momentum scattering is temperature-dependent, hence the damping of the coherent oscillation becomes more pronounced as the sample temperature is raised.

5.3 Spin dephasing in an external magnetic field

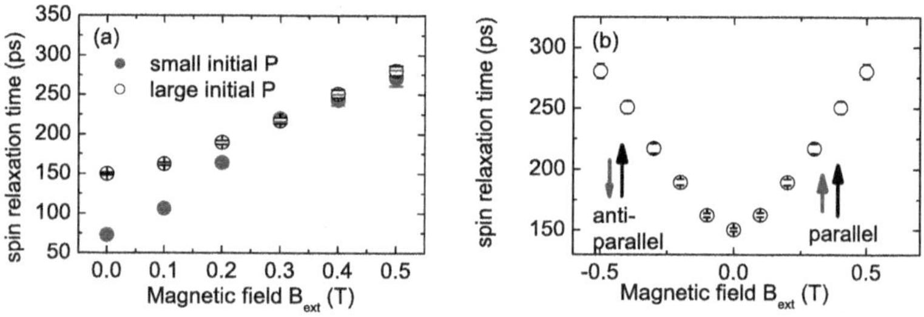

Fig. 3. (**a**) Spin dephasing times as a function of an external magnetic field perpendicular to the QW plane for small and large initial spin polarization. (**b**) Same as (**a**) for large initial spin polarization and both polarities of the external magnetic field.

Here, we present experiments performed in small external magnetic fields applied perpendicular to the quantum well plane. The field amplitude was kept below 0.5 Tesla, where effects due to Landau quantization may be neglected. In this regime, we observe a competition between the Hartree-Fock effective out-of-plane field, created by the interaction between spin-aligned electrons, and the external magnetic field. As Fig. 3(a) shows, the application of an external magnetic field monotonically increases the spin dephasing time for both small and large initial spin polarization. At zero and small external fields, the HF field dominates, leading to a significantly longer spin lifetime for large initial spin polarization. However, as the external field amplitude is increased, the spin lifetimes for large and small initial spin polarization start to merge and become almost identical for magnetic fields of 0.4 Tesla and above. This indicates that the mean value of the HF field for the large initial spin polarization is below 0.4 Tesla.

The spin polarization, which we create optically, defines a preferential direction along the growth axis. The external magnetic field can therefore be applied parallel or antiparallel to this preferential direction. Figure 3(b) shows that for both external field directions the spin lifetime increases monotonically, and the curve is symmetric in the external field B. This indicates that the external magnetic field cannot be used to compensate the effective Hartree-Fock field, since that is $\boldsymbol{k}$-dependent.

5.4 Probing magneto-anisotropy

The magneto-anisotropy in a [001]-grown GaAs QW leads to different spin lifetimes for electron spins aligned along different in-plane directions. In our experimental setup, however, we create an out-of-plane spin polarization by the circularly-polarized pump pulse, and the detection scheme using the linearly-polarized probe beam at near-normal incidence is only sensitive to the out-of-plane component of the spin polarization. Therefore, in order to probe the in-plane magneto-anisotropy, we use an in-plane magnetic field to force the spins to precess into the sample plane and back out again. By this means, the spin lifetimes we observe, represent an average of the out-of-plane and the different in-plane spin lifetimes within the sample and therefore allow us to infer the in-plane anisotropy. In the experiment, we use two sample pieces from the same wafer, which are mounted in the cryostat with their in-plane [110] axis either parallel or perpendicular to the applied in-plane magnetic field. Figure 4 shows the experimental results: (a) without an applied magnetic field, both sample pieces (black line and red circles) show the same spin lifetime $T_2^* = 113 \pm 1$ ps, as we expect due to symmetry reasons. (b) As an in-plane magnetic field of 1 Tesla is applied, the spin lifetimes become markedly different: if the magnetic field is applied along the in-plane [110] direction, the electron spins are forced to precess into the $[1\bar{1}0]$ direction. The spin lifetime average between the [001] and the $[1\bar{1}0]$ is $T^*_{2[1\bar{1}0]} = 127 \pm 1$ ps. If, however, the magnetic field is applied along $[1\bar{1}0]$, forcing the electron spins into the [110] direction, the averaged spin lifetime $T^*_{2[110]} = 204 \pm 2$ ps increases by about 60%. This is a clear indication of the in-plane magneto-anisotropy. From the magnetic-field dependence of the averaged spin lifetimes, both the Rashba and the Dresselhaus coefficients can be determined by comparing the experimental data to numerical many-body calculations [34, 35]. Here, we find values of $\beta = 1.38\,\mathrm{meV}\mathring{A}$ and $\alpha = 0.9\,\mathrm{meV}\mathring{A}$, yielding a ratio $\frac{\alpha}{\beta} = 0.65$. The symmetry of the CEF for this ratio is shown in Fig. 4 (c): the preferential direction of the CEF along the [110] direction is clearly visible. Using these values, an in-plane spin lifetime anisotropy of 60 to 1 can be inferred from the calculations. Thus, the spin lifetime for electron spins initially *aligned* along the [110] axis is estimated to be about 8 ns. We note that in order to observe the magneto-anisotropy by spin precession, the initial spin polarization has to be so high that the spin lifetime is long enough to allow for at least one precession cycle.

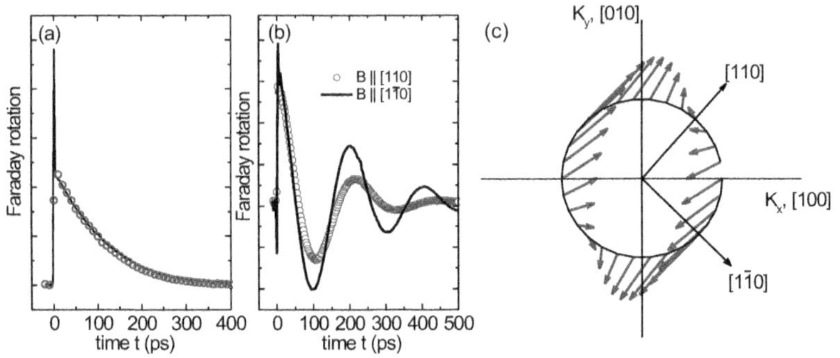

Fig. 4. (**a**) Time-resolved Faraday rotation traces for two orientations of the sample without external magnetic field. (**b**)TRFR traces at 1 Tesla in-plane magnetic field, for two relative orientations of the sample and the magnetic field. (**c**) Symmetry of the combined effective field for Rashba-Dresselhaus ratio $\alpha = 0.65\beta$.

5.5 Breakdown of magneto-anisotropy

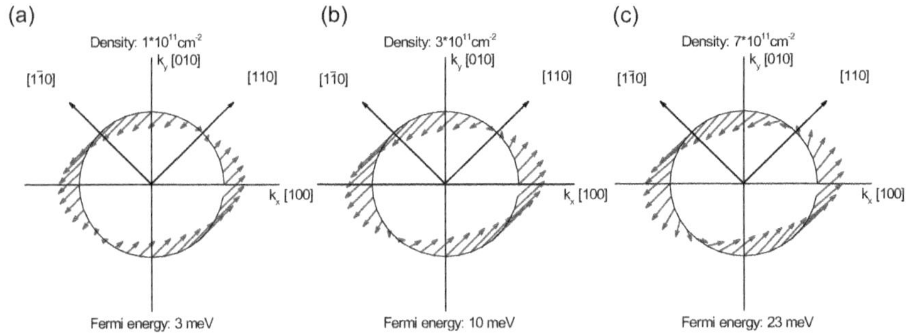

Fig. 5. Calculated symmetry of the combined effective field due to Rashba and *kubic* Dresselhaus terms. The carrier density and thus the Fermi energy is increased from (**a**) to (**c**).

The spin-dephasing anisotropy in [001]-grown QWs is an approximation based on the interference of the Rashba and the *linear* Dresselhaus spin-orbit fields. In this approximation, the vector sum of Rashba and Dresselhaus fields points along the [110] direction for $\alpha = \beta$, regardless of $\boldsymbol{k}$. As the carrier density in a 2DES is increased, the kubic Dresselhaus term has to be taken into account, and the symmetry of the combined effective field (CEF) changes. Here, we show density-dependent calculations of the symmetry of the CEF as a function of the carrier density. The calculations were performed for a 20 nm wide quantum well, and for the calculations the ratio of the Rashba and Dresselhaus terms was kept constant at $\alpha = \beta$. As Fig. 5(a) shows, for low density the orientation of the CEF remains almost perfectly aligned along the [110] direction. As soon as the density is increased to values more common for a 2DES, as in Fig. 5(b) and (c), a deviation from this orientation can be observed for a range of $\boldsymbol{k}$ values. In order to quantify the breakdown of the magneto-anisotropy, we calculate the average torque $\bar{\boldsymbol{\tau}}$ acting on an electron spin s pointing along the in-plane [110] or [1$\bar{1}$0] directions: $|\bar{\boldsymbol{\tau}}| = \sum_k |\boldsymbol{s} \times \boldsymbol{B}_{eff}(\boldsymbol{k})|$. Here, the sum is over all directions of $\boldsymbol{k}$ for $|\boldsymbol{k}| = k_F$. From these values, we can determine the (*dimensionless*) anisotropy $\bar{\boldsymbol{\tau}}_{[1\bar{1}0]}/\bar{\boldsymbol{\tau}}_{[110]}$ of the initial torque acting on electron spins pointing along [110] and [1$\bar{1}$0], and track its density-dependence. Figure 6 shows the results of this calculation

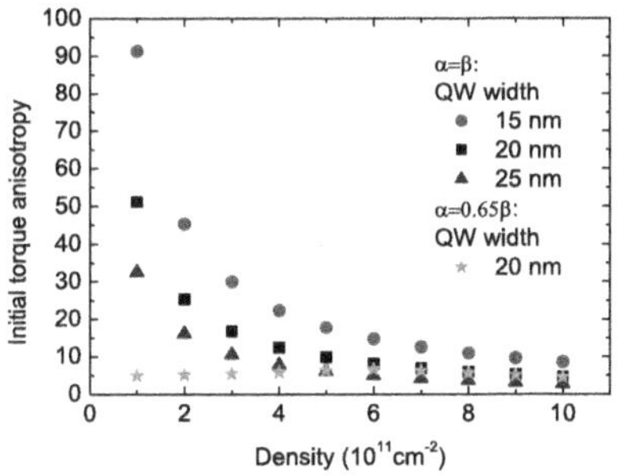

Fig. 6. Anisotropy of the initial torque acting on electron spins pointing along the in-plane [110] and [1$\bar{1}$0] directions as a function of electron density. Calculations for 15 nm (*red circles*), 20 nm (*black squares*), and 25 nm (*blue triangles*) wide QWs with equal Rashba and Dresselhaus terms ($\alpha = \beta$) are shown, as well as the values for a 20 nm wide QW with ($\alpha = 0.65\beta$) (*green stars*).

for 15–25 nm wide quantum wells with equal Rashba and Dresselhaus terms ($\alpha = \beta$). It can be clearly seen how the torque anisotropy is reduced by an order of magnitude as the density is increased from values common for low-doped 2DES samples ($n = 1 \times 10^{11}\,\mathrm{cm}^{-2}$) to highly-doped 2DES ($n = 8 - 10 \times 10^{11}\,\mathrm{cm}^{-2}$). For comparison, the same calculation was performed for a 20 nm wide QW with ($\alpha = 0.65\beta$), the ratio corresponding to the sample investigated in our experiments. There, the torque anisotropy is significantly lower than for ($\alpha = \beta$) and remains almost constant over a wide range of densities. We note that a direct experimental observation of the magneto-anisotropy breakdown will be difficult as tuning the carrier density by an external gate voltage will also change the Rashba/Dresselhaus ratio.

6 Summary

In conclusion, we have investigated the spin dephasing in a high-mobility 2DES as a function of the initial spin polarization, showing that due to electron–electron interaction, the spin dephasing is reduced as the initial spin polarization is increased. Additionally, we studied the spin dephasing in external magnetic fields applied perpendicular and parallel to the 2DES. Here, we observed a strong in-plane anisotropy of the spin dephasing due to interference between the Dresselhaus and Rashba spin-orbit fields. This anisotropy is strongest for equal Dresselhaus and Rashba fields, yet highly density-dependent, as the *kubic* Dresselhaus term changes the symmetry of the combined effective field.

7 Acknowledgements

The authors would like to thank J. Zhou, J.L. Cheng, J.H. Jiang and M.W. Wu for fruitful discussion. Financial support by the DFG via SPP 1285 and SFB 689 is gratefully acknowledged.

References

1. D. D. Awschalom, D. Loss, and N. Samarth, eds., *Semiconductor Spintronics and Quantum Computation*, Nanoscience and Technology (Springer, Berlin, 2002), and references therein.
2. I. Zutic, J. Fabian, and S. Das Sarma, Rev. Mod. Phys. **76**, 323 (2004), and references therein.
3. J. Fabian, A. Matos-Abiague, C. Ertler, Peter Stano, and I. Zutic, Acta Physica Slovaca **57**, 565 (2007), and references therein.
4. H. Ohno, A. Shen, F. Matsukura, A. Oiwa, A. Endo, S. Katsumoto, and Y. Iye, Appl. Phys. Lett. **69**, 363 (1996).
5. J. M. Kikkawa and D. D. Awschalom, Phys. Rev. Lett. **80**, 4313 (1998).
6. J. M. Kikkawa, I. P. Smorchkova, N. Samarth, and D. D. Awschalom, Science **277**, 1284 (1997).
7. S. A. Crooker, M. Furis, X. Lou, C. Adelmann, D. L. Smith, C. J. Palmstrøm, and P. A. Crowell, Science **309**, 2191 (2005).
8. J. M. Kikkawa and D. D. Awschalom, Nature **397**, 139 (1999).
9. N. P. Stern, D. W. Steuerman, S. Mack, A. C. Gossard, and D. D. Awschalom, Appl. Phys. Lett. **91**, 062109 (2007).
10. M. A. Brand, A. Malinowski, O. Z. Karimov, P. A. Marsden, R. T. Harley, A. J. Shields, D. Sanvitto, D. A. Ritchie, and M. Y. Simmons, Phys. Rev. Lett. **89**, 236601 (2002).
11. W. J. H. Leyland, G. H. John, R. T. Harley, M. M. Glazov, E. L. Ivchenko, D. A. Ritchie, I. Farrer, A. J. Shields, and M. Henini, Phys. Rev. B **75**, 165309 (2007).
12. Y. Ohno, R. Terauchi, T. Adachi, F. Matsukura, and H. Ohno, Phys. Rev. Lett. **83**, 4196 (1999).

13. M. W. Wu and M. Kuwata-Gonokami, Solid State Commun. **121**, 509 (2002).
14. S. Döhrmann, D. Hägele, J. Rudolph, M. Bichler, D. Schuh, and M. Oestreich, Phys. Rev. Lett. **93**, 147405 (2004).
15. O. Z. Karimov, G. H. John, R. T. Harley, W. H. Lau, M. E. Flatté, M. Henini, and R. Airey, Phys. Rev. Lett. **91**, 246601 (2003).
16. M. M. Glazov and E. L. Ivchenko, JETP Lett. **75**, 403 (2002).
17. M. Q. Weng and M. W. Wu, J. Appl. Phys. **93**, 410 (2003).
18. M. Q. Weng and M. W. Wu, Phys. Rev. B **68**, 075312 (2003).
19. M. Q. Weng, M. W. Wu, and L. Jiang, Phys. Rev. B **69**, 245320 (2004).
20. J. Zhou, J. L. Cheng, and M. W. Wu, Phys. Rev. B **75**, 045305 (2007).
21. M. Q. Weng and M. W. Wu, Phys. Rev. B **69**, 195318 (2004).
22. F. Meier and B. P. Zakharchenya, eds., *Optical Orientation* (Elsevier, Amsterdam, 1984).
23. B. Dareys, X. Marie, T. Amand, J. Barrau, Y. Shekun, I. Razdobreev, and R. Planel, Superlattices and Microstructures **13**, 353 (1993).
24. S. Pfalz, R. Winkler, T. Nowitzki, D. Reuter, A. D. Wieck, D. Hägele, and M. Oestreich, Phys. Rev. B **71**, 165305 (2005).
25. G. Dresselhaus, Phys. Rev. **100**, 580 (1955).
26. Y. A. Bychkov and E. I. Rashba, Pis'ma Zh. Éksp. Teor. Fiz. **39**, 66 (1984) [Sov. Phys. JEPT Lett. **39** 78 (1984)].
27. M. I. D'yakonov and V. I. Perel', Zh. Éksp. Teor. Fiz. **60**, 1954 (1971) [Sov. Phys. JEPT **33**, 1053 (1971)].
28. N. S. Averkiev and L. E. Golub, Phys. Rev. B **60**, 15582 (1999).
29. N. S. Averkiev, L. E. Golub, A. S. Gurevich, V. P. Evtikhiev, V. P. Kochereshko, A. V. Platonov, A. S. Shkolnik, and Yu. P. Efimov, Phys. Rev. B **74**, 033305 (2006).
30. B. Liu, H. Zhao, J. Wang, L. Liu, W. Wang, D. Chen, and H. Zhu, Appl. Phys. Lett. **90**, 112111 (2007).
31. D. Stich, T. Korn, R. Schulz, D. Schuh, W. Wegscheider, and C. Schüller, Physica E **40**, 1545 (2008).
32. D. Stich, J. Zhou, T. Korn, R. Schulz, D. Schuh, W. Wegscheider, M. W. Wu, and C. Schüller, Phys. Rev. Lett. **98**, 176401 (2007).
33. D. Stich, J. Zhou, T. Korn, R. Schuh, D. Schuh, W. Wegscheider, M. W. Wu, and C. Schüller, Phys. Rev. B **76** 205301 (2007).
34. D. Stich, J. H. Jiang, T. Korn, R. Schuh, D. Schuh, W. Wegscheider, M. W. Wu, and C. Schüller, Phys. Rev. B **76** 073309 (2007).
35. T. Korn, D. Stich, R. Schulz, D. Schuh, W. Wegscheider, and C. Schüller, Physica E **40**, 1542 (2008).

Magnetization Dynamics of Coupled Ferromagnetic-Antiferromagnetic Thin Films

Jeffrey McCord

Institute for Metallic Materials, IFW Dresden, Postfach 270116, 01171, Dresden, Germany
j.mccord@ifw-dresden.de

Abstract. The existence of mixed magnetic anisotropies and a systematic modification of the dynamic damping parameter in ferromagnet/antiferromagnet polycrystalline thin film systems as a function of antiferromagnetic layer thickness is demonstrated. Independent of the existence of exchange bias, using ultra-thin antiferromagnet layers, a controlled adjustment of static and dynamic magnetic properties of ferromagnetic thin films is achieved. Due to the combined increase of precessional frequency and effective damping parameter, a significant reduction in magnetic relaxation time is achieved. Simple relations based on the assumption of an interfacial antiferromagnetic layer contribution are derived. The addition of thin antiferromagnetic layers in contact to ferromagnetic thin films facilitates for controlled alteration of magnetic properties beyond the established manipulation of the intrinsic ferromagnetic material properties. The experimental findings are of great importance from a fundamental point of view and are significant for various applications based on ferromagnetic thin film technology.

1 Introduction

Exchange coupled magnetic thin films consisting of an antiferromagnetic (AF) and an adjacent ferromagnetic (F) layer are an essential part in todays spin electronic applications, as they are used as a part of magnetoresistive devices in current magnetic recording technology or magnetic random access memory (MRAM) cells. So far, the technological purpose of using F/AF systems is the magnetic stabilization of the F layer through the so-called exchange bias effect [1–3]. A variety of models have been developed to describe the exchange bias field [4–8], but recent experiments [9] provide direct experimental evidence for the existence of uncompensated interfacial spins being responsible for the described physical effect. Even though a fundamental understanding has been reached, the mechanism controlling the final spin structure at the interface is still not known.

The analysis of the magnetization dynamics [10–15] of F/AF systems provides an insight into the nature of the F/AF interaction that cannot be gained from other measurement techniques. Moreover, the understanding of magnetization dynamics and the possibility to influence the dynamic magnetic properties is also relevant for numerous spintronic applications [16–19]. This article focusses on the role of ultra-thin AF layers, in the thickness range below the onset of exchange bias, on the F material's effective magnetic properties. In the first part the relevant basics of magnetization dynamics are introduced. Then, the contemporary cognitions of the effects of AF layers on dynamics will be summarized, from which in the end the use of ultra-thin antiferromagnetic layers as an alternative route to adjust the effective static and dynamic magnetic material properties of ferromagnetic thin films is demonstrated.

2 Magnetization dynamics

The performance of magnetoelectronic devices on the time scale of a few nanoseconds rests on the dynamical response of the magnetization. The motion of magnetization $\boldsymbol{M}$ subjected to an effective magnetic field $\boldsymbol{H}_{\mathrm{eff}}$ is described by the phenomenological Landau-Lifshitz-Gilbert equation [20]

$$\frac{\mathrm{d}}{\mathrm{d}t}\boldsymbol{M} = -\gamma\mu_0\boldsymbol{M} \times \boldsymbol{H}_{\mathrm{eff}} + \frac{\alpha_{\mathrm{eff}}}{M_{\mathrm{s}}}\left(\boldsymbol{M} \times \frac{\mathrm{d}}{\mathrm{d}t}\boldsymbol{M}\right) , \tag{1}$$

where $\mu_0 = 4\pi \cdot 10^6$ Vs/Am is the vacuum permeability, M_{s} represents the saturation magnetization, and $\gamma = 1.76 \cdot 10^{11}\mathrm{T}^{-1}\mathrm{s}^{-1}$ the gyromagnetic ratio.

The first term in Eq. (1) describes the gyroscopic precession of the magnetization with a characteristic precession or resonance frequency f_{res}, scaling with the effective field $\boldsymbol{H}_{\mathrm{eff}}$ acting on $\boldsymbol{M}$. In the simplest case $\boldsymbol{H}_{\mathrm{eff}}$ is equal to the magnetic anisotropy field H_{k}. The dependence of f_{res} on H_{k} is given by the simplified Kittel equation [21]

$$f_{\mathrm{res}}^2 = \frac{\gamma^2\mu_0^2}{4\pi^2}M_{\mathrm{s}}H_{\mathrm{k}} . \tag{2}$$

Methods to change the induced magnetic anisotropy, and thus f_{res}, of polycrystalline films prepared by vacuum technology are limited to a few routes. The most practical way is to deposit the films in an applied magnetic saturation field or to perform a subsequent magnetic field annealing after the film deposition. In addition, the anisotropy of thin films can be significantly increased by inducing an additional anisotropy by off-axis film deposition [22, 23] or by changing the material's composition. The latter method potentially changes other magnetic properties like the magnetocrystalline anisotropy and in addition leads to a change of the magnetostriction constants.

The second term in Eq. (1) corresponds to the phenomenological damping torque, which is responsible for the reorientation of magnetization into the equilibrium state. The rate of relaxation is described by an effective damping

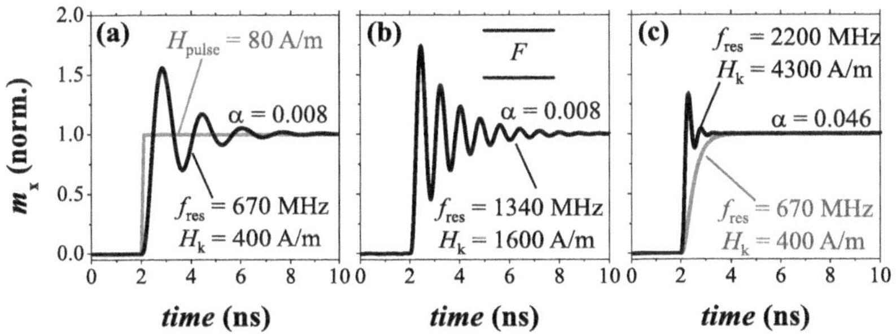

Fig. 1. Modeling of the magnetic response of a soft magnetic thin films to a fast field pulse with a rise-time of 100 ps. The evolution of the field pulse H_{pulse} is shown in (**a**). The values of H_{k} and the corresponding values of f_{res}, as well as the values of α used for the calculations are indicated. The saturation magnetization value of $\text{Ni}_{81}\text{Fe}_{19}$, $J_{\text{s}} = 1$ T, is used for all calculations.

parameter α. Correlated with α is a characteristic exponential decay time τ [24] of the amplitude of the precessional motion defined as

$$\tau = \frac{2}{\alpha \gamma \mu_0 M_s} \ . \tag{3}$$

Whereas for many potential applications in spintronics a small α is a precondition [25], for other applications a reasonably fast relaxation or settling time is favored [26, 27]. Methods to modify the magnetic relaxation time mostly rely on alloying the F films with rare earth or transition metal elements [16, 17, 28] or diluting the FM material [29].

In Fig. 1 the modeled dynamic magnetic response to a fast step-like excitation field H_{pulse} as derived from numerically solving Eq. (1) is shown. Both, anisotropy and/or damping are varied to illustrate the influence of both parameters on the magnetization response. With an increase in anisotropy field, f_{res} increases (Fig. 1(b)). However, only by additionally increasing α a congruent reduction in settling time is achieved. For high permeability materials the optimum condition of critically damped behavior [27] is obtained for

$$\alpha = \frac{1}{\sqrt{\mu_{\text{r}}}} \propto f_{\text{res}} \ , \tag{4}$$

where μ_{r} is the relative permeability. The change of magnetization response with f_{res} for a constant value of $\alpha = 0.046$ is shown in Fig. 1(c). As already mentioned above, exchange biased F/AF systems offer the opportunity to increase f_{res} and α_{eff}, which will be shown next. A special emphasis will be on the change of static and dynamic magnetic properties in polycrystalline $\text{Ni}_{81}\text{Fe}_{19}$-AF films with varying AF layer thickness (for details on sample preparation see [13–15]).

3 Statics and dynamics of exchange biased F/AF films

As a basic example for the diversity of the effects of F/AF thin films, the influence of AF layers on the static and dynamic magnetization behavior is reviewed, discussing the $Ni_{81}Fe_{19}/NiO$ and the $Ni_{81}Fe_{19}/Ir_{22}Mn_{78}$ system. The general statements are also valid for other material systems. Results of hysteresis loop measurements along the magnetic easy axis (EA) for a $Ni_{81}Fe_{19}$(30 nm)/NiO, respectively $Ni_{81}Fe_{19}$(40 nm)/IrMn, system with different AF thicknesses are summarized in Fig. 2. For the presented loop data, only for the thick NiO layer ($t_{\mathrm{NiO}} = 50$ nm, Fig. 2(c)) a loop shift, i.e. exchange bias, is observed. The overall dependency of H_{eb} and H_{c} for NiO and IrMn as AF material are displayed in (d) and (e). The difference in the onset of exchange bias with AF layer thickness in the IrMn- and the NiO-based bilayers is directly attributed to the difference in anisotropy energy of the AF materials, with $K_{\mathrm{u,IrMn}} = 3 \cdot 10^4 \mathrm{J/m^3}$ and $K_{\mathrm{u,NiO}} = 3 \cdot 10^3 \mathrm{J/m^3}$. In general, with the addition of the AF material, the EA coercivity increases relative to the pure F layer. For both materials, the maximum of H_{c} is around the onset of exchange bias. As no structural difference at the F/AF interface with variation of the AF thickness is expected for the investigated layers, the dependence is mainly related to the thermal stability of the AF layer scaling with its thickness t_{AF}.

Analyzing the hard axis (HA) magnetization behavior (Fig. 3) reveals a broader insight into the complicated situation of magnetic anisotropy in such systems. While the single F film displays a well defined loop shape, as expected for films with uniaxial anisotropy, a pronounced two-step reversal behavior is found for $t_{\mathrm{NiO}} = 5$ nm (Fig. 3(b)). Only a slight hint of a two-step magnetization process is visible in the corresponding nearly linear HA loop for $t_{\mathrm{NiO}} = 50$ nm (Fig. 3(c)). The total AF induced anisotropy field as derived from the HA curves, is much higher as anticipated from a combined contribution of induced F anisotropy field and exchange bias field H_{eb} [14]. This discrepancy can be described, introducing higher order uniaxial and cubic anisotropy terms [14, 30]. As macrostructural investigations revealed a clear $\langle 111 \rangle$ fiber texture in the films, contributions from the crystallinity of the NiO layer on the in-plane distribution of AF induced anisotropy can be ruled out. Moreover, it was shown that the direction of additional anisotropies [31] is not affected by the AF crystallinity, but only by the interfacial spin structure.

The shape of the hysteresis loops is modeled under the assumption of a simultaneous occurrence of AF-induced interfacial unidirectional, uniaxial, and cubic anisotropy terms. For the shown data, only the fractions of the unidirectional, uniaxial, and cubic anisotropy contribution is varied. The amount of coupling leading to exchange bias $\mathrm{f_{eb,AF}}$, uniaxial $\mathrm{f_{u,AF}}$, and cubic anisotropy $\mathrm{f_{c,AF}}$ is kept fixed with $\mathrm{f_{eb,AF}} + \mathrm{f_{u,AF}} + \mathrm{f_{c,AF}} = 1$. This is synonymous with the assumption of a constant exchange coupling J_{AF} across the F/AF interface. The AF induced energy contribution $e_{\mathrm{ani,AF}}$ acting on the F layer can then be written as [14]

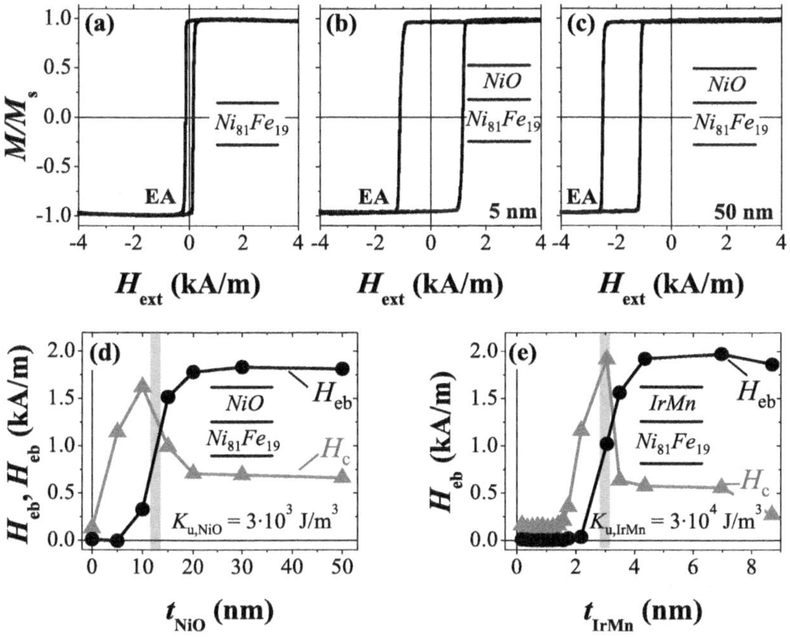

Fig. 2. Magnetization loops along the easy axis (EA) of a (**a**) single $Ni_{81}Fe_{19}$(30 nm) layer and of $Ni_{81}Fe_{19}$(30 nm)/NiO bi-layers with a NiO thickness (**b**) 5 nm, and (**c**) 50 nm. In (**d**) and (**e**) the thickness dependences of exchange bias field H_{eb} and coercivity field H_{c} for $Ni_{81}Fe_{19}$(30 nm)/NiO, respectively $Ni_{81}Fe_{19}$(40 nm)/$Ir_{22}Mn_{78}$ structures are displayed. The anisotropy constant $K_{\mathrm{u,AF}}$ of each AF material is indicated in (**d**) and (**e**).

$$e_{\mathrm{ani,AF}} = -\mathrm{f}_{\mathrm{eb,AF}}\, J_{\mathrm{AF}} \cos(\beta-\delta) + \frac{1}{2}\mathrm{f}_{\mathrm{u,AF}}\, J_{\mathrm{AF}} \sin^2(\beta-\delta) + \frac{1}{4}\mathrm{f}_{\mathrm{c,AF}}\, J_{\mathrm{AF}} \sin^2(\beta-\delta)\cos^2(\beta-\delta), \quad (5)$$

where δ is the direction of exchange anisotropy and β the direction of F magnetization. With the proposed phenomenological model for the AF contribution, the total energy density e_{total} of the system writes

$$e_{\mathrm{total}} = -H_{\mathrm{ext}}\, J_{\mathrm{s}}\, t_{\mathrm{F}} \cos(\varphi-\beta) + \frac{1}{2}\, K_{\mathrm{u,F}}\, t_{\mathrm{F}} \sin^2(\beta-\theta) + e_{\mathrm{ani,AF}}, \quad (6)$$

where the external field H_{ext} is aligned under an angle φ. The saturation polarization J_{s}, and the ferromagnetic layer thickness t_{F} combine to the Zeeman energy. The uniaxial anisotropy constant $K_{\mathrm{u,F}}$ of the F layer is aligned along the angle θ.

The results for $J_{\mathrm{AF}} = 10^{-5}$ J/m with equally contributing uniaxial and cubic AF terms, $\mathrm{f}_{\mathrm{k,AF}} = 0.5$ and $\mathrm{f}_{\mathrm{c,AF}} = 0.5$, are displayed in Fig. 3(e). The case of $\mathrm{f}_{\mathrm{eb,AF}} = 0.5$, $\mathrm{f}_{\mathrm{u,AF}} = 0.25$, and $\mathrm{f}_{\mathrm{c,AF}} = 0.25$, now with the inclusion

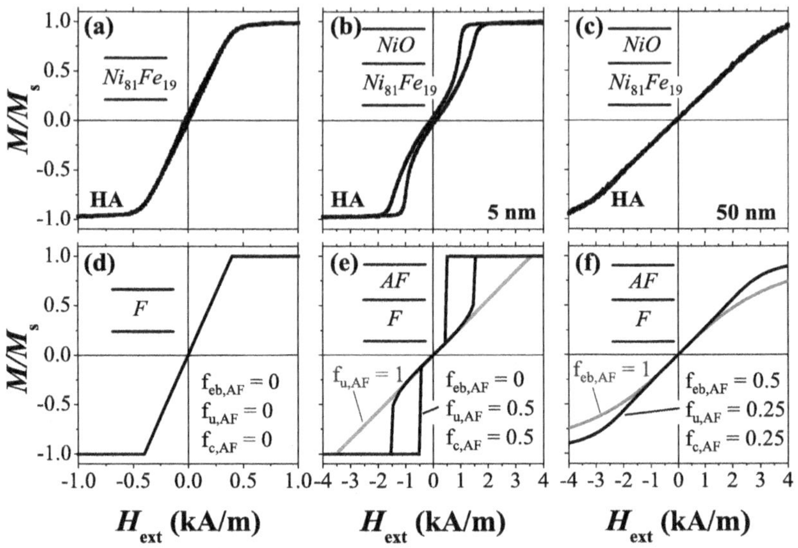

Fig. 3. Measured (**a**–**c**) and calculated (d–f) hard axis (HA) loops for the same sample set as in Fig. 2(a, b). Modeled HA magnetization loops for a constant coupling J_{AF} (**e**) without unidirectional, but a combination of uniaxial and unidirectional anisotropy and (**f**) a contribution from all three anisotropy terms. Fractions of $f_{eb,AF}$, $f_{u,AF}$, and $f_{c,AF}$ anisotropy contributions are indicated. Calculated loops for $f_{u,AF} = 1$ and $f_{eb,AF} = 1$ are included in (**e**) and (**f**) for comparison.

of unidirectional anisotropy, is shown in Fig. 3(f). A uniaxial anisotropy of $K_{u,F} = 200$ J/m^3 as derived from the F single layer measurements is used for all calculations. All features of the experiments, the two-step reversal seen in Fig. 3(b) and the extended linear regime shown in Fig. 3(c), are replicated.

An overview of the change of precessional frequency and the magnetic damping parameter with AF layer thickness for NiO, respectively IrMn, is given in Fig. 4. With the addition of the AF layer, even for thicknesses well below the onset of exchange bias, a shift of precessional frequency to higher values relative to the single layer $Ni_{81}Fe_{19}$ occurs. With increasing t_{NiO}, f_{res} first increases and then remains constant (Fig. 4(a)). A similar behavior is found for t_{IrMn} (Fig. 4(b)), but an additional peak of f_{res} below the onset of exchange bias becomes visible. The concurrent onset of H_c (Fig. 2) and f_{res} (Fig. 4) indicates that the increase in coercivity and in f_{res}, both occurring below the onset of exchange bias, originate from instability effects of the AF layer [13], which also reflects itself in the magnetic damping parameter α. Moreover, for the case of IrMn the onset of α takes place before the rise of f_{res} becomes visible. This behavior is interpreted as follows. At the onset of α the IrMn layer is already antiferromagnetic in character but displays superparamagnetic-like behavior with fluctuations of the AF magnetization, contributing to the enhanced magnetic damping [13] in the F/AF layer stack.

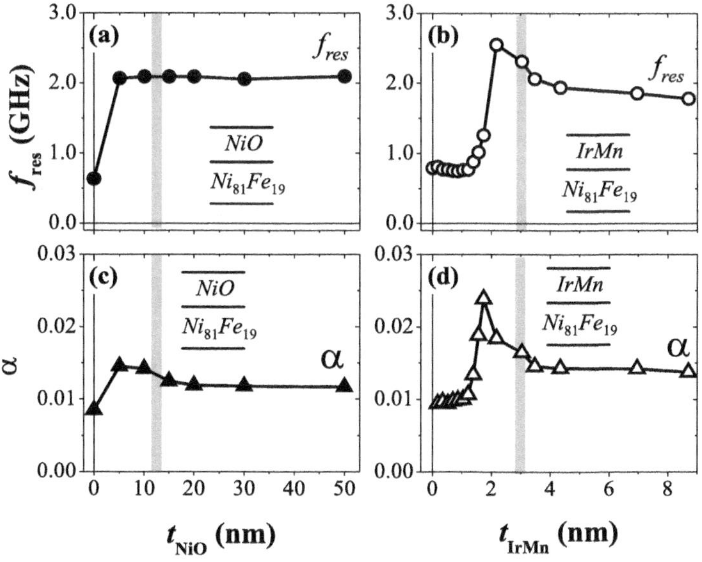

Fig. 4. Change of precessional frequency f_{res} (**a**, **b**) and damping parameter α (**c**, **d**) with AF layer thickness t_{NiO}, respectively t_{IrMn}. The position of onset of exchange bias is indicated (see. Fig. 2(d) and (e) for comparison).

Similar results are also reported for Co/CoO based systems [32]. With increasing IrMn thickness, the anisotropy energy of the AF then leads to a stabilization of the AF magnetization on the ns-time scale and, as a result f_{res} develops. With yet increasing IrMn thickness, the anisotropy stabilizes even more and an exchange bias field H_{eb} develops. Here, just the general influence of AF layer thickness in F/AF bi-layer stacks is described. Additional aspects and contributions from the interfacial coupling strength, e.g. modification of the interfacial structure with thermal treatment, are discussed in [13, 14].

In the following we will now exclusively focus on the systematics of effects below the onset of exchange bias in F/AF/F sandwich structures in the ultrathin IrMn thickness range.

4 F/AF/F structures below the onset of exchange bias

Representative magnetization loops from $Ni_{81}Fe_{19}/Ir_{19}Mn_{81}/Ni_{81}Fe_{19}$ sandwich structures with varying AF layer thicknesses, all below the onset of exchange-bias, are shown in Fig. 5. The scale of the magnetic field abscissa increases by one order of magnitude, comparing Fig. 5(a–d). By addition of the IrMn layer, the EA coercivity H_c increases constantly and peaks around $t_{IrMn} = 2.0$ nm. Only a very low exchange bias field H_{eb} is found for the thickest IrMn layer investigated here. Despite the change in structure and slight

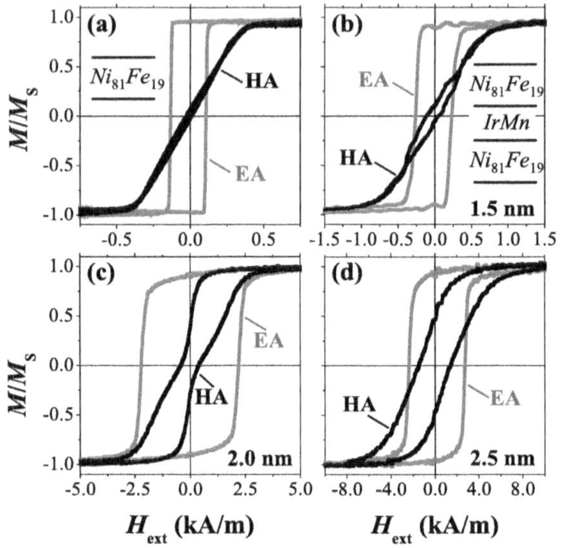

Fig. 5. (**a**–**d**) Easy axis (EA) and hard axis (HA) magnetization loops of $Ni_{81}Fe_{19}$(10nm)/$Ir_{19}Mn_{81}$/$Ni_{81}Fe_{19}$(10nm) films in the low IrMn thickness t_{IrMn} regime. The principle structure of the samples is sketched in (**a**) and (**b**).

difference in IrMn composition relative to the data presented above, the results are similar to Fig. 2(e). Moreover, a constricted loop shape as for the NiO example discussed before (see Fig. 3(b)) is found for $t_{\mathrm{IrMn}} = 2$ nm, demonstrating the existence of higher order anisotropy terms also for IrMn-based systems.

A summary of the resulting fields is given in Fig. 6. With the addition of a nanometer thin IrMn layer to the F sandwich, a strong uniaxial magnetic anisotropy is introduced. This effect occurs above a threshold AF layer thickness of $t_{\mathrm{IrMn}} = 1.5$ nm. Whereas H_{c} levels with increased AF layer thickness, the anisotropy field H_{k} evidences a steep dependence with t_{IrMn}, increasing by one order of magnitude relative to the initial F film value.

The magnetic responses to a fast field excitation, measured by microwave magnetometry [33], for a pure F and a F/AF/F sandwich structure in analogy to the computer experiments of Fig. 1 are displayed in Fig. 7. As can be seen, f_{res} and α have changed significantly with the addition of the AF layer. With small damping and anisotropy, as for the pure F layer, a delay in magnetization response together with visible oscillations of magnetization over several nanoseconds occurs. On the other hand, the F/AF/F layer's magnetization follows the excitation field almost instantaneously and settles much faster into the new equilibrium state. With increasing t_{IrMn} further, α decreases again but the precessional frequency stays rather constant.

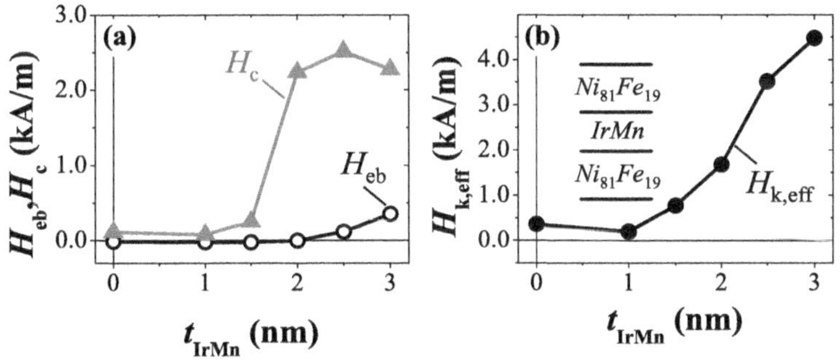

Fig. 6. (**a**) Easy axis coercivity H_c and exchange-bias field H_{eb} dependence on t_{IrMn} of $Ni_{81}Fe_{19}$(10 nm)/$Ir_{19}Mn_{81}$/$Ni_{81}Fe_{19}$(10 nm) sandwich structures. (**b**) Effective anisotropy $H_{k,eff}$ as derived from the HA loops.

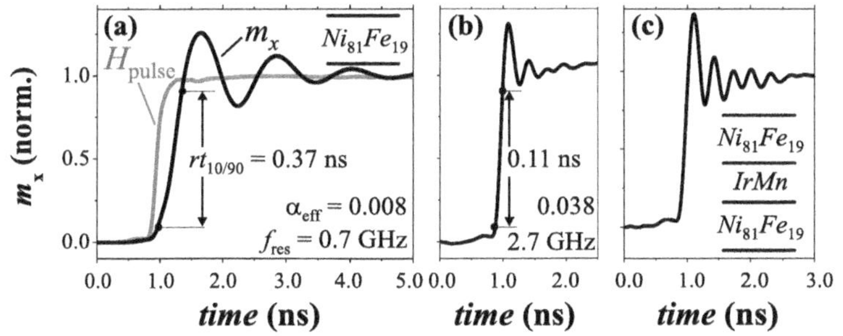

Fig. 7. Exemplary dynamic magnetization response of (a) a single $Ni_{81}Fe_{19}$(20 nm) layer and (**b, c**) $Ni_{81}Fe_{19}$(10 nm)/$Ir_{19}Mn_{81}$(2 nm)/$Ni_{81}Fe_{19}$(10 nm), respectively $Ni_{81}Fe_{19}$(10 nm)/$Ir_{19}Mn_{81}$(3 nm)/$Ni_{81}Fe_{19}$(10 nm) sample. Values of f_{res} and α are indicated in (**a**) and (**b**). The shape of the pulse field is underlaid in (**a**) and the values of 10/90-risetime $rt_{10/90}$ of magnetization response is shown in (**a**) and (**b**).

The overall dependence of f_{res} and α on t_{IrMn} is summarized in Fig. 8. The precessional frequency increases by a factor of 4 from $f_{res} \approx 0.7$ GHz to $f_{res} \approx 2.7$ GHz for the F/AF/F structures. No direct connection of the monotonically increasing H_k with AF layer thickness exists (compare Fig. 8 with Fig. 6). In contrast to H_k, f_{res} peaks at $t_{IrMn} \approx 2.5$ nm, but is not highly affected by possible AF layer thickness variations. Congruently, with the change of the precessional frequency, the damping parameter increases from $\alpha = 0.008$ to 0.04 exhibiting a maximum around $t_{IrMn} = 1.5$ nm. The data demonstrates that uniaxial magnetic anisotropy, f_{res}, and α_{eff} of F films are significantly enhanced by additions of thin AF layers. Comparing the results to the data of NiO- and IrMn-based F/AF structures (Sect. 3), the effects are much higher.

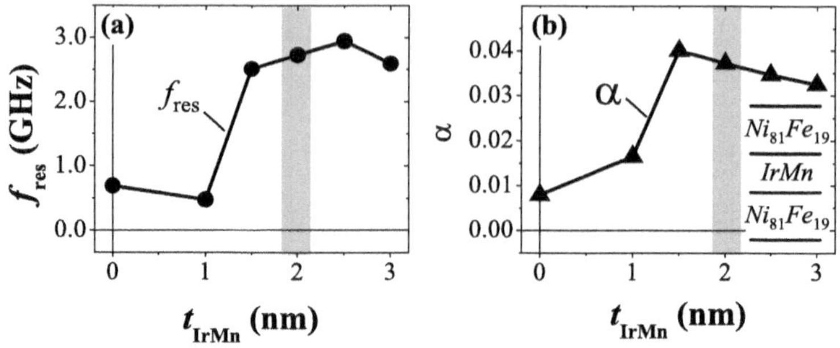

Fig. 8. (**a**) Change of precessional frequency f_{res} and (**b**) magnetic damping parameter α with t_{IrMn} for $Ni_{81}Fe_{19}(10\,nm)/Ir_{19}Mn_{81}/Ni_{81}Fe_{19}(10\,nm)$ sandwich structures.

In order to define general rules for the control of the relevant dynamic magnetic properties, the dependence of f_{res} and α for F/AF/F structures with now varying $Ni_{81}Fe_{19}$ layer thickness, but constant AF layer thickness, will be investigated. An AF layer thickness $t_{\mathrm{IrMn}} = 2.0$ nm was chosen for the experiments as in this AF thickness range, as indicated in Fig. 8, both characteristic dynamic values are close to their maximum. The variation of f_{res}^2 with the F layer thickness t_{NiFe} for different sets of F/AF/F stacks is displayed in Fig. 9(a). The precessional frequency is decreasing monotonically with the increase of t_{NiFe}. Over a wide range, the change with reciprocal F layer thickness $1/t_{\mathrm{NiFe}}$ follows a linear dependence of f_{res}^2 (Fig. 9(b)) in accordance with a modified Kittel relation (Eq. (2))

$$f_{\mathrm{res}}^2 = \frac{\gamma^2 \mu_0^2}{4\pi^2} M_{\mathrm{s}} \left(H_{\mathrm{k}} + J_{\mathrm{k,IrMn}} \cdot \frac{1}{t_{\mathrm{NiFe}}} \right) . \qquad (7)$$

The effective field H_{eff} (Eq. (1)) is composed of the F's uniaxial anisotropy field H_{k} and the AF induced uniaxial interfacial exchange coupling $J_{\mathrm{k,IrMn}}$, which together with t_{NiFe} describes the effective AF induced dynamic anisotropy field determining f_{res}. This dependency clearly proves the interfacial character of the dynamic anisotropy contribution. From the data in Fig. 9(b) together with Eq. (7)a $J_{\mathrm{k,AF}} \approx 1.15 \cdot 10^{-4}$ A is derived for the given the case of $t_{\mathrm{IrMn}} = 2$ nm.

Also the AF layer contribution of the damping parameter can be described assuming an interfacial AF contribution. The variation of α with t_{NiFe} is displayed in Fig. 10. A linear dependency with $1/t_{\mathrm{NiFe}}$ becomes obvious, which can be defined introducing an AF induced interfacial damping coupling parameter α_{IrMn}. Together with the F layer's damping contribution α_{NiFe}, the relaxation behavior is described by the simple relation

$$\alpha_{\mathrm{eff}} = \alpha_{\mathrm{NiFe}} + \alpha_{\mathrm{IrMn}} \cdot \frac{1}{t_{\mathrm{NiFe}}} . \qquad (8)$$

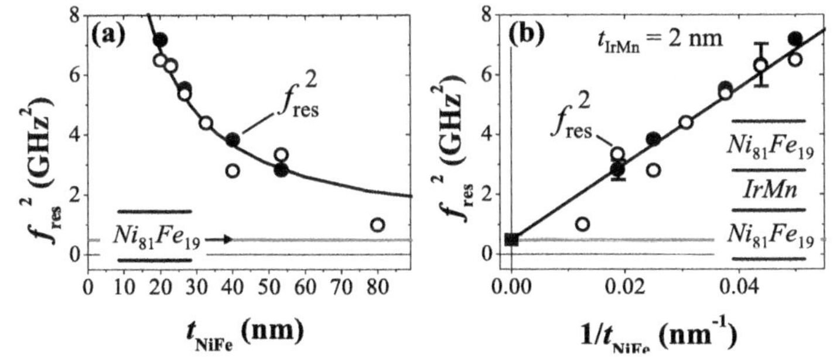

Fig. 9. (**a**) Dependence of the precessional frequency square f_{res}^2 on $Ni_{81}Fe_{19}$-thickness t_{NiFe} for $Ni_{81}Fe_{19}/Ir_{19}Mn_{81}(2\,nm)/Ni_{81}Fe_{19}$ sandwich structures. f_{res}^2 for the regular $Ni_{81}Fe_{19}$ films is added for comparison. (**b**) Variation of f_{res}^2 with $1/t_{NiFe}$. An antiferromagnetic contribution $J_{k,IrMn}$ in accordance with (Eq. (7)) is seen from the linear dependency.

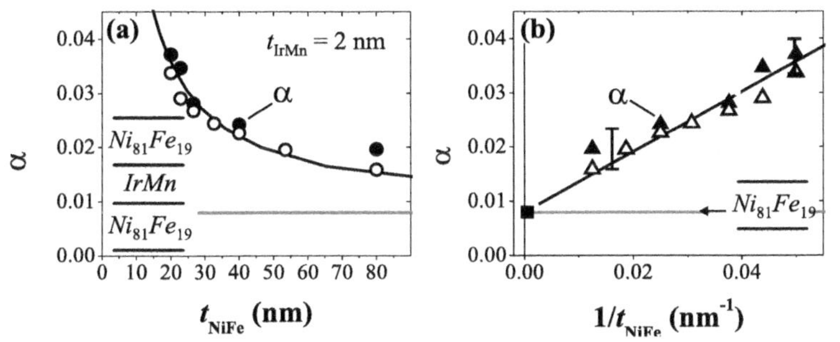

Fig. 10. (**a**) Dependence of α on the ferromagnetic layer thickness t_{NiFe} for different $Ni_{81}Fe_{19}/Ir_{19}Mn_{81}(2\,nm)/Ni_{81}Fe_{19}$ structures. (**b**) Variation of α with inverse ferromagnetic layer thickness $1/t_{NiFe}$. From the linear dependency in (b) an effective coupling damping parameter α_{AF} (Eq. (8)) is derived. See text for details.

For the given IrMn layer thickness a constant interfacial damping coefficient $\alpha_{AF} \approx 0.55 \cdot 10^{-9}$ m is found (Fig. 10(b)). This facilitates the controlled adjustment of α_{eff} by varying the F layer thickness.

As can be seen from the analysis presented above, the precessional frequency and the damping parameter are not varied independently. Nevertheless, due to the different dependence of f_{res} and α_{eff} with t_{NiFe}, almost critical damping conditions (Eq. (4)) are obtained and thus maximum high-frequency magnetization response can be achieved. Moreover, the relative contribution of both dynamic parameters could be altered by choosing different AF layer thicknesses than $t_{IrMn} = 2$ nm.

5 Conclusions

In summary, we evidence the controlled modification of static and dynamic F film properties by means of thin AF layer additions below the onset of exchange bias. The existence of higher order AF-induced anisotropy terms is not only of fundamental interest for the understanding of the exchange bias phenomena, but can also open new possible application for F/AF-layer film stacks. The effective induced uniaxial magnetic anisotropy and thus the precessional frequency, as well as the effective magnetic relaxation time can be adjusted over a wide range by changing both, the AF and F layer thickness. As shown, by dusting F layers with thin AF films, the quasi-static and high frequency response of F films can be changed substantially and optimized beyond the application of F layer doping or alloying. This establishes a novel method for the tailoring of ferromagnetic layer magnetic properties.

More details on the presented subject and additional references can be found in Refs. [13–15].

6 Acknowledgements

Special thanks go to C. Hamann for her continuous help and for proofreading the manuscript. I would like thank R. Kaltofen and R. Mattheis for the thin film deposition and their other numerous contributions to this work. I thank R. Schäfer for always helpful discussion and L. Schultz and O. G. Schmidt for continuous support. The initial code for the hysteresis loop modeling was developed by J. Dshemuchadse. Some of the measurements were performed and analyzed by Y. McCord. Funding through the Deutsche Forschungsgemeinschaft DFG is highly acknowledged.

References

1. W. H. Meiklejohn, C. P. Bean: New magnetic anisotropy, Phys. Rev. **102**, 1413–1414 (1956)
2. J. Nogués, I. K. Schuller: Exchange bias, J. Magn. Magn. Mat. **192**, 203–232 (1999)
3. F. Radu, H. Zabel: Exchange bias effect of ferro-/anti-ferromagnetic heterostructures, in H. Zabel, S. D. Bader (Eds.): *Magnetic Heterostructures, Advances and Perspectives in Spinstructures and Spintransport*, vol. 227, Springer Tracts in Modern Physics (Springer, 2008) pp. 97–183
4. A. P. Malozemoff: Mechanisms of exchange anisotropy, J. Appl. Phys. **63**, 3874–3879 (1988)
5. M. D. Stiles, R. D. McMichael: Coercivity in exchange-bias bilayers, Phys. Rev. B **63**, 064405 (2001)
6. N. C. Koon: Calculations of exchange bias in thin films with ferromagnetic/antiferromagnetic interfaces, Phys. Rev. Lett. **78**, 4865–4867 (1996)

7. U. Nowak, A. Misra, K. D. Usadel: Domain state model for exchange bias, J. Appl. Phys. **89**, 7269–7271 (2001)
8. D. Suess, M. Kirschner, T. Schrefl, J. Fidler, R. Stamps: Exchange bias of polycrystalline magnets with perfectly compensated interfaces, Phys. Rev. B **67**, 054419 (2003)
9. H. Ohldag, A. Scholl, F. Nolting, E. Arenholz, S. Maat, A. Young, M. Carey, J. Stöhr: Correlation between exchange bias and pinned interfacial spins, Phys. Rev. Lett. **91**, 017203 (2003)
10. R. D. McMichael, M. D. Stiles, P. J. Chen, W. F. Egelhoff: Ferromagnetic resonance linewidth in thin films coupled to NiO, J. Appl. Phys. **83**, 7037–7039 (1998)
11. P. Miltényi, M. Gryters, G. Güntherodt, J. Nogués, I. K. Schuller: Spin waves in exchange-biased Fe/FeF_2, Phys. Rev. B **59**, 3333 (1999)
12. S. Rezende, A. Azevedo, M. Lucena, F. Aguiar: Anomalous spin-wave damping in exchange-biased films, Phys. Rev. B **63**, 214418 (2001)
13. J. McCord, R. Mattheis, D. Elefant: Dynamic magnetic anisotropy at the onset of exchange bias: The NiFe/IrMn ferromagnet/antiferromagnet system, Phys. Rev. B **70**, 134418 (2004)
14. J. McCord, R. Kaltofen, T. Gemming, R. Hühne, L. Schultz: Aspects of static and dynamic magnetic anisotropy in $Ni_{81}Fe_{19}$-NiO films, Phys. Rev. B **75**, 94420 (2007)
15. J. McCord, R. Kaltofen, O. G. Schmidt, L. Schultz: Tuning of magnetization dynamics by ultrathin antiferromagnetic layers, Appl. Phys. Lett. **92**, 162506 (2008)
16. W. E. Bailey, P. Kabos, F. Mancoff, S. Russek: Control of magnetization dynamics in NiFe films through the use of rare-earth dopants, IEEE Tans. Magn. **37**, 1749 (2001)
17. S. G. Reidy, L. Cheng, W. E. Bailey: Dopants for independent control of precessional frequency and damping in $Ni_{81}Fe_{19}$ (50 nm) thin films, Appl. Phys. Lett. **82**, 1254–1256 (2004)
18. H. Song, L. Cheng, W. E. Bailey: Systematic control of high-speed damping in doped/undoped $Ni_{81}Fe_{19}$ bilayer thin films, J. Appl. Phys. **95**, 6592–6594 (2004)
19. Y. Guan, W. E. Bailey: Ferromagnetic relaxation in $(Ni_{81}Fe_{19})_{1-x}(Cu)_x$ thin films: Band filling at high Z, J. Appl. Phys. **101**, 09D104 (2007)
20. T. L. Gilbert: A lagrangian formulation of the gyromagnetic equation of the magnetization field, Phys. Rev. **100**, 1243 (1955)
21. C. Kittel: On the theory of ferromagnetic resonance absorption, Phys. Rev. **73**, 155–161 (1948)
22. D. O. Smith, M. S. Cohen, G. P. Weiss: Oblique-incidence anisotropy in evaporated permalloy films, J. Appl. Phys. **31**, 1755–1762 (1961)
23. R. D. McMichael, C. G. Lee, J. E. Bonevich, P. J. Chen, W. Miller, W. F. Egelhoff: Strong anisotropy in thin magnetic films deposited on obliquely sputtered Ta underlayers, J. Appl. Phys. **88**, 5296–5299 (2000)
24. G. Sandler, H. Bertram, T. J. Silva, T. Crawford: Determination of the magnetic damping constant in nife films, J. Appl. Phys. **85**, 5080–5082 (1999)
25. G. D. Fuchs, J. C. Sankey, V. S. Pribiag, L. Qian, P. M. Braganca, A. G. F. Garcia, E. M. Ryan, Z.-P. Li, O. Ozatay, D. C. Ralph, R. A. Buhrman: Spin-torque ferromagnetic resonance measurements of damping in nanomagnets, Appl. Phys. Lett. **91**, 062507 (2007)

26. N. Smith, P. Arnett: White-noise magnetization fluctuations in magnetoresistive heads, Appl. Phys. Lett. **78**, 1448–1450 (2001)
27. N. Sun, S. Wang, T. Silva, A. Kos: High-frequency behavior and damping of Fe-Co-N-based high-saturation soft magnetic films, IEEE Tans. Magn. **38**, 146–150 (2002)
28. J. O. Rantschler, R. D. McMichael, A. Castillo, A. J. Shapiro, W. F. Egelhoff, B. B. Maranville, D. Pulugutha, A. P. Chen, L. M. Conners: Effect of 3d, 4d, and 5d transition metal doping on damping in Permalloy thin films, J. Appl. Phys. **101**, 033911 (2007)
29. J. Fassbender, J. McCord: Control of saturation magnetization, anisotropy, and damping due to Ni implantation in thin $Ni_{81}Fe_{19}$ layers, Appl. Phys. Lett. **88**, 252501 (2006)
30. T. Ambrose, R. Sommer, C. Chien: Angular dependence of exchange coupling in ferromagnet/antiferromagnet bilayers, Phys. Rev. B **56**, 83–86 (1997)
31. C. Leighton, M. R. Fitzsimmons, P. Yashar, A. Hoffmann, J. Nogués, J. Dura, C. F. Majkrzak, I. K. Schuller: Two-stage magnetization reversal in exchange biased bilayers, Phys. Rev. Lett. **86**, 4394–4397 (2001)
32. R. D. McMichael, C. G. Lee, M. D. Stiles, F. G. Serpa, P. J. Chen: Exchange bias relaxation in CoO-biased films, J. Appl. Phys. **87**, 6406–6408 (2000)
33. A. B. Kos, T. J. Silva, P. Kabos: Pulsed inductive microwave magnetometer, Rev. Sci. Instr. **73**, 3563–3569 (2002)

Magnetic and Electronic Properties of Heusler Alloy Films Investigated by X-Ray Magnetic Circular Dichroism

Hans-Joachim Elmers[1], Andres Conca[1], Tobias Eichhorn[1], Andrei Gloskovskii[2], Kerstin Hild[1], Gerhard Jakob[1], Martin Jourdan[1], and Michael Kallmayer[1]

[1] Institut für Physik, Johannes Gutenberg-Universität, Staudingerweg 7, 55128 Mainz, Germany
elmers@uni-mainz.de

[2] Institut für Anorganische und Analytische Chemie, Johannes Gutenberg-Universität, Staudingerweg 9, 55128 Mainz, Germany

Abstract. We have investigated the magnetic properties of epitaxial Heusler alloy films using x-ray absorption spectroscopy (XAS) and x-ray magnetic circular dichroism (XMCD) in the transmission (TM) and in the surface sensitive total electron yield (TEY) mode. We have investigated Ni_2MnGa based shape memory alloys and half-metallic $Co_2Cr_{0.6}Fe_{0.4}Al$ films. Single crystalline $Ni_2MnGa(110)/Al_2O_3(11\bar{2}0)$ and $Ni_2MnGa(100)/MgO(100)$ films show a martensitic transition from a cubic high temperature phase to a martensitic low-temperature phase at 250–275 K as concluded from magnetometry and x-ray diffraction. The martensitic transition of this Heusler compound is shifted in films on Al_2O_3 to higher temperatures $T_m = 276$ K compared to the bulk value of 200 K. A remarkable change of the Ni x-ray absorption spectra occurs at T_m indicating specific changes of the electronic structure. The observed changes are in agreement with theoretical predictions. The orbital to spin momentum ratio of the Ni moment increases significantly on entering the martensite state thus explaining the macroscopic increase of magnetic anisotropy. The spin and orbital magnetic moments of $Co_2Cr_{0.6}Fe_{0.4}Al$ films are similar to values measured for the bulk materials of the corresponding compounds. Interface properties can severely deviate from the bulk properties. We have investigated the interfaces of $Co_2Cr_{0.6}Fe_{0.4}Al$ and Ni_2MnGa Heusler alloy films and Al cap layers. At elevated temperatures and at rough surfaces the deposited Al severely reacts with the surface of a Heusler alloy indicated by changes of the absorption spectra. Compositional deviations at the interface as detected by XAS can also severely influence magnetic interface properties. Micro-spectroscopy using photoemission electron microscopy reveals an Al surface reaction proceeding inhomogeneously with reaction nuclei separated on a micron length scale.

1 Introduction

Heusler alloys [1] have recently become a field of active research. The fabrication of magnetic Heusler compounds designed as materials for spintronic applications has made significant progress [2]. Heusler compounds based on the composition Co_2YZ (Y=3d-transition metal, Z=main group element) can be made as half-metals with a nearly complete spin polarization of the conduction electrons. A semiconducting energy gap for minority electrons occurs at the Fermi edge [3, 4]. The full spin polarization at the Fermi edge turns the Heusler compounds into exciting materials for next generation spin electronics [5] aiming at e.g. magnetic tunneling junctions, spin injection devices and new spin transport phenomena. Recent experimental progress confirms the large potential of these alloys in this field [6–12]. For a tunneling junction with two Co_2MnSi electrodes Sakuraba et al. [6] reported a tunneling magnetoresistance effect (TMR) of 580% at 4 K. A TMR device with $Co_2FeAl_{0.5}Si_{0.5}$ electrodes showed a TMR effect of 175% at room temperature [7]. The second important class of Heusler compounds are magnetic magnetic shape memory alloys based on the compound Ni_2MnGa. Ni_2MnGa -based Heusler compounds undergo a martensitic transformation from a Heusler $L2_1$ parent phase at high temperature to a lower symmetry martensite phase [13] at low temperature. In the martensitic state Ni_2MnGa revealed a huge magnetically induced length change of up to 10% [14, 15]. The stoichiometric compound Ni_2MnGa shows a ferromagnetic transition at $T_C = 376$ K and the martensitic transformation temperature at $T_m = 202$ K. However, T_m can be shifted for off-stoichiometric compounds to temperatures well above room temperature [14, 15]. Magnetic shape memory materials offer new functionalities in microdevices [16–18], e.g. micromotors and miniaturized linear actuators. For such applications, the fabrication of epitaxial thin films is an absolute necessity. Recently, epitaxial films of the Ni_2MnGa- derived compositions revealing a martensitic phase transition were prepared by DC-sputtering on $Al_2O_3(11\bar{2}0)$ and MgO(100) substrates [19, 20]. A magnetically induced reorientation of martensite variants in Ni_2MnGa films was reported in Refs. [21, 22].

Heusler compounds are magnetically complex materials with the magnetization distributed on different atoms. Hence, a quantitative information on the distribution of magnetization inside the material is of great interest. A full understanding of the magnetic properties can only be achieved using element-specific magnetometry. For the application of thin films it is of equal importance to determine interface properties. In addition to deviations of magnetic interface properties from the bulk properties that are due to the broken symmetry at the interface, Heusler compounds potentially show variations of the stoichiometry and the local order near interfaces. For the investigation of element-specific magnetic properties of Heusler alloys, x-ray magnetic circular dichroism (XMCD) has turned out to be a very useful tool [23, 24]. For XMCD the x-ray absorption measured by the total photoelectron yield (TEY), the electron mean free path of 2.5 nm limits the information depth and thus provides a surface sensitive information [25]. In contrast, a measurement of the transmitted x-ray intensity (TM) reveals an integrated information along the surface normal. For films of several tens of nanometer thickness this means practically a bulk information. The transmission signal can be measured via the substrate luminescence in Al_2O_3 and MgO, thus allowing for an investigation of expitaxial films [26, 27].

This article is organized as follows: We briefly introduce the experimental methods used here. We then report on results for Ni_2MnGa based films with a focus on magnetic bulk properties near the martensitic phase transition. Finally, the deviation of magnetic interface properties are discussed in comparison to the corresponding bulk properties of the films for both the Ni_2MnGa and $Co_2Cr_{0.6}Fe_{0.4}Al$ compounds.

2 Experimental

We prepared epitaxial Ni_2MnGa and $Co_2Cr_{0.6}Fe_{0.4}Al$ films of thicknesses between 50 nm and 100 nm using dc-sputtering onto $Al_2O_3(11\bar{2}0)$ and MgO(100) substrates at temperatures between 300 and 950 K. In order to prevent surface oxidation, we capped the films *in-situ* by metallic Al using sputter deposition at temperatures between 320 and 670 K. Details of film preparation and structural characterization are reported elsewhere [12, 19, 27–29].

We performed x-ray absorption spectroscopy (XAS) at the UE56/1-SGM beam-line at the synchrotron light source BESSY II (Berlin) with perpendicularly incident photons and an external magnetic field of 1.6 T applied parallel to the photon beam. The total photoelectron yield was measured by the sample current. The photon flux transmitted through the film was detected via x-ray luminescence in the substrate [26, 27].

Micro-spectroscopic measurements were performed at the soft x-ray beamline WERA at ANKA (Karlsruhe), combining XAS with photoemission electron microscopy (PEEM) [30].

3 Martensitic phase transition in Ni_2MnGa films

Figure 1 shows the incident-photon-flux-normalized TM XAS spectra of a Ni_2MnGa Heusler alloy film measured at room temperature, i.e. above the martensitic transition temperature T_m. The luminescence signal of the substrate $I^{\pm}$ is proportional to the transmitted x-ray intensity and proportional to the sensitivity function I_{ref} of the substrate. The absorption exponent $k = \mu d$ (absorption coefficient μ, thickness d) is calculated by $k^{\pm}(h\nu) = -ln[I^{\pm}(h\nu)/I_{\text{ref}}(h\nu)]$. $I_{\text{ref}}(h\nu)$ was determined by a corresponding measurement of the bare substrate crystal except a constant geometry factor and found to increase linearly with the photon energy. $I_{\text{ref}}(h\nu)$ was then normalized at

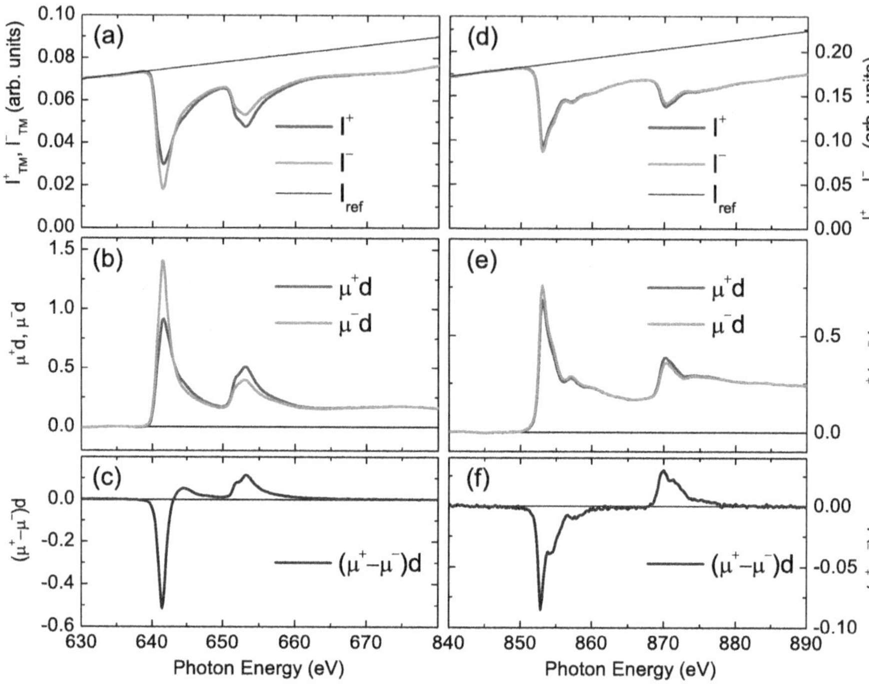

Fig. 1. Intensity of the luminescence light emerging from the substrate measured at the Mn (**a**) and Ni (**d**) $L_{2,3}$-edge of a Ni_2MnGa film on Al_2O_3 divided by the incident x-ray intensity. Corresponding absorption exponent $\mu d = -ln(I/I_{\mathrm{ref}})$ for film thickness d for Mn (**b**) and Ni (**e**). XMCD spectra, $(\mu^+ - \mu^-)d$, for Mn (**c**) and Ni (**f**) as calculated from the difference of absorption exponents for opposite magnetization directions.

the pre-edge. This normalization corresponds to an infinitely large penetration depth at the pre-edge and represents only a minor error with respect to the actual penetration depth of ca. 500 nm. In contrast to the XAS signal of pure Ni, the signal shows a satellite peak at about 4 eV above the L_3 edge. From the difference $k^+ - k^-$ measured for opposite magnetization direction with respect to the circular polarization we determine the effective spin and orbital magnetic moments using the sum rule analysis.

Variations of the x-ray absorption coefficient μ for different samples correspond to changes of the unoccupied local density of states function (LDOS). In Fig. 2 we show the Ni XAS, averaged for both magnetization directions, measured at various temperatures. We find a clear change of the intensity of the satellite peak A in the Ni L_3 absorption signal [indicated in Fig. 2] at T_m. This peak is well pronounced in the cubic phase above T_m and it is nearly suppressed in the tetragonally distorted phase below T_m. A similar change of the spectrum also occurs at the L_2 edge, however, with a larger lifetime broadening. The observed increase of the feature A at the phase transition

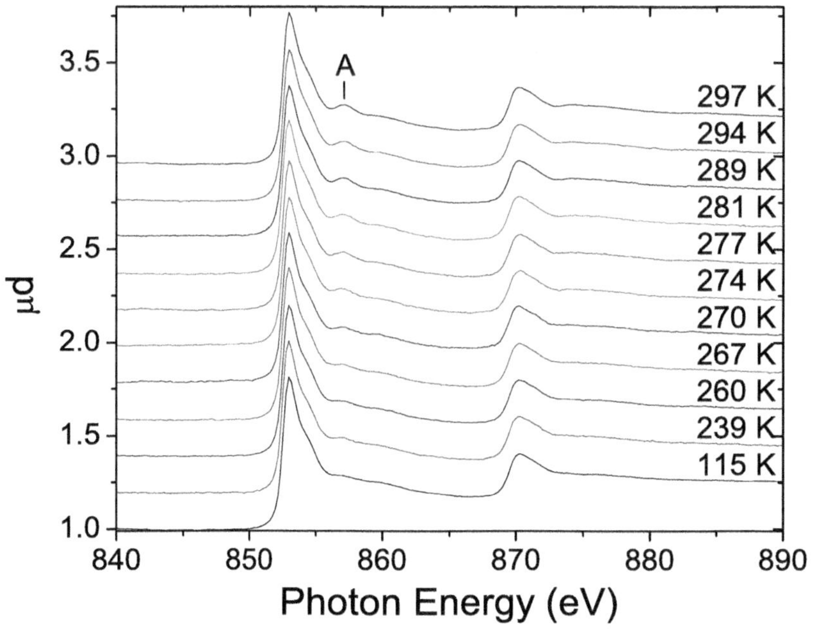

Fig. 2. Ni XAS, $\mu d = (\mu^+ + \mu^-)d/2$, measured at the indicated temperatures starting at 115 K. The spectra are shifted by a constant offset for clarity.

is in full agreement with the theoretical prediction by Ayuela et al. [31]. According to the calculations the peak originates from hybridized Ni $3d$ - Ga $2p$ states that are degenerated in the austenite state. Their degeneracy is lifted by the martensitic transition leading to a reduction of the density of states at the respective energy position.

The temperature dependence of the integrated intensity at peak A is shown in Fig. 3(a). A sharp increase occurs at the transition from the martensite to the austenite phase. Comparison to x-ray diffraction and magnetization data confirms that the observed changes in the XAS spectra are related to the martensitic phase transition [20].

In contrast to the Ni XAS changes in the Mn $L_{3,2}$ spectra at T_m are not detected. This is expected in the case of $c/a < 1$ where the Fermi level lies in a position at which there is almost an equal LDOS of majority and minority $t_{2}g$ states. For the case of $c/a > 1$ larger changes of the LDOS close to the Fermi level are expected because the atoms become closer in the $a - b$-plane. From this fact we might infer that the structural change with $c/a < 1$ is favored in agreement with other experiments.

The sum rule analysis [32, 33] results in independent values for the effective spin and orbital magnetic moments as shown in Fig. 3. The dipole term is neglected for the calculation of spin moments. We consider values of the number of d-holes, N_h, of 1.5 for Ni and 4.5 for Mn according to theoretical

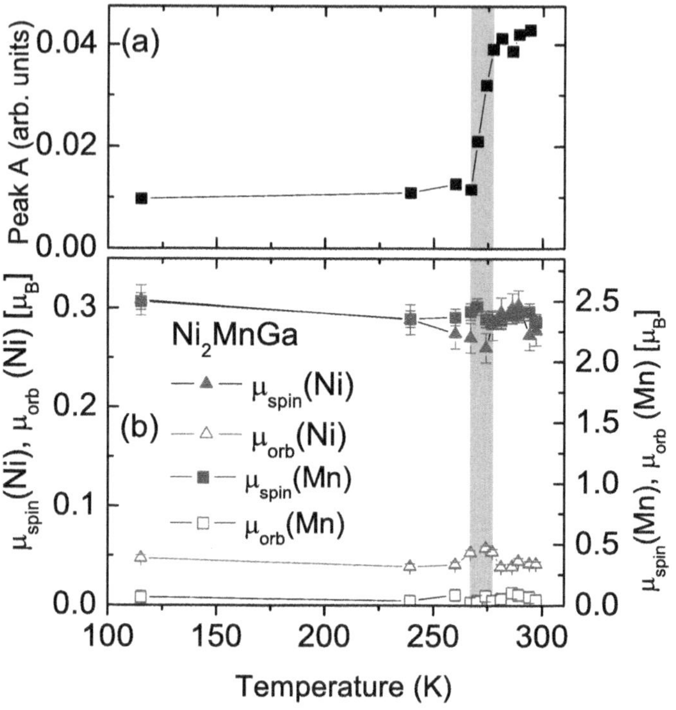

Fig. 3. (**a**) Intensity of the Ni XAS satellite peak A. The *gray area* indicates the temperature region of the martensitic phase transition. (**b**) Absolute values of the Mn and Ni spin and orbital magnetic moments.

calculations. For Mn a correction factor of 1.5 is taken into account because of the jj coupling. The ratio between Mn and Ni moment is roughly a factor of 9 in agreement with ab-initio theory [31, 34]. However, we could not confirm the predicted drop of the Ni moment at T_m. The cumulated magnetization is 3.5 $\mu_B/f.u.$ (extrapolated to $T = 0$ K), which is lower than the value of bulk samples (4 $\mu_B/f.u.$) [35]. We attribute this discrepancy to a residual atomic disorder which leads to deviations from the perfect $L2_1$ order.

The Mn spin moment shows a weak temperature dependence. In comparison to the Mn Moment, the Ni spin moment decreases with a slightly larger temperature dependence in the temperature region below and above T_m with a step-like increase at T_m. A larger temperature induced decrease of the Ni moment is in qualitative agreement with a theoretical prediction [36] based on a Heisenberg model. In a Heisenberg model a dominating nearest neighbor exchange constant $J_{\mathrm{Ni-Mn}}$ in combination with the different number of magnetic nearest neighbor atoms of Ni (4 Mn atoms) and Mn (8 Ni atoms) would lead to a stronger temperature dependence of m_{Ni} in comparison to m_{Mn} [36]. The observed difference is however much smaller than predicted which might be due to the itinerant character of the d-electron states. In contrast, a neutron diffraction experiment provided an indication of the opposite behavior of stronger temperature dependence of m_{Mn} [37] in agreement with previous XMCD results [20] for Ni_2MnGa films with a lower degree of $L2_1$ order and lower magnetization. The changes of the relative temperature dependencies with the degree of local atomic order reflect that subtle changes of local order in the Heusler compounds may provoke significant variations of macroscopic properties.

The Ni orbital moment of the film, as calculated from the transmission data, shows a peak-like increase at the phase transition exceeding 20% of the spin moment. The observed peak-like increase of the orbital moment might be explained by the large number of phase boundaries between martensitic and austenitic regions existing in the indicated temperature range. The reduced symmetry of lattice sites at these boundaries provokes an increase of the orbital moment similar e.g. to the case of free surfaces. The observed maximum orbital moment in the transition region might explain why the largest magnetic shape memory effects are observed close to T_m. The Mn orbital moment is almost zero. Therefore, the magnetic anisotropy is exclusively caused by the Ni magnetic moment.

4 Interface properties of Heusler compound films

Firstly, we focus on the interface properties of half-metallic Heusler compound films. XAS measured in transmission is directly compared to the simultaneously acquired TEY signal. The TEY spectra have been corrected for the self-absorption effect [25]. Figure 4 compares XAS for $Co_2Cr_{0.6}Fe_{0.4}Al$ films on $Al_2O_3(11\bar{2}0)$ capped with 4 nm Al at 480 K (sample I) and $Co_2Cr_{0.6}Fe_{0.4}Al$ films on MgO(100) capped with 4 nm Al at 320 K (sample II). The discrepancy between Co $L_{3,2}$-spectra measured in the surface sensitive TEY-mode and in the transmission mode for sample I indicates structural deviations at the interface of the Heusler alloy (see also Ref. [27]). The pronounced feature A appearing at 3.8 eV above the L_3 absorption maximum has been found for a number of Co_2YZ Heusler alloys and was attributed to a Co-3d – Al-2p hybridization state [24]. At first glance peak A appears smaller for the TEY – spectra, however, the difference of the TEY- and TM spectra of sample I reveals an additional peak B at 2.5 eV above the L_3 maximum (dashed red line in Fig. 4(a)). This peak is characteristic for the formation of a stoichiometric CoAl compound [38, 39]. The higher Al coordination for Co in CoAl in comparison to $Co_2Cr_{0.6}Fe_{0.4}Al$ shifts the hybridization state to lower energies. The paramagnetic properties of CoAl explain the decreased XMCD signal at the interface.

The implementation of a buffer layer on the MgO substrate and an additional annealing step leads to an increased atomic order of the $Co_2Cr_{0.6}Fe_{0.4}Al$ /MgO film (sample II) [29]. The increase of peak A (Fig. 4(a)) reflects the

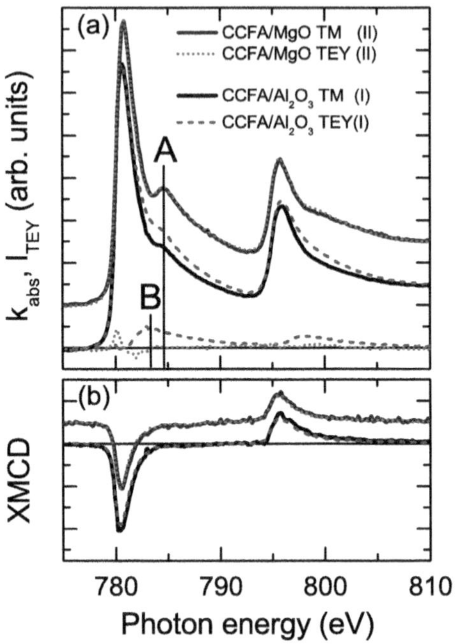

Fig. 4. (*Color online*) (**a**) XAS obtained at 300 K by total electron yield (TEY) and transmission (TM) for two $Co_2Cr_{0.6}Fe_{0.4}Al$ film samples. Sample I is a 110-nm-thick $Co_2Cr_{0.6}Fe_{0.4}Al$ Heusler alloy film grown on $Al_2O_3(11\bar{2}0)$ and capped by 6 nm Al at 480 K. Sample II is a 100-nm-thick $Co_2Cr_{0.6}Fe_{0.4}Al$ film grown on Fe/MgO(100) and capped by 4 nm Al at 320 K. XAS spectra shifted for clarity. The difference spectra of TEY and TM data for the same sample are indicated by the *dashed (red)* and *dotted (pink) lines*. (**b**) corresponding XMCD spectra.

change of the electronic structure due to the improved atomic order. No pronounced difference between TEY and TM appears above the L_3 maximum, signalizing that low-temperature deposition of Al avoids the formation of CoAl. The almost equal XMCD signals for sample II indicate that the Co magnetic moment at the interface and in the core of the film are similar except a small increase of the orbital magnetic moment at the interface. An interesting pronounced difference between TEY and TM XAS signals of sample II remains at the onset of the absorption maximum. This difference reflects a shift of the density-of-states maximum of minority states towards the Fermi edge at the Al-interface by 90 meV. The difference appears exclusively for highly ordered samples.

Figure 5 compares Ni XAS for Ni_2MnGa films on $Al_2O_3(11\bar{2}0)$ capped with 4 nm Al at 520 K (sample III) and 670 K (sample IV). The difference between TEY and TM comprises a decrease of the L_3 maximum and an increase of peak B appearing at 2.8 eV above the L_3 maximum with increasing deposition temperature for the Al capping layer. The position of peak B is characteristic

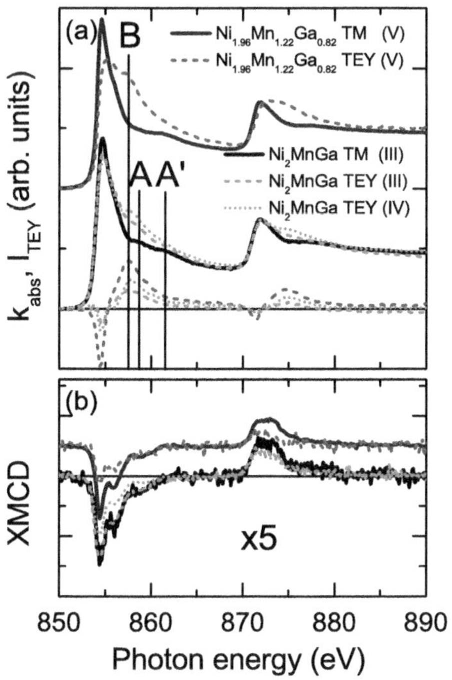

Fig. 5. (*Color online*) (**a**) XAS obtained at 70 K by total electron yield (TEY) and transmission (TM) for three Ni_2MnGa film samples. 100-nm-thick Ni_2MnGa films grown on $Al_2O_3(11\bar{2}0)$ are coated by Al at 520 K (sample III) and 670 K (sample IV). Sample V is a $Ni_{1.96}Mn_{1.22}Ga_{0.82}$ film on $Al_2O_3(11\bar{2}0)$ coated by Al at 640 K. Spectra shifted for clarity. The difference spectra of TEY and TM data for the same sample are indicated by the *blue* and *red dashed lines.* (**b**) corresponding XMCD spectra.

for the stoichiometric NiAl compound [38, 40–42] and can be well distinguished from peaks A and A' appearing in the case of well ordered Ni_2MnGa. Peak B is particularly pronounced for sample V with slightly changed composition $Ni_{1.96}Mn_{1.22}Ga_{0.82}$ and an Al capping deposited at 640 K. We attribute the increased NiAl formation in comparison to sample IV to an increased impact energy of Al atoms sputtered at a fivefold lower pressure in this case. The Ni XMCD signal decreases with increasing formation of NiAl precipitations because NiAl is paramagnetic.

From a series of PEEM images acquired with varying photon energy we calculate gray-level maps of the ratio $I(L_3)/I(B)$ of the L_3 maximum and the intensity of peak B indicating chemical contrast, and of the ratio $I(L_3)/I(L_2)$ indicating magnetic contrast (Peaks measured with respect to the pre-edge value).

The dark spots in Fig. 6(a) and (c) indicate areas with increased NiAl to Ni_2MnGa ratio. The prominent dark lines of high NiAl concentration indicated

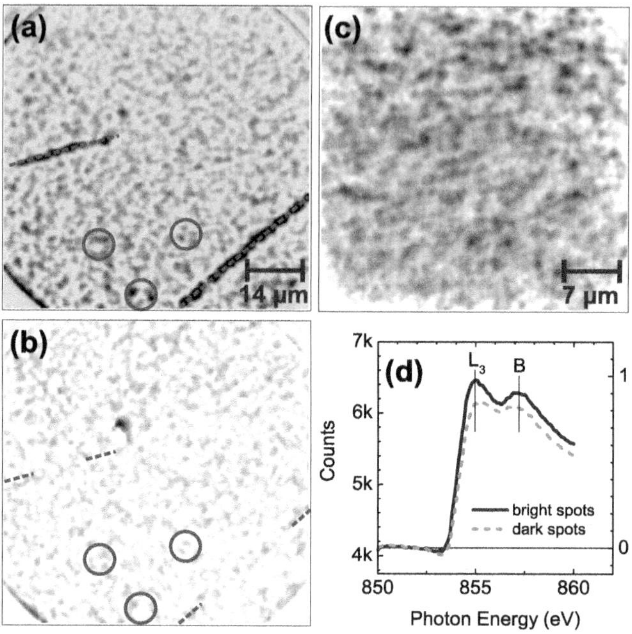

Fig. 6. (*Color online*) (**a**) *Gray-level image* of the ratio $I(L_3)/I(B)$ of the L_3 maximum and the intensity of peak B of sample V indicating the content of NiAl. (**b**) *Gray-level image* of the ratio $I(L_3)/I(L_2)$ of the L_3 and L_2 maximum indicating magnetic contrast obtained from the same PEEM series. The *red circles* mark identical areas in (**a**) and (**b**) and emphasize similar intensity variations. The single *black spot* is an image-processing artifact caused by a dust particle on the surface. (**c**) Chemical contrast as in (**a**) for the same sample with higher magnification. (**d**) Micro-spectra acquired in *dark spots* and surrounding areas of (**c**).

by the dashed lines in Fig. 6(a) are rare events probably caused by scratches on the substrate surface. The typical surface shows circular shaped NiAl nucleation centers which are randomly distributed with an average separation of a few micrometers. The high mobility of metallic Al deposited at elevated temperatures results in an inhomogeneous solid-state reaction. The micro-spectra averaged over the dark spots and the surrounding areas, respectively, reveal an overall deficiency of Ni in the dark, NiAl rich spots. This might be attributed to Al accumulation at these spots attenuating the Ni XAS signal or alternatively to a mass transport of Ni away from the NiAl nucleation centers.

The gray-level map of magnetic contrast shown in Fig. 6(b) reveals small magnetization variations ($< 2\%$ in comparison to 6% variation expected for domains with magnetization parallel and antiparallel to the photon beam), which are only in a few cases connected to compositional variations. The nearly independent domain structure indicates additional structural variations.

5 Summary

The results show, that besides the determination of element-specific magnetic moments, x-ray absorption spectroscopy provides information on the relation of structure and electron density-of-states. We have confirmed theoretically predicted variations of the electron density-of-states at the martensitic phase transition of Ni_2MnGa. Comparison of transmission and total electron yield data revealed deviations of structure and stoichiometry at buried interfaces of Heusler compounds. Using photoemission electron microscopy we gained laterally resolved information on the interface reactions.

6 Acknowledgements

The authors would like to thank for financial support from the Deutsche Forschungsgemeinschaft (Ja821/3-1 within SPP1239 and EL172/12-2, Ja821 /2-2, Jo404/4-1 within FG559) and S. Cramm for support at BESSY.

References

1. F. Heusler, W. Starck, and E. Haupt, Verh. Deutsch. Phys. Ges. **5**, 220 (1903)
2. C. Felser and B. Hillebrands, J. Phys. D: Appl. Phys. **40** (2007)
3. I. Galanakis, Ph. Mavrapoulos, and P. H. Dederichs, J. Phys. D: Appl. Phys. **39**, 765 (2006)
4. H. C. Kandpal, G. H. Fecher, and C. Felser, J. Phys. D: Appl. Phys. **40**, 1507 (2007)
5. J. M. D. Coey, M. Venkatesan, and M. A. Bari, Lecture Notes in Physics, vol. **595**, ed. C. Berthier, L. P. Levy, and G. Martinez (Springer, Heidelberg, 2002)
6. Y. Sakuraba, T. Miyakoshi, M. Oogane, Y. Ando, A. Sakuma, T. Miyazaki, and H. Kubota, Appl. Phys. Lett. **89**, 052508 (2006)
7. N. Tezuka, N. Ikeda, S. Sugimoto, and K. Inomata, Appl. Phys. Lett. **89**, 252508 (2006)
8. K. Inomata, S. Okamura, A. Miyazaki, M. Kikuchi, N. Tezuka, M. Wojcik, and E. Jedryka, J. Phys. D: Appl. Phys. **39**, 816 (2006)
9. D. Ebke, J. Schmalhorst, N. N. Liu, A. Thomas, G. Reiss, and A. Hütten, Appl. Phys. Lett. **89**, 162506 (2006)
10. T. Marukame, T. Ishikawa, S. Hakamata, K. Matsuda, T. Uemura, and M. Yamamoto, Appl. Phys. Lett. **90**, 012508 (2007)
11. A. D. Rata, H. Braak, D. E. Bürgler, and C. M. Schneider, Appl. Phys. Lett. **90**, 162512 (2007)
12. A. Conca, M. Jourdan, and H. Adrian, J. Phys. D: Appl. Phys. **40**, 1534 (2007)
13. P. J. Webster et al., Philos. Mag. B **49**, 295 (1984)
14. K. Ullakko, J. K. Huang, C. Kantner, R. C. O'Handley, and V. V. Kokorin, Appl. Phys. Lett. **69**, 1966 (1996)
15. A. Sozinov, A. A. Likhachev, N. Lanska, and K. Ullakko, Appl. Phys. Lett. **80**, 1746 (2002)

16. K. Bhattacharya and R. D. James, **53**, 307 (2005)
17. H. Morito, A. Fujita, K. Oikawa, K. Ishida, K. Fukamichi, and R. Kainuma, Appl. Phys. Lett. **90**, 062505 (2007)
18. Z. D. Han, D. H. Wang, C. L. Zhang, S. L. Tang, B. X. Gu, and Y. W. Du, Appl. Phys. Lett. **89**, 182507 (2006)
19. G. Jakob and H. J. Elmers, J. Magn. Magn. Mater. **310**, 2779 (2007)
20. G. Jakob, T. Eichhorn, M. Kallmayer, and H. J. Elmers, Phys. Rev. B **76**, 174407 (2007)
21. O. Heczko, M. Thomas, J. Buschbeck, L. Schultz L, and S. Fähler, Appl. Phys. Lett. **92**, 072502 (2008)
22. M. Thomas, O. Heczko, J. Buschbeck, U. K. Rössler, J. McCord, N. Scheerbaum, L. Schultz, and S. Fähler, New J. Phys. **10**, 023040 (2008)
23. J. Schmalhorst, S. Kämmerer, M. Sacher, G. Reiss, A. Hütten, and A. Scholl, Phys. Rev. B **70**, 024426 (2004)
24. H. J. Elmers, G. H. Fecher, D. Valdaitsev, S. A. Nepijko, A. Gloskovskii, G. Jakob, G. Schönhense, S. Wurmehl, T. Block, C. Felser, P.-C. Hsu, and W.-L. Tsai, Phys. Rev. B **67**, 104412 (2003)
25. R. Nakajima, J. Stöhr, and Y. U. Idzerda, Phys. Rev. B **59**, 6421 (1999)
26. M. Kallmayer, K. Hild, T. Eichhorn, H. Schneider, G. Jakob, A. Conca, M. Jourdan, H. J. Elmers, S. Schuppler, and P. Nagel, Appl. Phys. Lett. **91**, 192501 (2007)
27. M. Kallmayer, H. Schneider, G. Jakob, H. Elmers, K. Kroth, H. Kandpal, U. Stumm, and C. Cramm, Appl. Phys. Lett. **88**, 072506 (2006)
28. H. Schneider, G. Jakob, M. Kallmayer, H. J. Elmers, M. Cinchetti, B. Balke, S. Wurmehl, C. Felser, M. Aeschlimann, and H. Adrian, Phys. Rev. B **74**, 174426 (2006)
29. A. Conca, M. Jourdan, and H. Adrian, J. Crystal Growth **299**, 299 (2007)
30. Ch. Ziethen, O. Schmidt, G. K. L. Marx, G. Schönhense, R. Frömter, J. Gilles, J. Kirschner, C. M. Schneider, and O. Gröning, J. Electron Spectrosc. Relat. Phenom. **107**, 261 (2000)
31. A. Ayuela, J. Enkovaara, and R. M. Nieminen, J. Phys.: Condens. Matter **14**, 5325 (2002)
32. G. van der Laan and B. T. Thole, Phys. Rev. B **43**, 13401 (1991)
33. B. T. Thole, P. Carra, F. Sette, and G. van der Laan, Phys. Rev. Lett. **68**, 1943 (1992)
34. S. R. Barman, S. Banik, and A. Chakrabarti, Phys. Rev. B **72**, 184410 (2005)
35. V. V. Khovailo, V. Novosad, T. Takagi, D. A. Filippov, R. Z. Levitin, and A. N. Vasil'ev, Phys. Rev. B **70**, 174413 (2004)
36. M. Lezaic, Ph. Mavropoulos, J. Enkovaara, G. Bihlmayer, and S. Blügel, Phys. Rev. Lett. **97**, 026404 (2006)
37. Ch. Hordequin, E. Lelievre-Berna, and J. Pierre, Physica **234B–236B**, 602 (1997)
38. G. A. Botton, G. Y. Guo, W. M. Temmerman, and C. J. Humphreys, Phys. Rev. B **54**, 1682 (1996)
39. Y. S. Lee, C. H. Lee, K. W. Kim, H. J. Shin, and Y. P. Lee, J. Magn. Magn. Mater. **272–276**, 2151 (2004)
40. Y. K. Chang, K. P. Lin, W. F. Pong, M.-H. Tsai, H. H. Hsei, J. Y. Pieh, P. K. Tseng, J. F. Lee, and L. S. Hsu, J. Appl. Phys. **87** (2000)
41. D. A. Pankhurst, G. A. Botton, and C. J. Humphreys, Phys. Rev. B **63**, 205117 (2001)
42. L.-S. Hsu and A. I. Nesvizhskii, J. Phys. Chem. Solids **62**, 1103 (2001)

Coherent Spin Dynamics in Nanostructured Semiconductor-Ferromagnet Hybrids

Patric Hohage, Jörg Nannen, Simon Halm, and Gerd Bacher

Werkstoffe der Elektrotechnik and CeNIDE, Universität Duisburg-Essen, Bismarckstraße 81, 47057 Duisburg, Germany
gerd.bacher@uni-due.de

Abstract. We report on the ability to manipulate the coherent spin dynamics in a semiconductor by nanostructured ferromagnets. The fringe field of ferromagnetic wires causes an inhomogeneous magnetic field, which penetrates the underlying semiconductor. This allows us to locally change the frequency of electron as well as magnetic ion spin precession in an externally applied magnetic field, which is probed by time-resolved Kerr rotation. In Permalloy-GaAs hybrids, a fringe field induced frequency shift of the free electron spin precession of up to 10% is obtained at room temperature. For localized Mn^{2+} ion spins, the spatially inhomogeneous magnetic fringe field results in both a transient change of the precession frequency and a decrease of the dephasing time T_2^* of the spin ensemble in Co-CdZnMnSe hybrids. Model calculations are able to describe our findings.

1 Introduction

One key requirement for future spintronic device concepts is the ability to locally control the coherent evolution of spin states in a semiconductor (SC). Since the spin of a particle, like an electron, a hole or an impurity atom like Mn^{2+}, is accompanied by a magnetic moment, it can be manipulated by a magnetic field. On the one hand this can be an internal field resulting from an electric field or strain via spin orbit coupling [1–4] or from nuclear fields via hyperfine interaction [5, 6]. On the other hand nanostructured ferromagnets (FMs) on top of the SC do even allow one to generate a magnetic fringe field, which is varying locally on a micrometer or even sub-micrometer length scale. This can be used, e.g., for defining an energy landscape for spin states in magnetic semiconductors [7–10], where due to the efficient *sp-d* exchange interaction between carrier spins and the spins of the magnetic ions, fringe fields allow a pronounced spatial modulation of the carrier spin polarization even in the absence of an externally applied magnetic field [11, 12]. Moreover, magnetic fringe fields are utilized for coherent single electron spin control in a quantum dot [13] and for manipulating the frequency of coherent spin precession of both, magnetic ion spins [14, 15] and electron spins [16–19] in a SC on a micrometer length scale. Even an electrical manipulation of the coherent electron spin precession is achieved by dragging the electrons laterally within the inhomogeneous fringe field using an externally applied voltage [20].

There are several issues to be addressed when developing FM-SC hybrids for controlling coherent spin dynamics in a SC. First, patterns of nanostructured FMs with well-defined magnetization have to be designed in order to achieve a pronounced magnetic fringe field in the active semiconductor layer. Second, the impact of a spatially varying magnetic field on the coherent precession of an ensemble of free and localized spin states has to be evaluated, and third, the potential of high temperature operation up to room temperature is of interest. While, e.g., at low temperatures, spin coherence times of 100 ns and above have been obtained [21, 22], spin dephasing, e.g., due to the D'yakonov-Perel' mechanism, becomes more important if one increases the temperature. At room temperature, typical T_2^* times between less than 100 ps and a few ns, depending on the material system, have been found [23–28].

In this paper, we report on the coherent spin dynamics in FM-SC hybrids probed by time-resolved Kerr rotation. In Py-GaAs hybrids, we concentrate on the ability to define a frequency shift of the coherent electron spin precession by designing the geometry of a Py wire array and demonstrate fringe field induced manipulation of the coherent spin evolution up to room temperature. Hybrids of Co wires and CdZnMnSe/ZnSe quantum wells are studied in order to evaluate the impact of a locally varying magnetic field on the coherent precession of a localized Mn^{2+} spin ensemble.

2 Samples and experiment

In sample A, arrays with an extension of $(150\,\mu m)^2$ consisting of $t = 76$ nm thick Py ($Ni_{0.81}/Fe_{0.19}$) wires with a width of $w = 2\,\mu m$ and an interwire distance varying between $d = 0.25\,\mu m$ and $d = 8\,\mu m$ have been defined by electron beam lithography and lift-off technique on top of a GaAs heterostructure grown by molecular beam epitaxy (MBE). The active GaAs layer has a thickness of 100 nm and a nominal doping concentration of $n = 7 * 10^{15}$ cm^{-3} and is embedded between $Al_{0.3}Ga_{0.7}As$ barriers to prevent carrier diffusion to the substrate and the surface, respectively. Sample B consists of arrays of $t = 55$ nm thick Co wires capped by 5 nm of Cr to prevent oxidation on top of an MBE grown CdZnMnSe/ZnSe quantum well with a Mn concentration of $x_{Mn} = 0.08$ and a quantum well thickness of 18 monolayers. Width and period of the Co wires have been varied, while the length is $150\,\mu m$ in each case. In the SC heterostructures, thin top layers of 25 nm in both samples are used in order to achieve pronounced fringe fields from the FMs in the active area of the SC.

The time-resolved Kerr rotation (TRKR) measurements have been carried out in a continuous flow liquid He cryostat placed inside an electromagnet which provides magnetic fields up to 1.3 T in Voigt geometry, i.e. along the sample surface. By a circularly polarized pump pulse, electron-hole pairs with a well defined spin orientation perpendicular to the external magnetic field $\mathbf{B}_{\text{ext}}$ are generated. The spins start to process around the total magnetic field $\mathbf{B}_{\text{tot}}$ given by the sum of $\mathbf{B}_{\text{ext}}$ and the magnetic fringe field $\mathbf{B}_{\text{fr}}$ stemming from the FMs on top of the SC. The transient change of the Kerr rotation angle is measured by a linearly polarized probe pulse at time delay Δt. Both, pump and probe pulse are obtained from a tunable mode-locked Ti-Sapphire laser with a repetition rate of 76 MHz and a pulse width of 2 ps and our setup provides a spatial resolution of $< 6\,\mu$m. In case of the Py-GaAs hybrids, the fundamental wavelength of the laser was adjusted close to the GaAs bandgap and thus varied between 1.523 and 1.42 eV, depending on temperature, while for the Co-CdZnMnSe hybrids we used the second harmonic of the laser with an energy of about 2.7 eV according to the low temperature bandgap of CdZnMnSe. In order to improve the signal-to-noise ratio, a double modulation technique was used [23].

3 Free vs. localized spin precession in a semiconductor

In the left part of Fig. 1, typical examples of the TRKR signal obtained for the reference samples, i.e. SCs without FMs on top, are depicted. The measurements have been performed at $T = 5$ K and $B_{\text{ext}} = 0.9$ T. In the top part of the figure, the electron spin precession in GaAs is demonstrated for two different excitation energies E while in the bottom part, the precession of the Mn^{2+} ion spins in CdZnMnSe is shown. In each case, the transient Kerr angle $\theta(\Delta t)$ obtained experimentally follows the expression

$$\theta(\Delta t) = \theta_0 \cdot \exp(-\Delta t/T_2^*) \cos(\omega \Delta t + \phi) \tag{1}$$

where T_2^* is the dephasing time and $\omega = g\mu_B B_{\text{tot}}/\hbar$ the Larmor precession frequency of the spin ensemble with the g-factor g. A variety of information can be extracted from the results shown in Fig. 1. The electron precession can be described by Eq. (1) using $\phi \approx 0$ or $\phi \approx \pi$, respectively, while in case of Mn^{2+} spin precession, $\phi \approx \pi/2$ has to be used in order to describe the data. This can be easily understood by looking at the underlying mechanism starting the spin precession.

The spins of the optically excited electron-hole pairs are initially aligned perpendicular to $\mathbf{B}_{\text{ext}}$ before they start to precess. In case of CdZnMnSe, the spin dephasing of electrons and holes occurs rather rapidly within our time resolution. The fact that the Mn^{2+} spins are initially aligned parallel to $\mathbf{B}_{\text{ext}}$ and the coherent Mn^{2+} spin oscillation is induced via a coherent spin transfer from the optically generated, spin polarized carriers [29] results in a phase shift of $\phi = \pi/2$. By fitting the TRKR signal with Eq. (1), we get $g = 2.01$ and $T_2^* = 0.3$ ns.

In GaAs, the hole dephasing time is also rather short due to valence band mixing [30] and so only the electron spin precession contributes to the TRKR

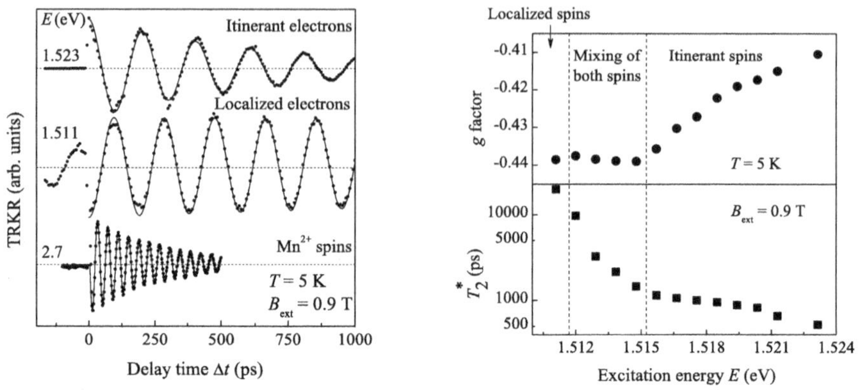

Fig. 1. *Left*: TRKR signal of electron spin precession in GaAs (*upper two curves*) and of Mn^{2+} spin precession in CdZnMnSe (*bottom curve*). In the *right part* of the figure, the g factor and the ensemble spin dephasing time T_2^* of electron spins in GaAs are depicted vs. excitation energy.

signal. For above bandgap excitation at $E = 1.523$ eV, free electrons are generated and precess around $\mathbf{B}_{\text{ext}}$ and from our data we extract $T_2^* = 0.5$ ns and $g = -0.41$. In case of below bandgap excitation (i.e. $E = 1.511$ eV), a significant increase of the T_2^* time up to more than 20 ns is obtained and we found $g = -0.44$.[1] This significant increase of T_2^* to values much longer than the recombination lifetime of the optically generated electron-hole pairs is attributed to the contribution of donor bound electrons in the regime of delocalized donor states [33]. As can be seen in Fig. 1, right, both the dephasing time T_2^* and the g factor depend strongly on the excitation energy: Localized electron spins are characterized by a much longer T_2^* time and a larger negative g factor than itinerant ones [23].

If at low temperatures the pump beam hits the GaAs layer not exactly perpendicular to the surface, nuclear spins of Ga and As will be polarized by the optically generated electrons via hyperfine interaction [34]. Vice versa, polarized nuclear spins act on the electrons like an effective magnetic field and modify the Larmor frequency [5]. In order to extract the impact of nuclear spins on the electron spin precession in GaAs, an angle of 15° against normal incidence was used for the pump beam. The pump pulse was switched on at Lab time = 0 and the transient variation of the oscillation frequency at $T = 3.1$ K and $B_{\text{ext}} = 885$ mT was monitored versus laboratory time.

[1] Note that the phase shift of π observed when changing the excitation energy from above bandgap to below bandgap excitation most likely results from the energy dependent sign of the Kerr angle [31, 32] and/or from the mixing of the Kerr and the Faraday effect due to an additional reflection of the probe beam at the back surface of the sample in case of below bandgap excitation.

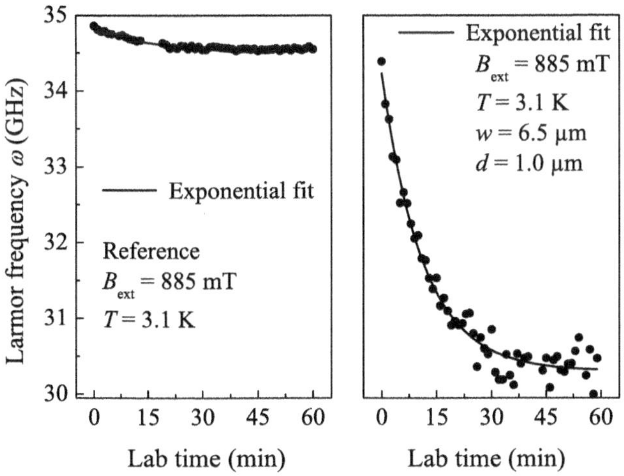

Fig. 2. Transient change of the Larmor frequency in GaAs (*left*) and in hybrids of Au wires (thickness 150 nm) on GaAs (*right*). The pump pulse was switched on at Lab time = 0. The *solid line* is a mono-exponential fit with a time constant of 11 min.

As can be seen in the left part of Fig. 2, a temporal variation of the Larmor frequency on the order of 1% is obtained. By a monoexponential fit, a time constant of 11 min is found which determines the build-up of the nuclear field. Surprisingly, the influence of the nuclear field increases by a factor of ≈ 10, if the same experiments are performed on hybrids consisting of metal stripes (in that case Au with $t = 150\,\mathrm{nm}$) on top of the GaAs heterostructure (see Fig. 2, right). No change of the characteristic time constant, however, is obtained. While the detailed mechanism of this strongly enhanced influence of nuclear fields on the coherent electron spin dynamics in hybrids is not yet clear, strain induced changes of the band structure [35] most likely contribute to this effect. In case of GaAs, our experiments show that for temperatures of 50 K and above, nuclear spin effects can be neglected.

4 Spin dynamics in ferromagnet-semiconductor hybrids

4.1 Hybrid design

From Eq. (1) it is obvious that by magnetic fringe fields $\mathbf{B}_{\mathrm{fr}}$ the total magnetic field $\mathbf{B}_{\mathrm{tot}}$ and thus the Larmor frequency can be easily modified. This results in a relative phase shift of the spin orientation after a certain time after generation. Let us assume that the FM wires are aligned along the y-direction and the external field is applied in x-direction, i.e. perpendicular to the wire orientation and parallel to the sample surface. Then, the magnetic fringe field of the FMs at the position of the active SC has components in both, x- and z-direction. We obtain $\mathbf{B}_{\mathrm{tot}} = (B_{\mathrm{ext}} + B_{\mathrm{fr,x}})\cdot\, \mathbf{x} + B_{\mathrm{fr,z}}\cdot\, \mathbf{z}$, where $B_{\mathrm{fr,x}}$ ($B_{\mathrm{fr,z}}$) are the components of $\mathbf{B}_{\mathrm{fr}}$ and $\mathbf{x}$ ($\mathbf{z}$) the unit vectors in x- and z-direction, respectively.

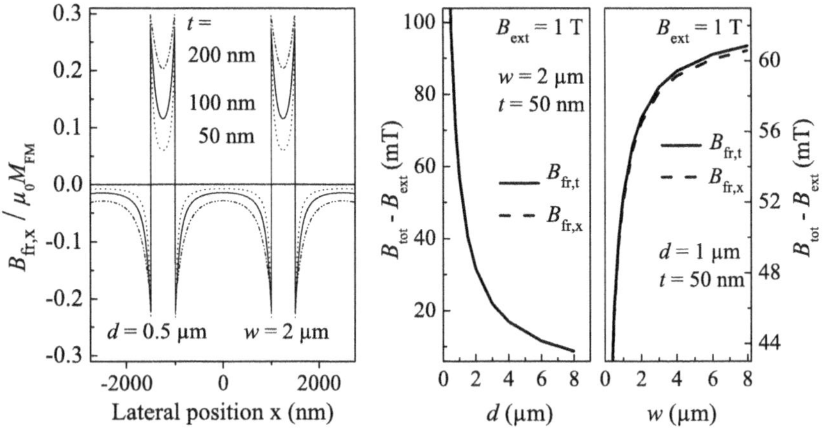

Fig. 3. *Left*: $B_{\mathrm{fr,x}}/(\mu_0 M_{\mathrm{FM}})$ versus lateral position x in a FM wire array. The wires are assumed to be oriented along the y-direction and magnetized along the x-direction. The fringe field is calculated on the SC surface for different values of t. *Right*: Contribution of the magnetic fringe field, $|\mathbf{B}_{\mathrm{tot}}| - |\mathbf{B}_{\mathrm{ext}}|$, versus gap distance d (*left*) and versus wire width w (*right*). The calculations labeled by $B_{\mathrm{fr,t}}$ consider the complete magnetic fringe field, $\mathbf{B}_{\mathrm{fr}}$, while the calculations labeled $B_{\mathrm{fr,x}}$ only consider the x-component of the magnetic fringe field.

On the left part of Fig. 3, we plot the ratio between $B_{\mathrm{fr,x}}$ and the magnetization of the FM, M_{FM}, versus lateral position x for an array of Py-wires with an interwire distance $d = 0.5\,\mu\mathrm{m}$ and a width of $w = 2\,\mu\mathrm{m}$. The fringe field is calculated directly at the SC surface and the FM thickness t has been varied between 50 and 200 nm. It is obvious that the fringe field is strongest in the vicinity of the wires and changes its sign going from beneath the wires to in between the wires. By increasing t, a strong enhancement of $B_{\mathrm{fr,x}}/(\mu_0 M_{\mathrm{FM}})$ is obtained and a remarkably high value of more than 20% can be expected in between the wires for $t = 200\,\mathrm{nm}$. As our TRKR measurements are performed from the top, we only probe the areas with positive values of the fringe field and because of our limited spatial resolution, we have to average over the whole area in between the wires.

It is quite instructive to discuss the dependence of the spatially averaged contribution of the magnetic fringe field, $|\mathbf{B}_{\text{tot}}| - |\mathbf{B}_{\text{ext}}|$, on the gap distance d and the wire width w in the example of Py wires fully magnetized perpendicular to the wire orientation (saturation magnetization $M_S = 1\ \text{T}/\mu_0$, external magnetic field $B_{\text{ext}} = 1$ T, see Fig. 3, right). With decreasing d, the contribution of the magnetic fringe field strongly increases and for $w = 2\,\mu\text{m}$, magnetic fringe fields in the order of 100 mT can be obtained for a Py thickness of $t = 50$ nm. Interestingly, $B_{\text{tot}} - B_{\text{ext}}$ is almost independent on whether the total magnetic fringe field, $\mathbf{B}_{\text{fr}}$, or purely the x-component of the fringe field, $B_{\text{fr,x}}$, is included into the calculation. This indicates that for the experiments discussed here, the z-component of the magnetic fringe field can be neglected in a good approximation. It is worth to note that the fringe field within the gap slightly increases with w but almost saturates, if w is on the order of 5 times the gap distance d. These calculations give us a hint how to optimize the fringe field induced spin manipulation in nanostructured FM-SC hybrids.

4.2 Spin precession in Py-GaAs hybrids up to room temperature

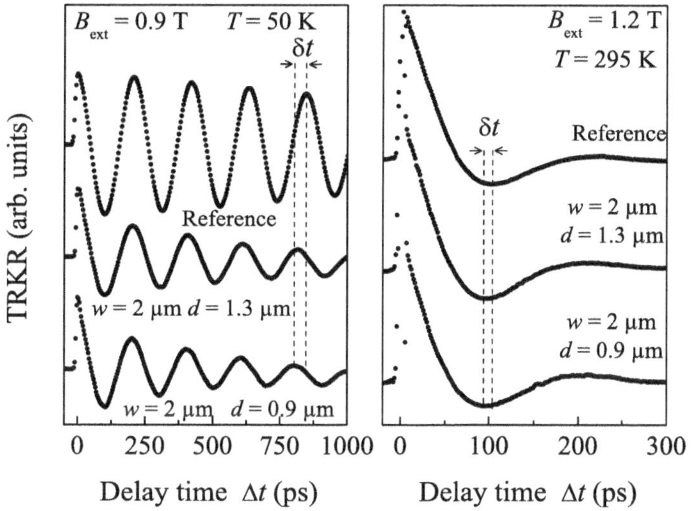

Fig. 4. Transient TRKR signal for different Py-GaAs hybrids compared to the measurement performed on the unpatterned GaAs reference layer. Experimental data at $T = 50\,\text{K}$ (*left*) and at room temperature (*right*) are depicted for comparison and δt indicates the phase shift of the electron spin orientation in the hybrids with $d = 0.9\,\mu\text{m}$ as compared to the reference at a certain time delay Δt.

In order to exclude any noticeable influence of nuclear spins on the coherent spin precession of electrons in Py-GaAs hybrids, the experiments have been performed between 50 K and room temperature. The wires have been oriented perpendicular to $\mathbf{B}_{\text{ext}}$ and $w = 2\,\mu\text{m}$ was kept constant while d was varied.

In Fig. 4, the TRKR signal of the electron spin precession in Py-GaAs hybrids with different values of d is compared to the results found for the GaAs reference sample. Measurements at $T = 50$ K (left) and at room temperature (right) are presented and the excitation energy has been adapted to the GaAs bandgap at each temperature. As expected, the increasing contribution of the magnetic fringe field results in an enhanced precession frequency in hybrids with small interwire distances d. Most important, this can be observed even at room temperature although dephasing due to the D'yakonov-Perel' mechanism results in a drastic reduction of T_2^* with increasing T [23]. Apparently not only the precession frequency but also T_2^* changes in hybrids as compared to the bulk reference. This can be most clearly seen at $T = 50$ K, where T_2^* decreases from about 2.5 ns for the unpatterned reference to 0.5 ns in case of the hybrids with $d = 0.5\,\mu\text{m}$. This is attributed to the impact of the laterally inhomogeneous magnetic fringe field and will be discussed in detail below.

From the data analysis using Eq. (1), we can extract the Larmor precession frequency of electrons in Py-GaAs hybrids and compare the results obtained for various wire geometries with fringe field calculations. For the calculations we perform a spatial averaging of the fringe field penetrating the active SC in the x-direction in order to account for the lateral inhomogeneous fringe field distribution of the Py wire array.

As can be seen in Fig. 5, a pronounced increase of the Larmor frequency with decreasing d is found and we reach a frequency change of $\approx 10\%$ for the smallest gap at room temperature. The solid line represents the result of our fringe field calculations assuming Py wires fully magnetized along $\mathbf{B}_{\text{ext}}$, i.e. perpendicular to the wire orientation, and considering the finite thickness of the cap layer above the active SC. A quite nice agreement between experiment and theory is obtained leading to the conclusion that the Larmor frequency of the electron spin precession can be systematically modified by magnetic fringe fields. In contrast, at $T = 50$ K, a clear deviation between experiment and theory can be seen, in particular for small values of d. A possible reason might be the impact of strain effects: Photoluminescence measurements performed on our hybrids at $T = 50$ K show a systematic reduction of the bandgap due to strain if one decreases d. As E was fixed, this results in an increase of excess energy and thus in a decrease of the g factor (see Fig. 1) and the precession frequency, respectively, in qualitative agreement to what is observed in experiment.

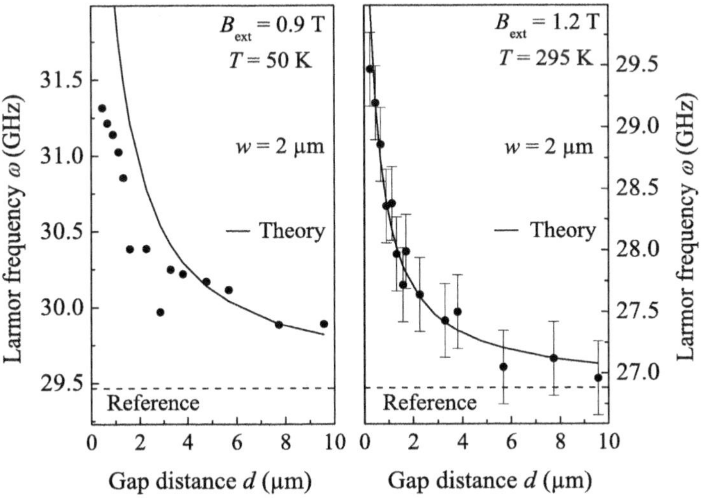

Fig. 5. Larmor frequency versus gap distance d for $w = 2\,\mu m$. Experimental data and theory are compared at $T = 50\,K$ (*left*) and at room temperature (*right*).

4.3 Localized spin dynamics in Co-CdZnMnSe hybrids

In contrast to free electron spins, the Mn^{2+} spin system is localized and characterized by a g factor, which is almost independent of the crystal structure and therefore on effects of strain, excitation energy or temperature. Thus, Mn^{2+} spins are ideally suited to study the impact of an inhomogeneous magnetic field on the coherent precession of a localized spin ensemble. For that purpose we investigated hybrids consisting of Co wires on top of a CdZnMnSe/ZnSe quantum well heterostructure (sample B) at a bath temperature of 2.3 K. Again, $\mathbf{B}_{ext}$ was oriented perpendicular to the wires and parallel to the sample surface.

In Fig. 6, left, the TRKR signal is depicted for a wire array with $w = 480$ nm and $d = 620$ nm for different external magnetic fields and compared to a reference measurement performed at $B_{ext} = 780$ mT on an unpatterned CdZnMnSe/ZnSe sample. Again, a pronounced frequency shift of the coherent spin precession due to the magnetic fringe field is obtained. This results, e.g. at $B_{ext} = 780$ mT, in a phase shift of π after about seven oscillation periods for the hybrid sample as compared to the reference. In addition, the hybrids show a clear reduction of the T_2^* time [14], similar to what was observed at low temperatures for the electron spins in Py-GaAs hybrids. A more in-depth analysis of the TRKR signal reveals an interesting detail: The precession frequency changes with delay time Δt. This is shown in the right part of Fig. 6, where the time-dependent precession frequency, determined by averaging over

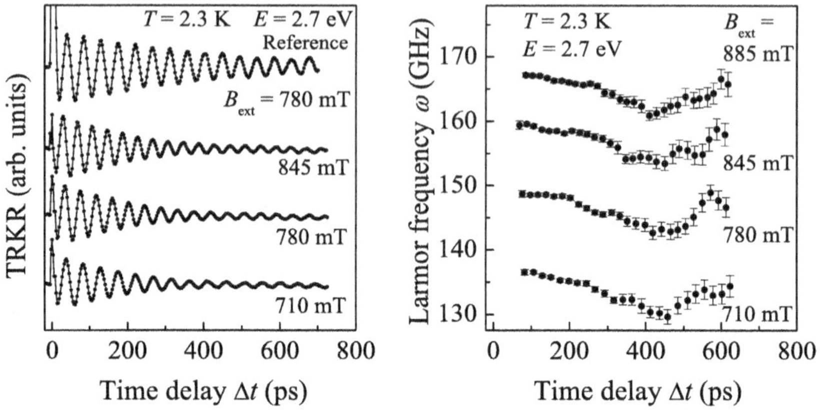

Fig. 6. *Left*: Transient TRKR signal of Co-CdZnMnSe hybrids with $w = 480$ nm and $d = 620$ nm for different external magnetic fields. For comparison the data obtained at $B_{\text{ext}} = 780$ mT on the unpatterned reference layer are included in the figure. *Right*: Larmor precession frequency ω versus delay time Δt for the same hybrids measured at different external fields.

3 adjacent oscillation periods, is depicted versus Δt. For any external magnetic field B_{ext}, the Larmor frequency first becomes smaller within the first few hundreds of ps, before increasing again and reaching almost its initial value.

In order to explain our experimental findings one should keep in mind that the magnetic fringe field is strongly inhomogeneous within the wire array (see Fig. 3). As the individual Mn^{2+} spins are localized, each of them experiences a different local magnetic field, which results in an individual precession frequency of each spin. For the Mn^{2+} spin ensemble one thus obtains

$$\theta(\Delta t) = \theta_0 \cdot \exp(-\Delta t/T_2^*) \sum_i \cos(\omega(x_i)\Delta t + \phi) \cos(\alpha_i) \tag{2}$$

where x_i is the lateral position of the i-th Mn^{2+} spin and α_i is the angle between the total magnetic field $\mathbf{B}_{\text{tot}}(x_i)$ and $\mathbf{B}_{\text{ext}}$ at each position x_i. Our calculations show that the magnetic fringe field of nanostructured FMs results in an asymmetric frequency distribution [14] with the consequence of a temporally varying precession frequency and a reduced effective T_2^* time for a localized spin ensemble exposed to a spatially inhomogeneous magnetic fringe field [36].

Typical results of our model calculations are shown in Fig. 7. As we know from energy dispersive X-ray spectroscopy that our Co wires are partially oxidized, we take the Co magnetization M_{FM} as a fitting parameter in order to account for the TRKR data. $M_{\text{FM}} = 980$ mT, which is about 54% of the saturation magnetization of pure Co, gives the best description of our data and both, the enhanced damping of the TRKR signal as well as the transient change of the Larmor precession frequency (see inset of Fig. 7) can reasonably well be described by our model.[2]

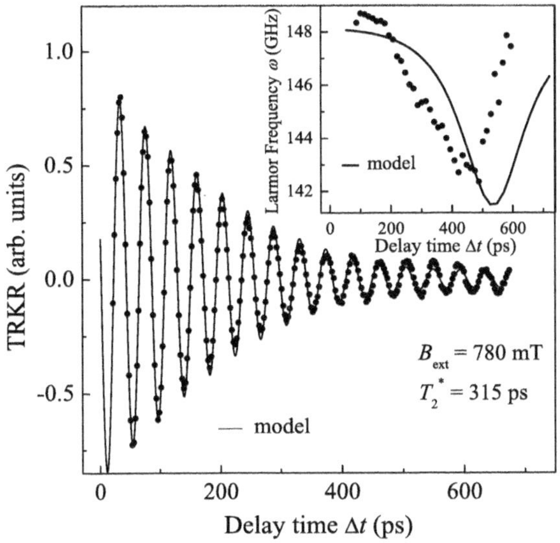

Fig. 7. Comparison of experimental data (*symbols*) and model calculations (*solid line*) of the TRKR signal of Mn^{2+} spin precession in Co-CdZnMnSe hybrids. In the inset, the transient change of the Larmor frequency obtained in experiment (*symbols*) and theory (*solid line*) is depicted.

5 Summary

In conclusion, we presented the coherent spin dynamics of localized and itinerant spins in semiconductor-ferromagnet hybrids. We demonstrate the ability to locally manipulate the Larmor frequency of coherent spin precession by nanostructured ferromagnets up to room temperature and point out the impact of laterally inhomogeneous fringe fields on the coherent dynamics of a localized spin ensemble. In fact, any lateral inhomogeneity, either in magnetic field, in electric field with spin orbit interaction or in g factor, which leads to an asymmetric frequency distribution of localized spins, will result in a transient frequency change and a reduction of T_2^* in the coherent precession of a spin ensemble.

[2] For an exact description it is necessary to include the non-perfect shapes of the FM, their domain structure and the non-homogeneous illumination profile of the laser. This has been omitted for the sake of simplicity.

6 Acknowledgements

We gratefully acknowledge the nanopatterning of the hybrids by F. Seifert, T. Kümmell, M. Wahle, S. F. Fischer and U. Kunze, the epitaxial growth of the semiconductors by J. Puls. F. Henneberger, D. Reuter and A. D. Wieck and the financial support of the Deutsche Forschungsgemeinschaft within the Sonderforschungsbereich 491 and the priority program SPP 1133.

References

1. S. Datta and B. Das, Appl. Phys. Lett. **56**, 665 (1990)
2. E. I. Rashba and Al. L. Efros, Phys. Rev. Lett. **91**, 126405 (2003)
3. L. Meier, G. Salis, I. Shorubalko, E. Gini, S. Schön, and K. Ensslin, Nat. Phys. **3**, 650 (2007)
4. Y. K. Kato, R. C. Myers, A. C. Gossard, and D. D. Awschalom, Nature **427**, 50 (2004)
5. R. K. Kawakami, Y. Kato, M. Hanson, I. Malajovich, J. M. Stephens, E. Johnston-Halperin, G. Salis, A. C. Gossard, and D. D. Awschalom, Science **294**, 131 (2001)
6. A. Greilich, A. Shabaev, D. R. Yakovlev, Al. L. Efros, I. A. Yugova, D. Reuter, A. D. Wieck, and M. Bayer, Science **317**, 1896 (2006)
7. J. Kossut, I. Yamakawa, A. Nakamura, G. Cywiński, K. Fronc, M. Czeczott, J. Wróbel, F. Kyrychenko, T. Wojtowicz, and S. Takeyama, Appl. Phys. Lett. **79**, 1789 (2001)
8. H. Schömig, A. Forchel, S. Halm, G. Bacher, J. Puls, and F. Henneberger, Appl. Phys. Lett. **84**, 2826 (2004)
9. H. Schömig, S. Halm, G. Bacher, A. Forchel, W. Kipferl, C. H. Back, J. Puls, and F. Henneberger, J. Appl. Phys. **95**, 7411 (2004)
10. P. Redliński, T. Wojtowicz, T. G. Rappoport, A. Libál, J. K. Furdyna, and B. Jankó, Phys. Rev. B **72**, 085209 (2005)
11. A. Murayama and M. Sakuma, Appl. Phys. Lett. **88**, 122504 (2006)
12. S. Halm, G. Bacher, E. Schuster, W. Keune, M. Sperl, J. Puls, and F. Henneberger, Appl. Phys. Lett. **90**, 051916 (2007)
13. Y. Tokura, W. G. van der Wiel, T. Obata, and S. Tarucha, Phys. Rev. Lett. **96**, 047202 (2006)
14. S. Halm, P. E. Hohage, J. Nannen, G. Bacher, J. Puls, F. Henneberger, Phys. Rev. B **77**, 121303(R) (2008)
15. S. Halm, P. E. Hohage, E. Neshataeva, F. Seifert, T. Kümmell, E. Schuster, W. Keune, M. Sperl, Y.-H. Fan, J. Puls, F. Henneberger, and G. Bacher, Phys. Stat. Sol A **204**, 191 (2007)
16. P. E. Hohage, F. Seifert, T. Kümmell, G. Bacher, D. Reuter, and A. D. Wieck, Phys. Stat. Sol C **3**, 4346 (2006)
17. P. E. Hohage, J. Nannen, S. Halm, G. Bacher, M. Wahle, S. F. Fischer, U. Kunze, D. Reuter, and A. D. Wieck, Appl. Phys. Lett., **92**, 241920 (2008)
18. L. Meier, G. Salis, C. Ellenberger, K. Ensslin, and E. Gini, Appl. Phys. Lett. **88**, 172501 (2006)
19. L. Meier, G. Salis, N. Moll, C. Ellenberger, I. Shorubalko, U. Wahlen, K. Ensslin, and E. Gini, Appl. Phys. Lett. **91**, 162507 (2007)

20. L. Meier, G. Salis, C. Ellenberger, E. Gini, and K. Ensslin, Phys. Rev. B **74**, 245318 (2006)
21. J. M. Kikkawa and D. D. Awschalom, Phys. Rev. Lett. **80**, 4313 (1998)
22. M. Römer, J. Hübner, and M. Oestreich, Rev. Sci. Instrum. **78**, 103903 (2007)
23. P. E. Hohage, G. Bacher, D. Reuter, and A. D. Wieck, Appl. Phys. Lett. **89**, 231101 (2006)
24. J. M. Kikkawa, I. P. Smorchkova, N. Samarth, and D. D. Awschalom, Science **277**, 1284 (1997)
25. G. Salis, Y. Kato, K. Ensslin, D. C. Driscoll, A. C. Gossard, and D. D. Awschalom, Nature **414**, 619 (2001)
26. J. A. Gupta, D. D. Awschalom, X. Peng, and A. P. Alivisatos, Phys. Rev. B **59**, R10421 (1999)
27. S. Ghosh, V. Sih, W. H. Lau, D. D. Awschalom, S.-Y. Bae, S. Wang, S. Vaidya, and G. Chapline, Appl. Phys. Lett. **86**, 232507 (2005)
28. B. Beschoten, E. Johnston-Halperin, D. K. Young, M. Poggio, J. E. Grimaldi, S. Keller, S. P. DenBaars, U. K. Mishra, E. L. Hu, and D. D. Awschalom, Phys. Rev. B **63**, 121202 (2001)
29. S. A. Crooker, J. J. Baumberg, F. Flack, N. Samarth, and D. D. Awschalom, Phys. Rev. Lett. **77**, 2814 (1996)
30. M. Oestreich, S. Hallstein, A. P. Heberle, K. Eberl, E. Bauser, and W. W. Rühle, Phys. Rev. B **53**, 7911 (1996)
31. A. V. Kimel, F. Bentivegna, V. N. Gridnev, V. V. Pavlov, R. V. Pisarev, and Th. Rasing, Phys. Rev. B **63**, 235201 (2001)
32. Z. Chen, R. Bratschitsch, and S. T. Cundiff, Opt. Lett. **30**, 2320 (2005)
33. K. V. Kavokin, Phys. Rev. B **64**, 075305 (2001)
34. G. Lampel, Phys. Rev. Lett. **20**, 491 (1968)
35. H. Knotz, A. W. Holleitner, J. Stephens, R. C. Myers, and D. D. Awschalom, Appl. Phys. Lett. **88**, 241918 (2006)
36. S. Halm, P. E. Hohage, J. Nannen, E. Neshataeva, L. Schneider, G. Bacher, Y. Fan, J. Puls, and F. Henneberger, J. Phys. D: Appl. Phys. **41**, 164007 (2008)

Part IV

Organic Materials and Water

Coupling of Paramagnetic Biomolecules to Ferromagnetic Surfaces

Heiko Wende

Fachbereich Physik and Center for Nanointegration (CeNIDE), Universität Duisburg-Essen, Lotharstraße 1, 47048 Duisburg, Germany
heiko.wende@uni-due.de

Abstract. Ferromagnetic coupling of a monolayer of Fe-porphyrin molecules to ferromagnetic films is revealed by X-ray magnetic circular dichroism measurements. By means of thickness- and temperature-dependent investigations the range and the strength of the coupling to Ni and Co films is analyzed. Element-specific hysteresis curves demonstrate that the Fe-spins in the molecules can be switched by reversing the magnetization of the ferromagnetic films.

1 Introduction

The fundamental understanding of the interaction of magnetic molecules with surfaces is essential to realize the vision of molecular spintronics (see e.g. [1–3]). Various works have been conducted on ferromagnetic molecules like Mn_{12} (see e.g. [4] and references therein). Here we are following a different route: To study the important molecule-surface interaction we chose purely paramagnetic biomolecules – namely Fe-porphyrins. The reason is that nature made these molecules stable enough that they can even be sublimated under UHV conditions. A schematic representation of the octaethylporphyrin-Fe(III)-chloride molecule sublimated here is given in Fig. 1(a). Thereby the molecules can be deposited in the monolayer-regime on surfaces of single crystals. Furthermore, the molecules can be ordered geometrically by this procedure. A further goal is to manipulate the Fe-spin in this paramagnetic porphyrin molecule which is vital for the possible use in molecular spintronics. Here, we show that the spin direction can be modified by the coupling to ferromagnetic films which were epitaxially grown on a Cu(100) single crystal (schematic picture given in Fig. 1(b)). We study the geometric ordering and the magnetic properties of the molecules including the coupling by X-ray absorption spectroscopy (XAS). The adsorption is characterized by angular-dependent near-edge X-ray absorption spectra (NEXAFS) and the element-specific magnetic properties are investigated by X-ray magnetic circular dichroism spectroscopy (XMCD) [5, 6]. The XMCD technique even allows for the measurement of element-specific hysteresis curves. Thereby the magnetic coupling can be studied in greater detail. To achieve a deeper insight into the nature of the coupling Ni as well as Co films were used as the ferromagnetic substrates for the Fe-porphyrins. The ferromagnetic coupling of the Fe-spin in the molecule to the substrate can be directly concluded from the sign of the XMCD spectra as exemplarily given for the $L_{2,3}$-edges of Fe and the Ni film substrate shown in Fig. 1(c). Furthermore, the choice of the film thickness determines the easy axis of magnetization and thereby we analyzed the coupling to an in-plane magnetized film (5 ML Co) and an out-of-plane magnetized film (15 ML Ni) [7].

2 Experimental details

The samples were prepared and measured in situ under UHV conditions (base pressure $p = 2.0 \times 10^{-10}$ mbar). The Cu(100) single crystal was cleaned by Ar^+-sputtering and the ultrathin Co and Ni films were prepared by e-beam evaporation (for details see [6]). The 2,3,7,8,12,13,17,18-octaethylporphyrin-Fe(III) chloride (Fe-OEP-Cl) molecules were sublimated at 485 K and adsorbed on the epitaxial films at room temperature. In the past experimental works utilizing the XMCD technique have also been performed on Mn-porphyrins [8]. However, we showed in earlier works that the XMCD analysis by means of the so-called sum rules is questionable for light $3d$ elements [9]. This is the reason for choosing the heavier 3d element Fe here. The thickness of the porphyrin molecules was cross-checked by the signal-to-background ratio (edge-jump) at the X-ray absorption edges of the elements in the molecule (Fe, N and C labeled in Fig. 1(a)). The XAS spectra were measured at the beamline UE56-2/PGM2 of BESSY, the synchrotron radiation facility in Berlin, Germany, utilizing the electron yield detection mode. By means of angular-dependent NEXAFS spectra at the N K-edge we confirmed that the Fe-porphyrins are basically lying flat on both ferromagnetic substrates [5] which paves the way for the coupling of the molecules to the films. Furthermore, the fine structures at the N K-edge reveal that the sublimated Fe-porphyrins exhibit the same structures as reported for similar Zn-tetraphenylporphyrin molecules [10]. This demonstrates that the basic porphyrin-structure is not disintegrated neither by the preparation process nor by radiation damage. However, the isotropic spectra at the Fe $L_{2,3}$-edges indicate a divalent state [6] which hints at the disconnection of the Cl atom from the molecule in the adsorbed state.

3 Results and discussion

One fundamental question is in which way the paramagnetic molecules interact with ferromagnetic surfaces. Is there a magnetic coupling? To answer this question we investigated a coverage of 1 ML Fe-OEP on a 15 ML Ni film on a

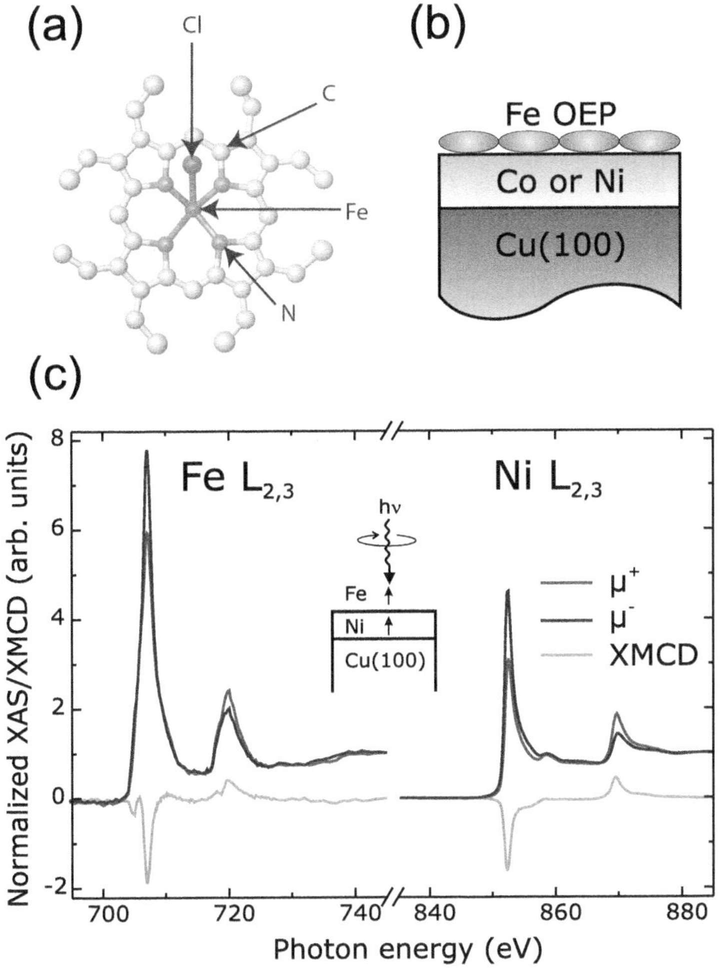

Fig. 1. Schematic representations of (**a**) the Fe-octaethylporphyrin(OEP)-chloride molecule and (**b**) the sample setup with the sublimated molecules on the epitaxially grown ferromagnetic films. (**c**) As an example for the element specificity of the XAS technique the normalized X-ray absorption coefficients for *right* and *left* circularly polarized X-rays $\mu^+(E)$ (*red*) and $\mu^-(E)$ (*blue*) are presented together with the corresponding XMCD signal at the Fe and Ni $L_{2,3}$-edges for a monolayer (ML) of Fe-OEP on a 15 ML Ni film (300 K, 10 mT) (Fe-spectra taken from Ref. [5]). The inset depicts the orientation of the sample to the incident X-rays. The *arrows* for Fe and the Ni film show the alignment of the spins.

Cu(100) single crystal. One advantage of the XMCD technique is the element specificity. Thereby, the magnetism of the molecules can be studied separately from the magnetism of the underlying metallic film. The XAS spectra for right and left circularly polarized X-rays, are presented at the Fe and Ni $L_{2,3}$-edges

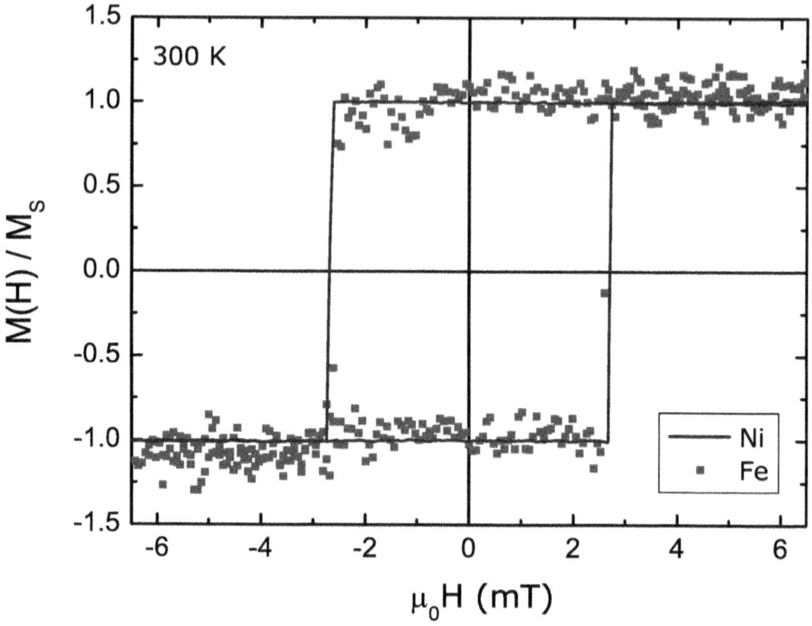

Fig. 2. Element specific hysteresis curves of the porphyrin Fe atom (*filled squares*) and the Ni film (*full line*) recorded at the L_3-edge XMCD maximum of Fe OEP on Ni/Cu(100) at 300 K (figure taken from Ref. [5]).

in Fig. 1(c). A clear difference between $\mu^+(E)$ and $\mu^-(E)$ – the X-ray magnetic circular dichroism (XMCD) – is revealed at the absorption edges of the two elements. This demonstrates that both the Fe and Ni spins are ordered. The data are recorded at 300 K and an applied magnetic field of 10 mT. At these conditions a pure paramagnet could not exhibit such a high ordering. Hence, a magnetic coupling of the molecules to the ferromagnetic film must exist. This coupling is of ferromagnetic nature since the XMCD spectra exhibit the same sign at the respective edges. The nature of this coupling phenomenon is revealed by density functional theory calculations (GGA+U) which were carried out in the group of O. Eriksson (Uppsala University, Sweden) [5]. These calculations show that the magnetic coupling is due a 90° indirect exchange mechanism via the N-atoms. No direct hybridization of the Fe-atoms in the molecules with the ferromagnetic film is found. In a next step we analyze the element specific hysteresis curves shown in Fig. 2. These are determined by measuring the XMCD difference at the Fe and Ni L_3-edges as a function of the applied field. Obviously the Fe magnetization directly follows the one of the Ni film. This reveals that the magnetic moment of the Fe atom in the molecule can be switched by reversing the magnetization of the underlying film. This might open a new route for molecular spintronics by measuring the spin-dependent transport through the molecule. From Fig. 2 it is apparent

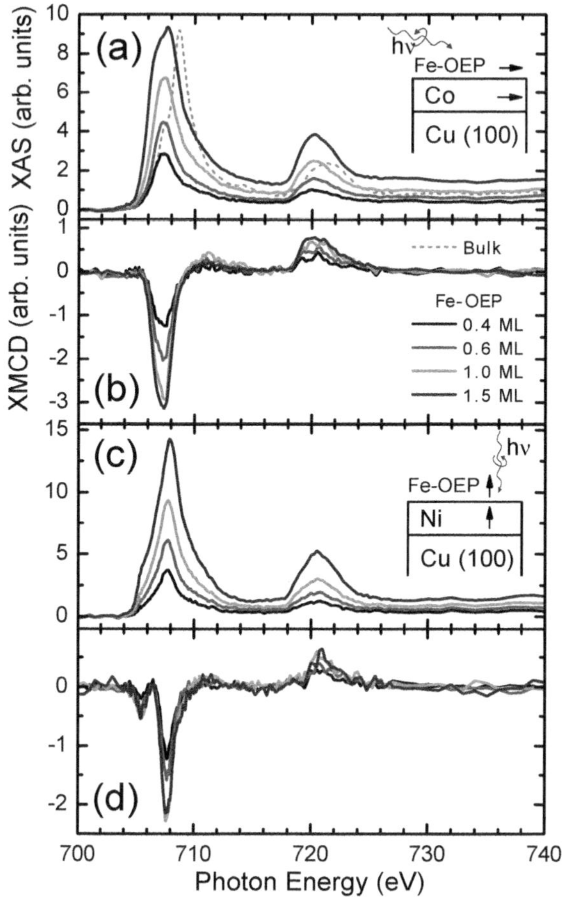

Fig. 3. Thickness dependence of the Fe $L_{2,3}$-edge XAS and XMCD spectra of Fe-porphyrin molecules on 5 ML Co/Cu(100) (**a, b**) and on 15 ML Ni/Cu(100) (**c, d**) (300 K, 10 mT). *Panel* (**a**) includes the XAS data (*dashed line*) of Fe-OEP-Cl bulk (figure taken from Ref. [6]).

that the data at the Fe-edge are more noisy than the one at the Ni edge. The simple reason is the amount of material being probed: Because of the size of the porphyrin molecules the coverage of 1 ML Fe-OEP molecules corresponds to an effective Fe coverage in the 1/100 ML regime which has to be compared to 15 ML of Ni. The clear XMCD signal at the Fe $L_{2,3}$ edges demonstrates the extraordinary sensitivity of the technique.

After revealing the coupling of the Fe spins to the ferromagnetic substrate the next questions concern the range and the strength of the coupling. Therefore, we analyzed the thickness dependence of the magnetic properties of the molecules on two substrates, namely a 15 ML Ni film and a 5 ML Co film both epitaxially grown on Cu(100) (Fig. 3). Since the easy axis of magnetization of the Co film is in-plane whereas the one of the Ni film is out-of-plane as already discussed above, the measurements were carried out at grazing and normal X-ray incidence, respectively. The results are presented in Fig. 3. They show that the Fe-porphyrin molecules are also magnetically ordered by a coupling to the Co film. A prominent difference in the spectral shape of the XMCD between the two substrates can be detected in Fig. 3: For the Ni case a double-structure can be seen clearly at the L_3 edge and weaker at the L_2 edge. However, for the Co substrate only a relatively broad contribution in the Fe-XMCD can be identified at the $L_{2,3}$-edges. In our earlier works we could show that these distinctions originate from the different Fe $3d$-orbitals being probed at the respective X-ray incidences [6]. Next we turn to the thickness dependence of the spectra. It should be noted that the XAS data are normalized to the pre-edge regime only. Therefore, the signal-to-background ratio (edge-jump) of the XAS spectra and the fine structures increase with the molecular coverage – the X-ray absorption is simply stronger for a larger amount of material. Interestingly, the behavior of the XMCD intensity is different: There is a clear increase for coverages from 0.4 to 1.0 ML. However, the size of the XMCD for 1.5 ML is basically identical to the one of 1 ML for both substrates. This shows that the additional 0.5 ML does not contribute any magnetic signal. From these results it can be concluded that only the first molecular monolayer couples magnetically to the ferromagnetic films and already the second layer is decoupled from the substrate.

Furthermore, we study the coupling strength for the two substrates by temperature-dependent measurements. These are presented in Fig. 4. For the Ni substrate case the XMCD spectrum is taken at room temperature, then the sample is cooled down, measured and warmed up again to room temperature to control the reproducibility of the measurements. For the Co-substrate case hardly a change in the XMCD intensity is detected when cooling down to 85 K from room temperature. In contrast, on the Ni film the Fe-XMCD signal increases by a factor of two when cooling from RT to 70 K. For a conclusive interpretation the XMCD temperature dependence of the ferromagnetic film was determined in parallel [6] which is a strength of the element specificity of the XMCD technique. These measurements reveal that the Co magnetization is reduced by about 5% when warming up from 85 K to RT and the Ni magnetization decreases by about 15% when annealing from 70 K to RT. Hence, there is only a minor temperature dependence of the underlying ferromagnetic films in the investigated temperature regime. Therefore, the stronger temperature dependence of the Fe-XMCD for the Ni substrate in comparison to the Co substrate reveals a clearly weaker coupling of the Fe-spins in the molecule to Ni than to Co. Describing the temperature-dependent reduction by means of the Brillouin function a coupling strength of 20 and 70 meV is determined for the Ni and the Co case, respectively [6]. Finally, we determine the effective Fe spin moment at 70 K by applying the XMCD sum rules. This yields the magnetic moment per atom per d-hole. We obtain $\mu_S^{\text{eff}}(\text{Fe}) = 0.65 \pm 0.1$/hole and $0.45 \pm 0.1\ \mu_B$/hole for molecules deposited on Co and Ni films, respectively [6].

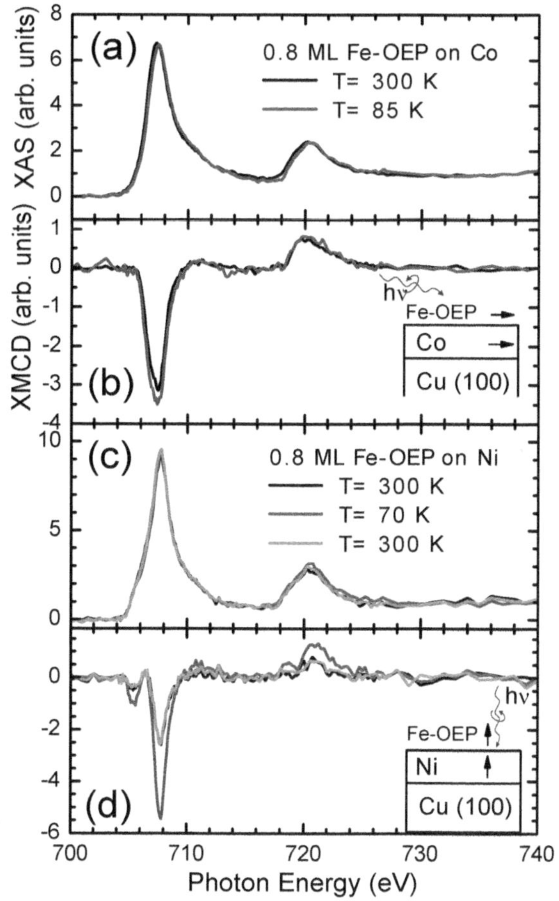

Fig. 4. Fe $L_{2,3}$-edge XAS and XMCD spectra for 0.8 ML Fe-OEP molecules on the Co substrate recorded at 300 K and 85 K (**a, b**) and on Ni (**c, d**) at 300, 70 and 300 K (figure taken from Ref. [6]).

As mentioned above the XAS measurements indicate that Fe is in a divalent state. Hence, the thumber of d-holes is about $n_h \approx 4$ and therefore the effective Fe spin-moment is about $\sim 2\,\mu_B$. This is consistent with the intermediate spin state of $S = 1$ which was determined by the GGA+U calculations [5]. The ratio of the orbital to effective spin moment is about μ_L/μ_S^{eff} ~5% to 10%, which shows that the orbital moment is nearly quenched.

4 Summary

By means of X-ray absorption spectroscopy it is shown that a monolayer of Fe-porphyrin molecules couples ferromagnetically to Ni and Co films which are epitaxially grown on Cu(100). The range and the strength of this coupling is studied by thickness- and temperature-dependent measurements. For both substrates the first molecular monolayer only couples to the substrate. We determine a clearly weaker coupling strength to the Ni film ($E_C = 20$ meV) in contrast to the Co film ($E_C = 70$ meV). The possibility to switch the Fe-spin in the molecule by reversing the magnetization of the substrate film might open new prospects for molecular spintronics in the future by measuring the spin-dependent transport through the molecule.

5 Acknowledgements

M. Bernien, J. Luo, C. Weis, N. Ponpandian, J. Kurde, J. Miguel, M. Piantek, X. Xu, Ph. Eckhold, W. Kuch, K. Baberschke, and P. Srivastava are acknowledged for their various experimental contributions to this work. We thank P.M. Panchmatia, B. Sanyal, P.M. Oppeneer, and O. Eriksson (Uppsala University, Sweden) for their enlightening DFT calculations which revealed the coupling mechanism. F. Senf, B. Zada, and W. Mahler are acknowledged for their support during the measurements. This work is supported by BMBF (05 KS4 KEB/5) and DFG (SFB 658, Heisenberg Programm) grants.

References

1. J.V. Barth, G. Costantini, and K. Kern: Nature **437**, 671 (2005)
2. L. Grill, M. Dyer, L. Lafferentz, M. Persson, M.V. Peters, and S. Hecht: Nat. Nanotechnol. **2**, 687 (2007)
3. T. Yokoyama, S. Yokoyama, T. Kamikado, Y. Okuno, and S. Mashiko: Nature **413**, 619 (2001)
4. J. Gómez-Segura, J. Veciana, and D. Ruiz-Molina: Chem. Commun. 3699 (2007)
5. H. Wende, M. Bernien, J. Luo, C. Sorg, N. Ponpandian, J. Kurde, J. Miguel, M. Piantek, X. Xu, Ph. Eckhold, W. Kuch, K. Baberschke, P.M. Panchmatia, B. Sanyal, P.M. Oppeneer, and O. Eriksson: Nat. Mater. **6**, 516 (2007)
6. M. Bernien, X. Xu, J. Miguel, M. Piantek, Ph. Eckhold, J. Luo, J. Kurde, W. Kuch, K. Baberschke, H. Wende, and P. Srivastava: Phys. Rev. B **76**, 214406 (2007)
7. K. Baberschke: Lecture Notes in Physics **580**, 27 (2001)
8. A. Scheybal, T. Ramsvik, R. Bertschinger, M. Putero, F. Nolting, and T.A. Jung: Chem. Phys. Lett. **411**, 214 (2005)
9. H. Wende: Rep. Prog. Phys. **67**, 2105 (2004)
10. S. Narioka, H. Ishii, Y. Ouchi, T. Yokoyama, T. Ohta, and K. Seki: J. Phys. Chem. **99** 1332 (1995)

Band Alignment in Organic Materials

F. Flores[1], J. Ortega[1] and H. Vázquez[2]

[1] Departamento de Física Teórica de la Materia Condensada, Universidad Autónoma de Madrid, 28049 Spain
[2] DTU-Nanotech, Technical University of Denmark, DTU, DK-2800 Lyngby, Denmark

Abstract. Band alignment at metal/organic (MO) and organic/organic (OO) interfaces is discussed within a unified Induced Density of Interface States (IDIS) model, which incorporates most of the effects that can be expected to operate at weakly interacting organic interfaces: compression of the metal electron tails due to Pauli repulsion, orientation of molecular dipoles and electron charge transfer between the two media. This last mechanism tends to align the Charge Neutrality Level (CNL) of the organic material and the metal Fermi level (E_F): electron charge transfer reduces the initial misalignment between the CNL and the metal work function ($\phi_M - CNL$) to $S(\phi_M - CNL)$, where S is the interface screening parameter which is shown to also screen the 'Pauli' and molecular interface dipoles. Results for several Au/organic and organic/organic interfaces are presented and discussed.

PACS numbers: 79.60.Jv, 79.60.Dp, 73.40.Gk, 73.20.-r

1 Introduction

New electronic devices, like OFETs, OLEDs and photovoltaic cells, based on organic materials have already started to appear in the market. The growing field of organic and spin-based electronics relies on the use of organic conjugated molecules and polymers as active components in multilayer device applications [1]. The performance of these organic devices depends crucially on the different energy barriers that control carrier injection into, and transport between, different layers. These energy barriers are key quantities for organic thin film devices and are determined by the relative positions of molecular levels across metal-organic (MO) and organic-organic semiconductor (OO) interfaces [1–4].

Figure 1 shows the energy diagram for an organic light emitting diode (OLED) with three interfaces. In this device, electrons and holes are injected across MO interfaces into the two organic films of the system; these charges are collected at the OO interface, where electron-hole recombination creates the emitted photon. The performance of this device depends on how electrons and holes are injected into the organic materials and on the way those charges are collected and recombine at the OO interface: this is determined by the barriers at the different MO and OO interfaces. Since the barriers are one of the most important physical parameters defining device behaviour, understanding interface mechanisms and predicting energy level alignment is therefore highly relevant to engineering new devices and designing new functionalities.

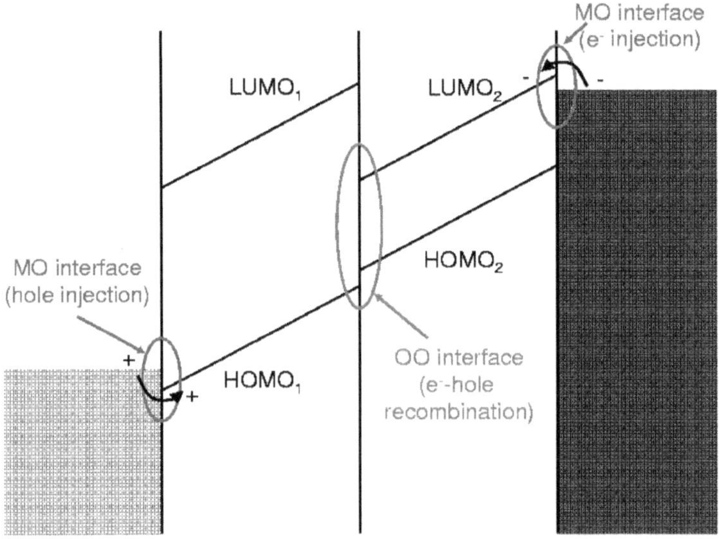

Fig. 1. Schematic energy level diagram for an organic light-emitting diode.

Let us first consider MO interfaces. Molecular level alignment at MO junctions has been extensively investigated over the last decade [5–8]. Vacuum evaporation of molecular films on clean metal surfaces, which are expected to form an intimate MO interface, has been shown experimentally [9, 10] to form interfaces that depart from the Schottky-Mott limit (where the vacuum level alignment rule is used), because of the substantial interface dipole they exhibit. Figure 2 shows the energy diagram for such an interface with and without the interface dipole; several mechanisms have been proposed to explain the formation of these dipoles and the MO injection barriers: chemical reaction and the formation of gap states in the organic material [8, 11–14]; orientation of molecular dipoles [15, 16]; or compression of the metal electron tails at the MO interface due to Pauli repulsion [13, 17–19]. It has also been suggested [20, 21], that an additional important mechanism is the tendency of the Charge Neutrality Level (CNL) of the organic material to align with the metal Fermi level (or with the CNL of the other organic material at OO interfaces [23]); this mechanism is associated with the rearrangement of charge at the interface, and this is why the model is called the Induced Density of Interface States (IDIS) model. More recently, this model has been extended [19, 24] to include Pauli repulsion and intrinsic molecular dipoles.

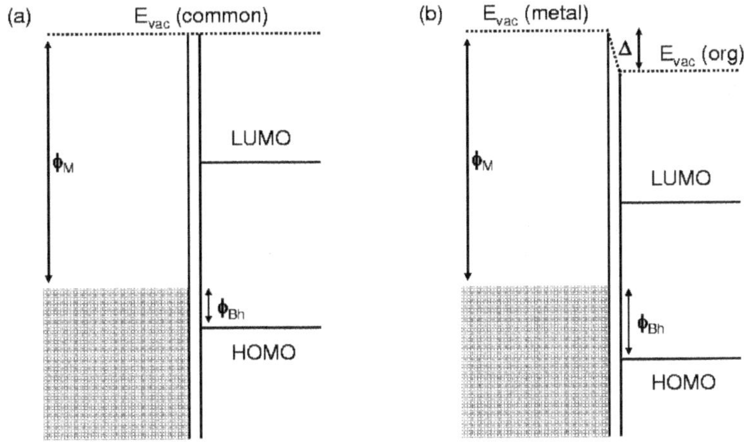

Fig. 2. Band alignment at a MO junction in the absence (*left*) or presence (*right*) of an interface dipole Δ.

Within the IDIS model, the organic CNL is a kind of material electronegativity which defines the direction in which electronic charge is transferred to (or from) other materials. This concept is closely related to Pauling´s electronegativity as shown in the next table taken from 'The Nature of the Chemical Bond' by Linus Pauling [25].

Table 1. Pauling's electronegativity and its relation to $(I + A)/2$. All energies are in eV.

	I	A	$(I + A)/2$	Electronegativity
F	17.4	3.6	10.5	4.0
Cl	13.0	3.8	8.4	3.0
Br	11.9	3.5	7.7	2.8
H	13.6	0.8	7.2	2.1
Li	5.4	0.0	2.7	1.0
Na	5.2	0.0	2.6	0.9

Table 1 illustrates with a few examples how Pauling's atomic electronegativity is closely related to $(I+A)/2$, the average of the first ionization level and electron affinity, a quantity that should be a measure of the electron attraction by the neutral atom. While it can be intuitively understood how the donor or acceptor character of an organic molecule is directly related to its HOMO and LUMO positions, the organic CNL generalizes this concept to an organic material and takes into account not only its HOMO and LUMO levels but the whole energy spectrum of the system. At a MO interface, within the IDIS model electron charge should flow between the two materials depending on the sign of $(\phi_M - CNL)$, in such a way that a dipole, Δ, will be induced at the interface. The effect of this dipole is to screen $(\phi_M - CNL)$ to $S_{MO}(\phi_M - CNL)$, where S_{MO} is a screening parameter $(0 < S < 1)$ that depends on the interface properties; the interface dipole is then given by: $(1 - S_{MO})(\phi_M - CNL)$. At an OO interface [11], the relevant energy is $(CNL_1 - CNL_2)_{ini}$ and the screening parameter S_{OO} can be approximated by $\frac{1}{2}(\frac{1}{\epsilon_1} + \frac{1}{\epsilon_2})$, where ϵ_1 and ϵ_2 are the static dielectric constants of both media.

2 The CNL in the IDIS model

Let us first discuss how the CNL position of the organic material is calculated, by considering an organic molecule (such as, say, PTCDA or CuPc, see Fig. 3) physisorbed on an unreactive metal [26], say Au: this is the case we consider in this paper. Notice that, at variance with chemisorptive interfaces, where metal-organic bonds exist, DFT is not well-suited for this case, where there is a weak chemical interaction between a noble metal like Au and a closed-shell organic molecule separated a distance $\sim 3 - 3.5$ Å; in particular, standard (i.e. LDA or GGA) DFT calculations are not very accurate since, for instance, HOMO-LUMO gaps or image potential effects are not described properly. We therefore start with a DFT calculation of the isolated molecule and include in a first step many-body corrections to the molecular energy levels; then the interaction between metal and molecule is considered by projecting the metal states onto the organic ones, using a self-energy which acts on the molecular Hamiltonian [20]. An important characteristic of DFT methods lies in the underestimation of the organic molecular gap, whereby the DFT HOMO-LUMO difference is significantly smaller than the measured transport gap. When analysing charge transfer and band alignment at interfaces, this obviously represents an important limitation that needs to be addressed. We do so by introducing many-body corrections (as detailed in reference [22]) to the position of the molecular energy levels; these provide a (surprisingly good) description of the Ionization and Affinity levels corresponding to the (positively- or negatively-) charged molecule, a situation which is the one measured in photoemission experiments. These corrections are introduced as a scissors operator acting on the molecular Hamiltonian, shifting each DFT eigenenergy by an amount which depends on the energy associated with adding an electron (empty states) or a hole (occupied states) to that particular molecular orbital. These corrections compensate for the self-interaction that is included in conventional (i.e. LDA or GGA) functionals, and increase the molecular gap, in the direction of what is observed experimentally. This is shown in Fig. 4 for the case of PTCBI, which compares the LDA electronic levels (top) with the spectrum after the level corrections outlined in the text have been introduced (bottom).

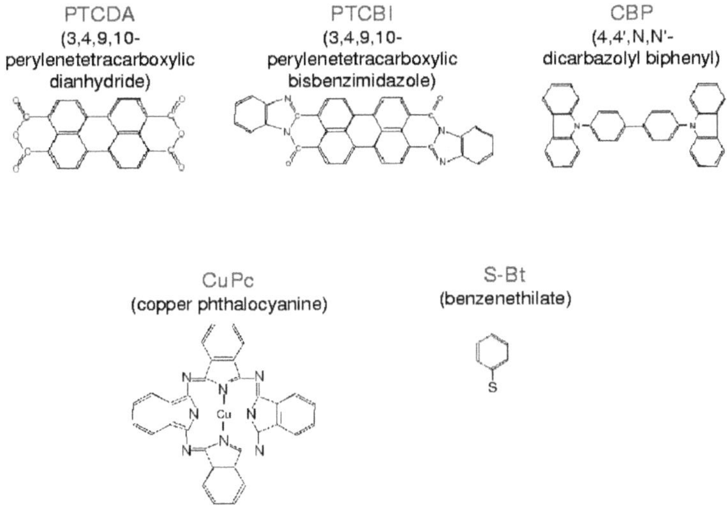

Fig. 3. Low-weight organic semiconductors that are discussed in this paper.

The CNL position is then calculated by introducing the interaction between the molecule and the metal using perturbation theory; in this calculation, the molecular energy levels (including corrections due to many-body as well as image charge effects) are broadened by their interaction with the metal, and an 'induced density of states' is created in the molecular energy gap. The organic CNL is calculated by integrating its total density of states up to the charge of the neutral molecule. It should be stressed that the CNL position is calculated by using the DFT energy spectrum after these many-body corrections, as well as the reduction of the gap by image charge effects have been introduced. In this way, we approximate the relevant contributions when a charge is added to the molecule at the interface, and reproduce the transport gap, a situation which a simple DFT calculation for a weakly-interaction interface would not describe satisfactorily.

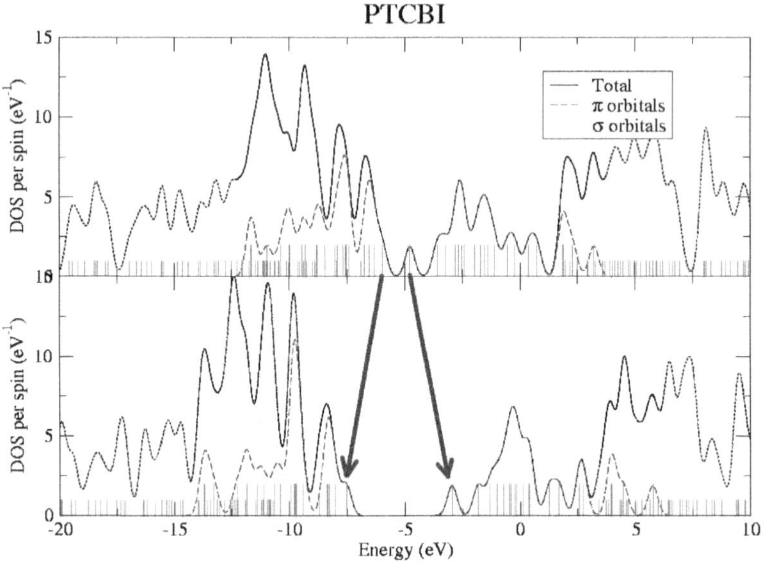

Fig. 4. Electronic levels for PTCBI calculated using LDA (*top*) and introducing the corrections to molecular level positions outlined in the text (*bottom*). The *arrows* show the shift of the HOMO and LUMO.

3 The IDIS model at MO interfaces

To gain some insight into the alignment, before discussing the Unified IDIS model, we first consider an organic molecule on a metal, characterised respectively by CNL and ϕ_M. To try to equalise this initial potential difference, a dipole is induced across the interface, whose magnitude is calculated according to the IDIS theory as

$$\Delta^{IDIS} = \frac{4\pi e^2 D(E_F)d}{A}\left(\phi_M - CNL - \Delta^{IDIS}\right) \quad (1)$$

In this equation, Δ^{IDIS} is proportional to the potential offset or misalignment, which in turn depends on the dipole induced Δ^{IDIS}. The screening parameter, S, can be defined in terms of the density of states at the Fermi level, D(E_F), the area per molecule, A, and the distance, d, between the molecule and the metal [27]:

$$S_{MO} = \frac{1}{1 + 4\pi e^2 D(E_F)d/A} \quad (2)$$

Solving for Δ^{IDIS}, and using Eq. (2), we get [28]:

$$\Delta^{IDIS} = (1 - S_{MO})(\phi_M - CNL) \quad (3)$$

which is the central IDIS equation for MO interfaces.

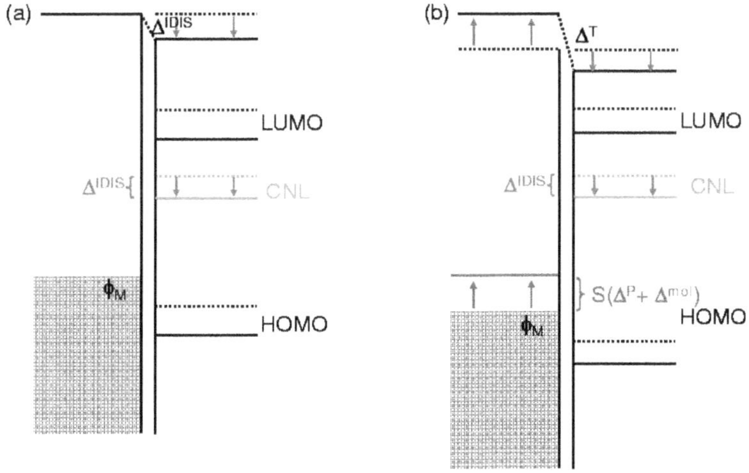

Fig. 5. Energy level alignment within the IDIS formalism.

This analysis can be extended to include Pauli (exchange) effects and molecular permanent dipoles at the interface. The Pauli repulsion can be described by the 'pillow' dipole, Δ^P, which is associated with the compression of the metal electron tails at the interface [19]. We calculate the charge rearrangement due to the orthogonalization of the organic and metal orbitals by expanding the organic-metal many-body interactions up to second order in the organic-metal overlap; this rearrangement corresponds to the 'push-back' effect experienced by the tail of the metal wave function and changes the value of ϕ_M, so that it can be described by an effective dipole at the interface [19]:

$$\Delta^P = \frac{4\pi}{A} \sum_{i \in molec, i' \in metal} -(n_i + n_{i'})\, S_{ii'} \int \vec{\Delta} r' \phi_{i'} \phi_i \; + \; (n_i - n_{i'})\, S_{ii'} \frac{\vec{d}}{4}. \quad (4)$$

The effect of having molecules with a permanent dipolar moment, P^0, with a non-zero component perpendicular to the interface can also be taken into account as a modification of the metal work-function; within the Topping model [29] this effect is described by means of the dipole Δ^{mol}:

$$\Delta^{mol} = \frac{\Delta^0}{1+\alpha} \quad (5)$$

where $\Delta^0 = 4\pi P^0/A$, A being the area per organic molecule, and $\alpha = 2\chi/a^3$ $(A = \pi a^2)$, χ being the molecular susceptibility. This molecular dipole is screened both by the surrounding organic material (through its polarizability α) and, if it is localised at the MO interface, by the metal electrons too.

A similar analysis as before yields

$$\Delta_{MO}^{T} = S_{MO}(\Delta^{P} + \Delta^{mol}) + (1 - S_{MO})(\phi_{M} - CNL) \tag{6}$$

This result is physically very intuitive: the 'pillow' and the molecular dipoles (if it is located at the interface) have been screened by the surface polarizability to $S\Delta^{P}$ and $S\Delta^{mol}$, respectively. The total interface dipole now incorporates the effects of the 'pillow' and the molecular dipole.

The magnitudes of the Pauli ('pillow') and molecular permanent dipoles are obtained from respective separate first-principles calculations for the isolated molecule (to obtain P^{0}) and according to Eq. (4) for the Pauli-pillow case.

4 IDIS model for OO interfaces

One of the advantages of the IDIS model is its description of MO and OO junctions within the same formalism. The analysis of molecular band offsets within DFT requires careful interpretation due to the large intermolecular distances and weak interaction.

Within our formulation [23], the alignment at OO interfaces is determined by the tendency of the CNLs of both materials to align (as determined from their initial CNL offset), and follows eqn 6 with the simplification that some terms are now negligible: the Pauli (pillow) term, for instance, does not appear because of the similar size of the orbitals in both organic materials.

The screening parameter for organic heterojunctions, S_{OO}, can in this case be calculated as [23, 31]:

$$S_{OO} = \frac{1}{2}\left(\frac{1}{\epsilon_1} + \frac{1}{\epsilon_2}\right) \tag{7}$$

which assumes that the potential drop takes place at mid-distance between both organics, whose contributions are proportional to the inverse of their static dielectric functions. We note in passing that this equation reduces to the MO case by considering $\epsilon_{metal} \to \infty$, so that $\frac{1}{2\epsilon_{org}}$ provides an estimate for S_{MO}.

Thus, the alignment equation for organic-organic interfaces is

$$\Delta_{OO}^{T} = S_{OO}\Delta^{mol} + (1 - S_{OO})\,(CNL_1 - CNL_2) \tag{8}$$

5 Results and discussion

This approach has been applied to different MO and OO interfaces. Since a direct DFT calculation will not yield an accurate description, our approach for weakly-interacting interfaces is to project the interaction at the interface onto the molecular Hamiltonian, and then calculate the CNL position and the total dipole using the above equations.

It should be stressed that the CNL position does not necessarily coincide with the Fermi level of the junction; instead, it is a useful energy marker measuring 'organic electronegativity' and indicating the magnitude and direction of charge transfer. The final (mis)alignment between the CNL and the Fermi level depends on their initial offset and on how efficiently potential differences are screened at the interface (S parameter).

Equations (2) and (6), and (7) and (8) have been applied to several weakly-interacting interfaces [19–21, 23, 24, 30]; Table 2 shows the results for junctions between Au and PTCDA, PTCBI, CBP and CuPc, and between Cu and a full monolayer of benzenethiolate (S-Bt). The table details the different contributions (see Eq. 6) to the interface dipole, as well as a comparison with experiment of the the total dipole and hole injection barriers.

Table 2. Results for several MO interfaces [19–21, 23, 24].

	CNL	S (th.)	S (exp)	$S\Delta^P$	$S(\phi_M - CNL)$	$S\Delta^{mol}$	Δ^T (th.)	Δ^T (exp)
Au/PTCDA	−4.8	0.16	∼0	0.12	0.25	–	0.39	∼0.25
Au/PTCBI	−4.4	0.16	∼0	0.17	0.50	–	0.67	0.4
Au/CBP	−4.05	0.50	∼0.6	0.21	0.43	–	0.64	0.5
Au/CuPc	−3.8	0.30	∼0.25	0.22	1.05	–	1.27	∼1.2
Cu/S-Bt	−3.75	0.28	–	0.11	0.26	0.63	1.00	1.00

The first four cases exhibit no molecular dipole in the direction perpendicular to the interface (PTCDA, for instance has polar C-O bonds, but the molecule is adsorbed flat on Au [26]). For these cases, the IDIS (charge transfer) term is seen to yield the strongest contribution to the dipole. The Pauli-pillow term also represent an important contribution, but not as strong; this explains why our early analysis where the Pauli-pillow term was not included [20, 21] already resulted in a good agreement with experiment (in fact, it provides a better agreement than the combined IDIS-Pauli expression!)

The Cu/S-Bt interface, on the other hand, illustrates a case where there is a strong molecular dipole. In agreement with previous studies [15, 16, 32], we find the effect of molecular dipoles to be dominant; in our case it contributes 0.63 eV to a total dipole of 1.0 eV for a full monolayer. This description was also used to determine the Bt orientation on the surface as a function of coverage [24].

Organic heterojunctions, on the other hand, show a weak interaction and the energy level alignment at these interfaces is characterised by exhibiting no interface dipole (vacuum level rule) in the majority of cases, while at the same time the large dipoles that have been measured at several interfaces constitute a few significant exceptions. The fact that the vacuum level rule is followed in most cases is consistent with the unreactive nature of the interface and the large bandgap of organic semiconductors. As for the exceptions, large ($\lesssim$0.5 eV) dipoles have been measured [33], which are not easy to rationalise in terms of the either the HOMO or LUMO relative positions, yet clearly appeal to the 'donor' or 'acceptor' character of the organic materials.

When applied to OO interfaces, our model yields good agreement with the observed interfacial dipoles [23]: the direction of the dipole is always correctly predicted, and its magnitude is in good agreement with the experimental values (see Fig. 6), within 0.15 eV of the measured dipoles. Notice the large values of S_{OO} in Fig. 6; this points in the direction of poor screening and is consistent with experimental observation of small or zero interface dipole ($S_{OO} = 1$ would correspond to vacuum level alignment). Thus, the initial offset between the CNL positions is not efficiently screened once the junction is formed.

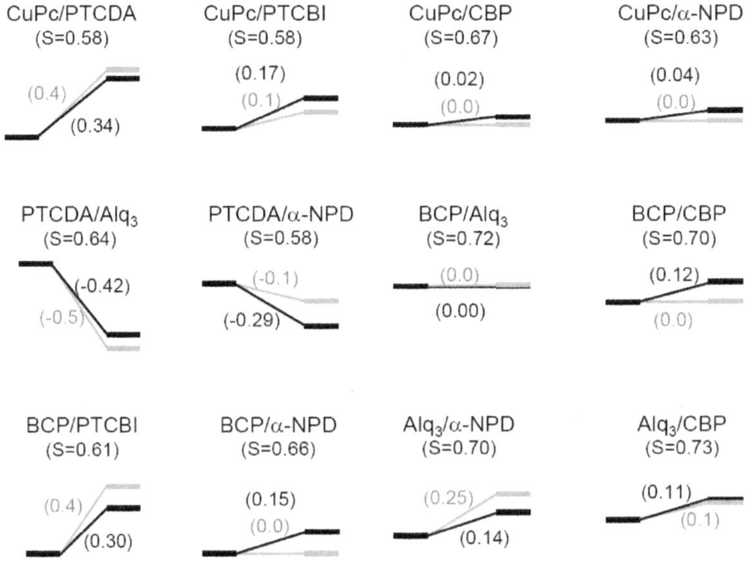

Fig. 6. Molecular band offsets at several organic heterojunctions. The figures give the theoretical [34] (*black*) and experimental [33] (*grey*) values for the interface dipole, as well as the calculated values of S_{OO}. Notice that the vacuum alignment rule is followed in most cases, but large dipoles are observed at some interfaces.

Recent experimental results [35] have shown disagreement with the predictions of the IDIS approach; instead, the authors propose the alignment to be governed by polarons, which pin the Fermi level. While we believe their model to be correct, at the same time we think it does not contradict our approach: notice that the alignment described in their work would correspond to $S_{OO} = 1$ in our model, with the molecular level positions 'renormalised' by polaron formation. Additionally, let us mention that the molecular levels at organic heterojunctions are probably broadened much less than in the proximity of a metal; however, we have verified the CNL position to be almost unsensitive to the level broadenings, so that our analysis holds for OO interfaces.

Thus, when considering molecular band offsets of low-weight organic molecules (though perhaps not those involving polaron formation [35]), we believe our approach in terms of the CNL to be very useful. The puzzling observation of vacuum level alignment at most heterojunctions, coexisting with large dipoles in a few interfaces can be easily understood within the CNL picture: only interfaces with very large initial CNL offsets (such as those involving PTCDA or PTCBI) will give rise to large dipoles at the interface. This allows for the understanding and prediction of molecular band offsets of organic semiconductors in a general and intuitive manner.

6 Conclusions

Although this approach has been very successful in understanding the role played by different mechanisms to create the surface dipoles and the interface barriers at MO and OO interfaces, there is a need to go beyond this model and to apply improved DFT methods to calculate these systems. The main problem in undertaking this program is associated with the limitations discussed above for LDA or GGA functionals: the crucial point to realize is that a full LDA (GGA) calculation for these weakly-interacting interfaces would not allow an accurate description of the charge transfer at the interface, since LDA (GGA) results yield organic molecular transport energy gaps that can be underestimated by several eVs. This serious limitation points to the need for going beyond standard DFT methods in order to obtain an accurate description of organic semiconductor interfaces. This is still an open problem to be addressed in the near future.

7 Acknowledgements

We acknowledge financial support from the Spanish CICYT under projects NAN-2004-09183-C10-07 and MAT2007-60966, the Comunidad de Madrid CAM under contract 0505/MAT/0303, and the Danish Research Council (Forskningsrådet for Teknologi og Produktion).

References

1. Conjugated Polymer and Molecular Interfaces, edited by W. R. Salaneck, K. Seki, A. Kahn, and J. J. Pireaux (Marcel Dekker, New York, 2001).
2. W.R. Salaneck, S. Stafström and J.L.Brédas, Conjugated Polymer Surfaces and Interfaces: Electronic and Chemical Structure of Interfaces for Polymer Light emitting Diodes, Cambridge University Press, Cambridge (1996).
3. N. Koch, Chem. Phys. Chem. **8** 1438 (2007).
4. J. C. Scott, J. Vac. Sci. Technol. A **21** 521 (2003).

5. C. Shen, A. Kahn, and I. G. Hill: in Conjugated Polymer and Molecular Interfaces, edited by W. R. Salaneck, K. Seki, A. Kahn, and J. J. Pireaux (Marcel Dekker, New York, 2001), pp. 293–350.
6. C. Shen, A. Kahn, and I. G. Hill, in Conjugated Polymer and Molecular Interfaces, edited by W. R. Salaneck, K. Seki, A. Kahn, and J. J. Pireaux (Marcel Dekker, New York, 2001), pp. 351–400.
7. D. Cahen, A. Kahn, and E. Umbach, Mater. Today **8** 32 (2005).
8. H. Ishii, K. Sugiyama, E. Ito, and K. Seki, Adv. Mater. (Weinheim, Ger.) **11** 605 (1999).
9. S. Narioka, H. Ishii, D. Yoshimura, M. Sei, Y. Ouchi, K. Seki, S. Hasegawa, T. Miyazaki, Y. Harima and K. Yamashita, Appl. Phys. Lett. **67** 1899 (1995).
10. I. G. Hill, A. Rajagopal, A. Kahn and Y. Hu, Appl. Phys. Lett. **73** 662 (1998).
11. I. G. Hill, J. Schwartz, and A. Kahn, Org. Electron. **1** 5 (2000).
12. C. Shen and A. Kahn, Org. Electron. **2** 89 (2001).
13. X. Crispin, V. Geskin, A. Crispin, J. Cornil, R. Lazzaroni, W. R. Salaneck and J. L. Brédas, J. Am. Chem. Soc. **124** 8131 (2002).
14. M. Eremtchenko, D. Bauer, J. A. Schaefer and F. S. Tautz, New J. Phys. **6** 4 (2004).
15. S. Kera, Y. Yabuuchi, H. Yamane, H. Setoyama, K. K. Okudaira, A. Kahn and N. Ueno, Phys. Rev. B **70** 085304 (2004).
16. M. Knupfer and G. Paasch, J. Vac. Sci. Tech. A **23** 1072 (2005).
17. P. S. Bagus, V. Staemmler, and C. Wöll, Phys. Rev. Lett. **89** 096104 (2002).
18. G. Witte, S. Lukas, P. S. Bagus and C. Wöll, Appl. Phys. Lett. **87** 263502 (2005).
19. H. Vázquez, Y. J. Dappe, J. Ortega and F. Flores, J. Chem. Phys., J. Chem. Phys. **126** 144703 (2007).
20. H. Vázquez, R. Oszwaldowski, P. Pou, J. Ortega, R. Pérez, F. Flores, and A. Kahn, Europhys. Lett. **65** 802 (2004).
21. H. Vázquez, F. Flores, R. Oszwaldowski, J. Ortega, R. Pérez, and A. Kahn, Appl. Surf. Sci. **234** 108 (2004).
22. R. Oszwaldowski, H. Vázquez, P. Pou, J. Ortega, R. Pérez and F. Flores, J. Phys.: Condens. Matter 15, S2665 (2003).
23. H. Vázquez, W. Gao, F. Flores and A. Kahn, Phys. Rev. B **71** 041306(R) (2005).
24. M. G. Betti, A. Kanjilal, C. Mariani, H. Vázquez, Y. J. Dappe, J. Ortega and F. Flores, Phys. Rev. Lett. **100** 027601 (2008).
25. L. Pauling, The Nature of the Chemical Bond (Cornell University Press, Ithaca, NY, 1960), 3rd edition.
26. P. Fenter, F. Schreiber, L. Zhou, P. Eisenberger and S. R. Forrest, Phys. Rev. B **56**, 3046 (1997).
27. S. M. Sze. Physics of Semiconductor Devices (John Wiley, New York, 1981).
28. C. Tejedor, F. Flores and E. Louis, J. Phys. C: Solid State Phys. **10** 2163 (1977).
29. J. Topping, Proc. R. Soc. London, Ser. **A 114**, 67 (1927).
30. O. V. Molodtsova, M. Grobosch, M. Knupfer and V. Yu. Aristov, Appl. Phys. Lett. **91** 244103 (2007).
31. J. Tersoff, Phys. Rev. B **32** R6968 (1985).
32. S. Yanagisawa, Y. Morikawa, Chem. Phys. Lett. **420** 523 (2006).
33. I.G. Hill, D. Milliron, J. Schwartz and A. Kahn, Appl. Surf. Sci. **166** 354 (2000).
34. H. Vázquez, F. Flores and A. Kahn, Org. Electron. **8** 241 (2007).
35. S. Braun, W. Osikowicz, Y. Wang and W.R. Salaneck, Org. Electron. **8** 14 (2007).

Organometallic Nanojunctions Probed by Different Chemistries: Thermo-, Photo-, and Mechano-Chemistry

Martin Konôpka[1], Robert Turanský[1], Nikos L. Doltsinis[2,3], Dominik Marx[2], and Ivan Štich[4]

[1] Center for Computational Materials Science, Department of Physics, Slovak University of Technology (FEI STU), 81219 Bratislava, Slovakia
{martin.konopka,robert.turansky}@stuba.sk

[2] Lehrstuhl für Theoretische Chemie, Ruhr–Universität Bochum, 44780 Bochum, Germany
dominik.marx@theochem.ruhr-uni-bochum.de

[3] Department of Physics, King's College London, London WC2R 2LS, United Kingdom
nikos.doltsinis@kcl.ac.uk

[4] Institute of Physics, Slovak Academy of Sciences, 84511 Bratislava, Slovakia
ivan.stich@savba.sk

Abstract. Based on *ab-initio* simulations, three different types of chemistry, namely thermo-, photo-, and mechano-chemistry are compared for organometallic nanojunctions. In the first part we provide the first direct comparison of mechanical *versus* thermal activation of bond breaking. Study of thiolate/copper interfaces provides evidence for vastly different reaction pathways and product classes. This is understood in terms of directional mechanical manipulation of coordination numbers and system fluctuations in the process of mechanical activation. In the second part mechanically and opto-mechanically controlled azobenzene (AB) switch based on AB-gold break-junction have been studied. It was found that both *cis*→*trans* and *trans*→*cis* mechanically driven switchings in the lowest singlet state are possible. Bidirectional optical switching of mechanically strained AB through first excited singlet state was also predicted, provided that the length of the molecule is adjusted towards the target isomer equilibrium length. The simulations reveal the paramount importance played by mechanical activation for this class of systems.

1 Introduction

The rapidly developing field of molecular electronics is directed towards single-molecule devices which are based on specific responses of certain molecules. The aim is to build electrically, optically and mechanically driven active components, switches, sensors or data storage media [1–4] based on organic molecules embedded into junctions. Such junctions will be exposed to thermal, electrical, optical, and mechanical activation which may trigger "chemical reactions" in the junction. Different ways of activating chemical reactions such as thermochemistry, electrochemistry or photochemistry are well understood. In contrast, and despite pioneering studies done a long time ago [5], the field of mechanically induced chemistry, or "mechanochemistry", still remains largely unexplored [6–9] when it comes to a nanoscale understanding [10–19]. Clearly this is at odds with the intriguing current possibilities to use mechanical, nano-Newton forces as a *tool* to induce, alter, or control chemical reactions by manipulating molecules [20–22]. This is typically achieved within setups like atomic force/scanning tunnelling microscopy (AFM/STM) or mechanically-controllable break junctions (MCB) [23] using either single molecules covalently attached to metallic tips or metal surfaces coated with molecular monolayers [24–27]. As mentioned above, there is an increasing technological interest into such heterogeneous molecule/metal systems. Typical system classes used are thiolated polyenes or aromatic rings that are covalently anchored via sulphur bridges to noble metal leads. Another vast application field is the coating and functionalisation of metal surfaces using similar thiolates to make the strong covalent bonds across the molecule/metal interface, e.g. using self-assembled monolayer (SAM) techniques [28, 29]. Crucial to applications is to understand and to control those factors that control the chemistry in such molecule/metal junctions and eventually lead to wear, fatigue, and degradation. These are obviously very general questions which, however, can only be addressed in a meaningful way by analysing the behavior of well-controlled specific but representative systems. Along the lines of these thoughts, using *ab initio* simulations we study two different kind of systems:

- mechanochemical stability and decomposition of interfaces and junctions between thiolates (here: ethylthiolate CH_3–CH_2–S, see insets in Figs. 1, 2) and extended copper surfaces and compare them to the traditional thermochemical behavior;
- mechanically and opto-mechanically controlled azobenzene (AB) switch based on AB-gold break-junction (see Fig. 3).

The former system directly probes and compares thermochemistry and mechanochemistry, while the latter system compares photo- and mechano-chemistry.

2 Thermo- and Mechano-chemistry of copper-ethylthiolate junctions

Heterogeneous nano-structured systems consist of bonds that differ greatly in strength. If subject to sufficiently strong thermal fluctuations the weakest bond is expected to be affected most and thus to break first. Does this also hold in the realm of mechanochemistry, i.e. would the very same bond(s) break if the chemical transformation is induced by an external pulling force?

A short account of this work was published elsewhere [30]. The reason we chose to study thiolate/copper junctions is because, compared to their gold counterparts [13, 14, 31], thiolate/copper contacts [32–35] are expected to display a rich disintegration behavior, including fragmentation of intramolecular covalent bonds.

2.1 Models and methods

To compare the mechanochemical behavior to the thermal decomposition as closely as possible, two distinct simulation protocols were applied. Successive static pulling of a single attached molecule mimics the effect of mechanical strain, whereas *ab initio* MD [36] at elevated temperatures probes the thermal pathway. The surface is modelled by a c(4×4) supercell of Cu(111) with four CH_3–CH_2–S anchored thus as to concur with experiment [34, 35]; a six-layer slab was used with the bottom Cu layer fixed. Mechanical pulling was simulated by constraining the terminal C atom of the thiolate, i.e. the CH_3– group, and increasing its distance relative to the fixed Cu atoms by increments of 0.2–0.8 Å, which defines the extension D, all the way up to fragmentation while carrying out full structure relaxation at each step. Thermochemical decomposition was achieved by increasing the temperature of the nuclei (while keeping the bottom layer fixed) from 300 K up to fragmentation (typically at $\approx$ 1800 K; note that such a high temperature was used to speed up rates into the range accessible to *ab initio* MD). All calculations were carried out using Perdew–Burke–Ernzerhof (PBE) functional [37], ultrasoft pseudopotentials at a 25 Ry plane wave cutoff, and `CPMD` [36, 38] (see [32, 33] for technical details).

2.2 Results

As a first step, the thermochemical decomposition of the above-defined thiolate/copper interface is studied. Most interestingly, heating up this interface gradually leads to thermally induced cleavage of covalent C–S bonds. This breaks apart the molecules and leads to desorption of hydrocarbons while sulphur remains bound to the surface as depicted in Fig. 1 for one out of the four molecules (see leftmost inset in Fig. 2 for a view of the SAM). This result not only reproduces the well-known behavior observed in experiments [34, 35], but is also consistent with energetic considerations on cluster models [32]. What happens in the mechanochemically driven process? To achieve this end, one of the four anchored molecules is pulled out of the SAM by applying uniaxially an external force perpendicular to the surface using the protocol introduced above, thus mimicking AFM/STM/MCB-type setups. The structural changes are monitored by the number of S–Cu bonds, i.e. the coordination number n_{S-Cu}, defined as the number of S–Cu distances that are within 2.75 Å around an S. Initially, n_{S-Cu} is found

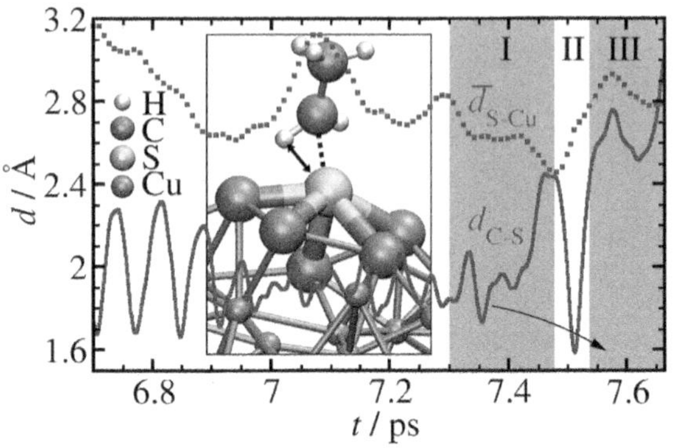

Fig. 1. Thermally induced decomposition of the $(CH_3CH_2S)_4/c(4{\times}4)Cu(111)$ surface obtained by *ab initio* MD. C–S distance d_{C-S} (*solid line*) and average S–Cu distance $\bar{d}_{S-Cu}$ (*dotted line*; obtained at each time-step as the average of the five shortest S–Cu distances d_{S-Cu}) are shown shortly before fragmentation. I, II, and III label the three key stages of the bond rupture process as discussed in the text. The snapshot shows the configuration at $t \approx 7.45$ ps where CH_3–CH_2 is about to detach while S remains on the surface (black *dotted line* and *arrow* symbolise the broken C–S bond and the repulsive H/S interaction as described in the text, respectively, while thicker bonds highlight the increased fivefold S coordination with Cu, $n_{S-Cu} \approx 5$, compared to $n_{S-Cu} \approx 3$ at equilibrium).

to be systematically reduced from $\sim$ 3 at equilibrium conditions [33] to $n_{S-Cu} = 1$ by applying forces of the order of 1–2 nN (see Fig. 2). The tensile stress is relieved by a sequence of isomerisation transformations such that a single Cu atom attached to the thiolate molecule by the remaining S–Cu bond is pulled out of the SAM so that finally the Cu–Cu bond breaks. At this stage it is concluded that thermochemistry and mechanochemistry lead to *vastly different reaction scenarios*, including different pathways and different products for both system classes representing typical molecule/metal interfaces and junctions. The next step consists in *understanding why* the thiolate C–S bond easily breaks by the action of thermal fluctuations both experimentally and in the simulation, while for the same system, it strengthens and becomes very stiff under the influence of an external pulling force.

We thus characterise the C–S bond in terms of the "vertical fragmentation energy" $E^{\mathrm{frag}}_{\mathrm{C-S}}$ as a measure of its strength, which is defined for a system A–B by $E^{\mathrm{frag}}_{\mathrm{A-B}} = E(A) + E(B) - E(A\text{–}B)$ without allowing any structural relaxation of the A and B fragments; thus it is not a binding or dissociation energy. These energies are compiled in Table 1 for prototypical configurations sampled from simulations (equilibrium results included for comparison). Two

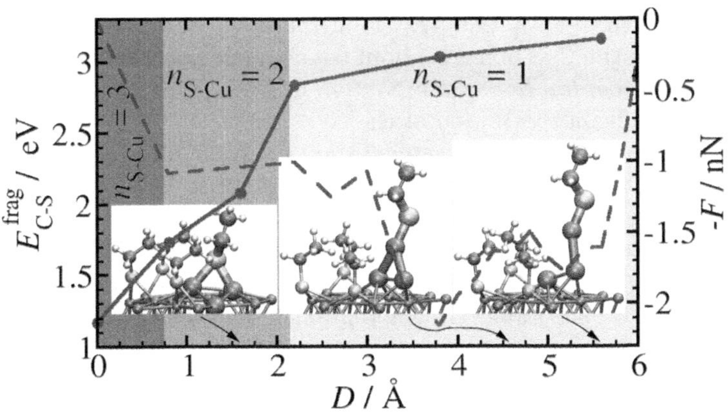

Fig. 2. Mechanically induced decomposition of the $(CH_3CH_2S)_4$/c(4×4)Cu(111) surface obtained by mechanically pulling out one of the four thiolates. Vertical fragmentation energy for C–S bond breaking E^{frag}_{C-S} (*solid line, left axis*) and applied pulling force $-F$ (*dashed line, right axis*) as a function of extension D; note that use of the minus sign follows convention. The insets show representative configurations at extensions D indicated by the arrows (the important bonds are highlighted by thick bonds, atom symbols as defined in Fig. 1). The reduction of the S coordination number n_{S-Cu} from 3 to 2 to 1 is highlighted by the *shaded areas.*

of the configurations were chosen from the thermochemical C–S bond rupture simulation in order to demonstrate the strong effect of different S–Cu coordinations. One of them exhibits a very low coordination, $n_{S-Cu} = 1$, while in the other configuration close to bond breaking the coordination number is greatly increased. Although both structures are characterised by the *same*

Table 1. Vertical fragmentation energies (in eV, see text for definition) for breaking the C–S and C–C bonds of thiolate on the Cu(111) surface. Data was obtained for the initial equilibrium structures (see "E" subcolumns) and along the thermochemical (T) and mechanochemical (M) pathways. Within each of the two setups, C–S bonds lengths d_{C-S} in T structures are the same, whereas n_{S-Cu} coordinations are very different. The T configurations have been sampled at ≈ 6.82 and 7.45 ps (where $n_{S-Cu} = 1$ and 5, respectively, see Fig. 1). The M structures correspond to $D \approx 5.6$ (see Fig. 2).

	E	T	T	M
n_{S-Cu}	3	5	1	1
d_{C-S}	1.87	2.31	2.31	1.88
d_{C-C}	1.52	1.61	1.39	1.56
E^{frag}_{C-S}	1.2	0.3	1.1	3.2
E^{frag}_{C-C}	4.3	4.0	4.1	4.5

C–S bond length $d_{C-S} = 2.31$, E^{frag}_{C-S} in the over-coordinated situation, just before breaking the C–S bond, is about 0.8 eV smaller. This demonstrates that *transient over-coordination* of the S atom due to strong thermal fluctuations leads to a significant *weakening* of the C–S bond.

In stark contrast, such a fragmentation analysis applied to the mechanochemical pathway shows that exerting the *pulling force* very significantly *strengthens* the very same C–S bond by ~ 2 eV upon increasing D (see the solid line in Fig. 2). Concurrent to strengthening the C–S bond, also the C–C bond gets stronger compared to the thermal case (i.e. $E^{frag}_{C-C} = 4.5$ eV *versus* 4.0–4.1 eV; see Table 1), which implies that the entire molecule, CH_3–CH_2–S, is affected by the mechanical stress. These differences can again be traced back to different coordination numbers. Hence, the manipulation of the coordination number of the metal/molecule contact is the key factor which governs the strength or weakness of the intramolecular C–S bond, and thus the stability of the interface with respect to the degradation of the molecular coating.

At this stage it is clear that coordination number changes are a necessary ingredient in determining the decomposition pathway, but this alone is not sufficient. Further analysis of thermal fragmentation displayed in Fig. 1 shows that several key processes must happen *simultaneously*. In phase I, thermal fluctuations lead to a transient *decrease* of the average S–Cu distance, down to $\bar{d}_{S-Cu} \approx 2.4$ Å, and thus to an *increase* of the coordination number n_{S-Cu} up to ≈ 5 (see highlighted S–Cu bonds in snapshot) which drains charge from the C–S bond thus destabilising it. In phase II, the C–S bond, which is already largely weakened due to process I, is further destabilised by fluctuations of the –CH_2-group leading to strongly repulsive interactions of one of its two H's with S (see black arrow in Fig. 1). This short-lived process is responsible for the ultimate C–S bond weakening. The destabilisation can be estimated in retrospect by allowing only these two H's to relax while keeping all other atoms frozen: E^{frag}_{C-S} *increases* from 0.3 to 0.8 eV thus *strengthening* the C–S bond as a result of *reducing* these repulsive H/S interactions. Finally, absorption of these strong vibrations (seen in the analysis of C and S momenta, not shown) eventually induces rupture of the strongly destabilised C–S bond leading to the breaking apart of the molecule in phase III. As a result, S remains chemisorbed on the surface (see highlighted S–Cu bonds in Fig. 1) whereas CH_3–CH_2 desorbs.

Given these insights into thermal decomposition it should be clear that the sequence of processes I $\rightarrow$ II $\rightarrow$ III in Fig. 1 is severely counteracted if an uniaxial pulling force is applied. The reason is most transparently demonstrated by the systematically *decreased* coordination number n_{S-Cu} from 3 to 1 by virtue of the externally applied directional force. This systematically disfavours any fluctuation-induced increase of n_{S-Cu} which has been shown to be a necessary ingredient to break the C–S bond. Mechanochemical action on the surface produces a *monoatomic* S/Cu contact junction that is characterised by a constant C–S bond distance of about 1.9 Å throughout pulling and a vertical fragmentation energy of ≈ 3 eV implying a significant C–S bond *strengthening* induced by the pulling (see solid line in the $n_{S-Cu} = 1$ regime of Fig. 2 and Table 1). This effect is responsible for the stiff response to the mechanically applied pulling force that leads to the extraction of Cu out of the surface but not to cleavage of the molecule itself.

From the analysis above it is clear that two completely *different reaction pathways and product classes* were observed in mechanochemical *versus* thermochemical decomposition.

3 Mechanically and opto-mechanically controlled azobenzene (AB) switch based on AB-gold break-junction.

A lot of attention has been focused on azobenzene (AB) which is a molecule capable of optical switching [39]. Its two different conformations — *trans* (TAB) and *cis*(CAB) — can be optically switched using laser light of appropriate wavelength [40]. The *trans*→*cis* isomerisation is accompanied by a significant decrease in the length of the molecule (by ≈ 2.4 Å). AB can be used as a molecular engine [41] driven by optical pulses. In molecular–electronics applications AB is a subsystem of a larger device. For this purpose the molecule is functionalised by sulphur atoms which serve as very convenient bridges between carbons and metals like gold (also copper at low temperatures, see recent computational studies [32, 33]). Hence the anchoring is done by functionalising the AB molecule by the *thiolate* bonds between sulphurs and phenyl rings forming dithioazobenzene (DAB), see Figs. 3 and 4. Embedding AB into junctions introduces new complexities such as the modification of mechanical, electronic and chemical properties of AB with possible modification of the isomerisation mechanisms. Indeed, experiments on anchored AB [42] and related photochromic molecules [43] indicate that two-way photo-switching of these systems may not be straightforward.

The key issues raised by embedding AB in a mechanically controllable break-junction (MCB) [23] are as follows. (1) In which way does the anchoring to the gold tips alter the electronic structure of AB and hence its physical properties? (2) In which way can the optical switching probabilities be manipulated by applying mechanical strain? (3) Can DAB be switched by purely mechanical means? (4) Can this switching be bidirectional, i.e. both *cis*→*trans* and *trans*→*cis*? (5) What is the interplay between optical [44, 45] and mechanical [30] switching forces?

3.1 Models and methods

We model both mechanical and opto-mechanical isomerisation of a DAB bridge connected to two gold electrodes [46]. Mechanical *cis*→*trans* switching

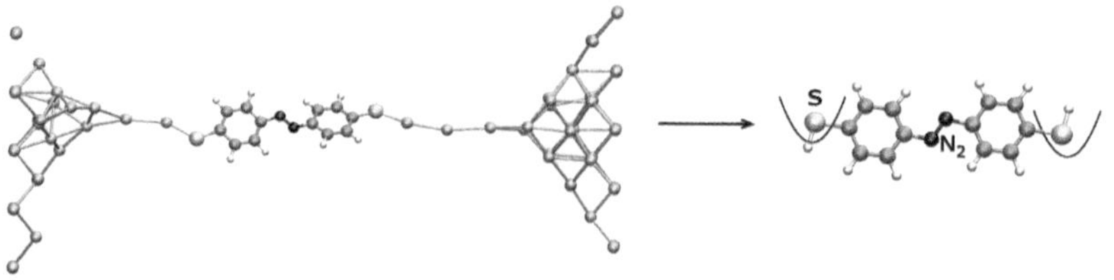

Fig. 3. A model of azobenzene junction between gold tips used in mechanical switching simulations and the schematics of the model with harmonic restraints applied on the sulphur atoms and used for the opto-mechanical simulations. The chosen configuration with the tips corresponds to extension $D = 15.2$ Å of the system described up to 8.0 Å by Fig. 4.

has been modelled using two differently shaped tips each composed of 40–60 Au atoms. These tips, interconnected through periodic boundary conditions, are large enough to allow for elastic/plastic deformations arising during mechanical manipulation. We performed ground state DFT calculations employing the Perdew–Burke–Ernzerhof (PBE) functional [36], a 19-electron (semicore) pseudopotential for Au, and an all-electron representation for the other species. The calculations were carried out using the DMol3 code [47] with a DNP (double numerical plus polarisation) basis set, and a 0.005 Ha electronic smearing. To obtain a starting point for the DFT calculations, the gold tips were first prepared using empirical interatomic potentials [48] by a MCB [23] type of procedure. A gold rod coupled to two gold (110) plates was generated. The system was then heated and let to evolve under Langevin dynamics followed by a relatively fast pulling process up to the point of breakage of the Au nanojunction. The tips so formed were then allowed to evolve at low temperatures (5 K) for a few picoseconds. All remaining modelling was performed using DFT techniques: the tips resulting from the dynamics were relaxed, their separation adjusted so as to accommodate a relaxed CDAB or TDAB (stripping the hydrogen atoms off the terminating SH groups). The junction was then relaxed by DFT geometry optimisation at different constrained inter-tip distances, typically in steps of 0.2 Å. Straining of the junction in static manner was continued until rupture in the case of pulling and until occurrence of reisomerisation in the case of compression. The pulling simulations have been applied mainly to *cis* isomer (which is the shorter one in equilibrium) while the compressing strain was exerted in most cases on *trans* isomers. Several realisations of the junction have been modelled, differing by the model tips used and/or sulphur bonding site on the tip.

To study opto-mechanical switching in the S_1 excited state we employed the generalised restricted open-shell Kohn–Sham (gROKS) method [49, 50], which has been successfully applied to the AB chromophore previously [45]. In the present scenario, the electronic structure of AB is considerably complicated by the presence of the gold electrodes. In order to disentangle mechanical and optical/chemical effects, we adopted a simplified model to study opto-mechanical switching, in which the mechanical action of the tips was modelled by harmonic restraining potentials in which the hydrogen–terminated sulphur atoms of DAB are allowed to move during dynamical evolution in the S_1 excited state, see Fig. 3. One choice of the force constant was 0.167 a.u., corresponding to the value of the Au–S molecule. The other value was twice of this, in order to model harder metal-sulphur bonding. Our model neglects chemical effects of gold on AB electronic structure. However, the chemical effects have been found not to change our model qualitatively. In particular, chemical quenching of the $n \rightarrow \pi^*$ (HOMO $\rightarrow$ LUMO) excitation on the nitrogen atoms due bonding to the tips was not observed [42]. The model was found qualitatively robust also against the choice of the force constants. Varying the distance between the two restraining centres mimics variations of a tip distance of an STM/AFM apparatus. Straining of the system is continued until rupture of the molecule and, in case of the *trans* conformation, we studied also its compression. The excited state simulations were performed with the CPMD code [37, 38] using a plane-wave basis set truncated at 70 Ry, Goedecker pseudopotentials [51, 52], and the PBE functional [36]. Comparative ground state calculations were carried in the same setup. We do not perform a non-adiabatic simulation but rather generate a complete set of ground-state energies for configurations visited in the S_1 state, see Sect. 3.4. Based on a comparison with experimental results,[†] we have added a uniform shift of 0.7 eV to all gROKS (S_1) energies, see the values in Table 2 to be discussed below.

3.2 Results

3.3 Mechanical switch

We have found that both *cis*→*trans* and *trans*→*cis* conformation changes in the S_0 electronic state are possible if appropriate mechanical manipulation is applied. *Cis*→*trans* isomerisation occurred under an external pulling action while for the opposite isomerisation direction a compressing force was used. Basically this behaviour is consistent with the different equilibrium lengths of the two isomers which in terms of sulphur–sulphur distances are about 9 and 12.6 Å for *cis* and *trans* DAB, respectively. Typical example of our simulation results of mechanical *cis*→*trans* switching induced by the external pulling force is shown in Fig. 4. The quantities are plotted against the extension D which is defined as the displacement of the fixed Au atoms plane from its equilibrium position. In the case without tips D is just the displacement of the pulled sulphur atom. The results indicate that mechanical *cis*→*trans* switching in the lowest singlet is easily feasible requiring fairly modest forces

[†]The gROKS $n \rightarrow \pi^*$ excitation energies are calculated to be 2.21 eV for *trans* AB (2.13 eV for *cis* AB) compared to 2.82 eV (2.91 eV) from gas-phase experiments [53] and 2.84 eV (3.0 eV) from correlated Coupled Cluster calculations [54]. Similar errors are expected for DAB isomers.

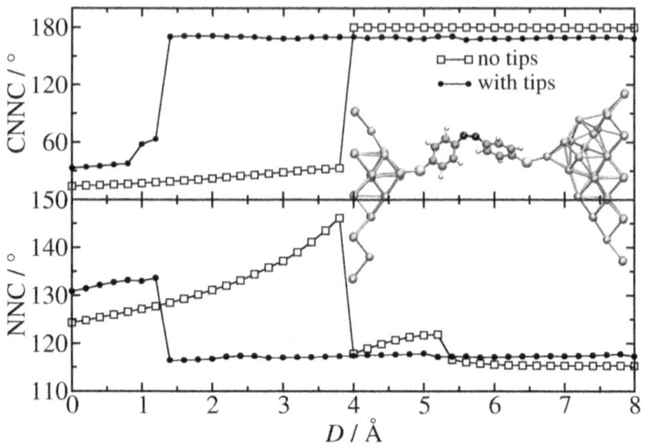

Fig. 4. Simulation of the mechanical *cis*→*trans* switching. The system with tips was simulated using 130 Au atoms in the supercell. *Upper panel*: CNNC dihedral angle. *Lower panel*: Average NNC bond angle. The extension D is defined as the displacement of the fixed Au atoms plane from the equilibrium value. For the system without tips the displacement of the pulled sulphur atom is considered as D. The inset shows the starting configuration which is the *cis*-equilibrium one. See also the *left-hand* side of Fig. 3 where the stretched configuration (*trans*-conformation) at $D = 15.2$ Å is drawn in the same zoom as the inset. All the results have been calculated using numerical localised basis sets and the DMol3 code [47].

(∼1 nN) irrespective of the details of the tips used. Isomerisation proceeds primarily via torsion about the N=N bond as can be seen from the changes in the CNNC dihedral angle. The dihedral angle is a soft degree-of-freedom, and therefore the switching proceeds on a flat part of the potential energy surface (PES) and changes only weakly with the model tips used. However, after the *cis*→*trans* isomerisation has occurred, further force application to the *trans* isomer proceeds on much steeper part of the PES with a number of elastic/plastic tip(s) deformations. The *trans* isomer presents a relatively stiff structure without any significant structural changes during the pulling. Fragmentation of the junction occurs by breaking an Au–Au bond in the monatomic gold nanowire(s) pulled out of the tip(s); see Fig. 3 with the configuration one simulation step ($\Delta D = 0.2$ Å) before the gold nanowire at the right-hand sulphur atom breaks. Starting the mechanical treatment from the *trans* isomer (not shown) reproduces the behaviour observed after the *cis*→*trans* switching corroborating the fact, that the *trans* isomer is rather stiff and the gold tips rather ductile. The mechanical effect of gold tips on both molecular isomerisation and breaking of the junction can be appreciated by comparison to the same processes but generated using hard distance constraints on the two S atoms instead of gold tips, see "no tips" plots on Fig. 4. Clearly, the system without the tips is much stiffer due to the missing elastic degrees of freedom in the tips and thus the PES is much steeper and the force required for isomerisation much larger (∼3 nN). Qualitatively, the strain–induced processes with and without tips have several common features, e.g. *cis*→*trans* isomerisation at fairly small extensions (at ≈ 1.5/5 Å). There are, however, some important differences between the scenarios with and without tips. First, without tips breakage occurs at much smaller extensions, ≈ 6 Å, compared to 9/16 Å with tips. This is an obvious consequence of the tips deformations accounted to the extension D. Moreover, without tips the isomerisation takes place primarily via the *inversion mechanism*, i.e. by change of NNC bond angle, and not via the (mostly) rotation mechanism observed with the tips. In both cases however the isomerisation mechanisms are neither pure rotation nor pure inversion but rather a mixture of both.

Our simulations have shown that, contrary to intuition, the reverse mechanical *trans*→*cis* switching is also possible [55]. That is the process of TDAB compression in the S_0 singlet state can turn the molecule into the *cis* conformation. Again, we have examined the process on the two levels of modelling, with just dithioazobenzene alone and with the molecule attached to small gold tips of 40 atoms together (18 of them fixed). In both cases we observed the switching. The switching without tips occurred at a compression corresponding to sulphur–sulphur distance $d^{\mathrm{sw}}_{\mathrm{S-S}} = 5.6$ Å. On the other hand, for azobenzene attached to the gold tips, the isomer was changed at $d^{\mathrm{sw}}_{\mathrm{S-S}} = 7.4$ Å. The minimum of energy for the structure with the small tips has an inter-sulphur distance corresponding to a partially compressed AB molecule which leads to difficulties when comparing the results in terms of the quantity D. The minimum-energy structure with the tips is a result of the tip preparation procedure and accompanying structural changes of Au–Au bonds in the tips. The compression regime of AB attached between gold tips is still under investigation.

3.4 Opto-mechanical switch

The protocol of the dynamical simulations is as follows: (1) First dynamics of DAB in its S_0 ground-state at a given stretch for given isomer is simulated at 300 K for 1 ps. (2) A laser pulse applied in an experiment is modelled by a vertical $S_0 \rightarrow S_1$ electronic excitation. (3) System dynamics in the S_1 state is followed for 0.5 ps. The S_0 energies for the excited state trajectories are simultaneously generated and the $S_1 - S_0$ energy separation, crucial for non-adiabatic relaxation, calculated. Steps (1–3) are repeated for several stretch values for both *cis* and *trans* isomers. If not otherwise said the results shown corresponds to the restraining parameter $\kappa = 0.167$ a.u.

Figure 5 shows a typical example of time evolution of the CNNC dihedral angle leading up to and after photo-excitation of the *cis* and *trans* isomers at different tip separations. (The tips are modelled by the harmonic restraints in this case.) Extensions 0.0/4.5 Å for the *cis* and −2.0/1.6 Å for the *trans* isomer have been studied, several representative cases are shown in the figure. Comparisons of "no tips" vs. "tips" plots on Fig. 4 indicate that the model

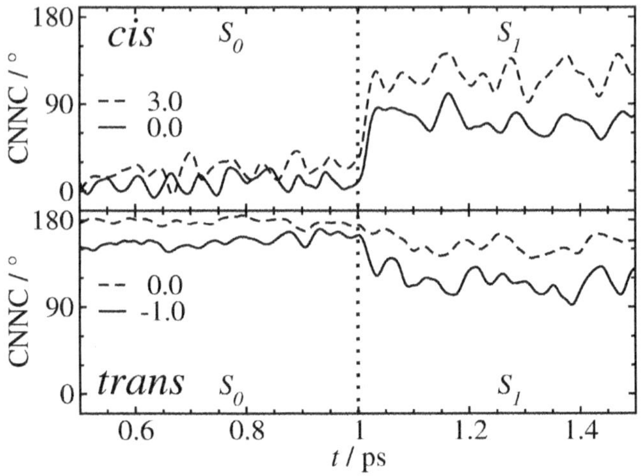

Fig. 5. Simulation of the opto-mechanical switching. Time evolution of the CNNC dihedral angles. Interval [0, 1] ps (second half shown)presents the equilibration stage in the S_0 electronic state. Interval [1, 1.5] ps corresponds to the excited S_1 state dynamics. The *dotted vertical line* represents the instant of the vertical $S_0 \rightarrow S_1$ excitation. *Upper graph*: evolution starting from the *cis* conformation. *Lower graph*: evolution from the *trans* conformation. Extension D (different values distinguished by different lines) measures for each of the isomers the distance difference between the harmonic restraining potential centres and the equilibrium isomer length (8.6 Å for *cis* and 12.6 Å for *trans* structure). The restraining parameter was $\kappa = 0.167$ a.u.

with restraints qualitatively reproduces mechanical effects of the model with the tips only for small extensions D. In Fig. 5 an abrupt response to the vertical $S_0 \rightarrow S_1$ excitation at time $t = 1$ ps can be seen, in particular for the *cis* isomer which undergoes changes in the CNNC dihedral by about 90/150° within 10 fs. These CNNC dihedral changes are the larger the larger is the tip separation (Fig. 5). The amplitude of the rotational relaxation of the *trans* isomer is much less pronounced although still very fast. An extremely fast partial (15/20°) opening of NNC angles takes effect especially at larger stretches (not shown in figures) for the *trans* isomer. For negative stretches the CNNC variations upon photo-excitation are considerably larger since the DAB molecule is allowed sufficient freedom to relax to the excited state global minimum rotamer structure [56]).

So far we have discussed the dynamical evolution in the excited state. However, photoisomerisation is only complete after the chromophore has relaxed back to the electronic ground state and its excess kinetic energy has dispersed into the environment. An important criterion controlling non-radiative decay is the $S_1 - S_0$ "vertical" energy gap, $E_{\rm deexc}$. In order to extract trends for the efficiency of non-radiative decay pathways at the different elongations, we computed the time averages $\langle E_{\rm deexc} \rangle$ over the 0.5 ps excited-state trajectories for simulations starting from both isomers, see Table 2. For the *cis* isomer the

Table 2. Average de-excitation energies $\langle E_{\mathrm{deexc}} \rangle$ in eV's for the set of simulations shown in Fig. 5. The legends *cis*/*trans* denote the *initial* conformations of the isomers. For the definition of the extension D see Fig. 5 and its caption.

D/Å	−1.0	0.0	3.0
cis	–	1.00	1.24
trans	1.19	1.75	–

gap size is increased when the extension of 3 Å is applied. The pronounced feature of the *trans*-DAB is the fall of $\langle E_{\mathrm{deexc}} \rangle$ in the compression regime. The effect of extension/compression on $\langle E_{\mathrm{deexc}} \rangle$ can qualitatively be understood as to follow from profiles of the S_1 and S_0 PESs [56, 57]; the average de-excitation gap is a function of the molecule length and has its minimum in between equilibrium *cis* and *trans* lengths. These trends depend only weakly on the force constant of the restraining potentials [55].

Analysis of time evolution of E_{deexc} shows that the gap is smallest for CNNC dihedrals around 100°. This value is in agreement with recent quantum chemical calculations [56]. The CNNC dihedral is thus the major structural parameter governing the de-excitation gap and consequently the probability of $S_1 \leftrightarrow S_0$ switch.

The compressive regime in *trans*→*cis* switching is important also in view of counteracting the large length mismatch between the *trans* and *cis* isomers which is in equilibrium about 4 Å measured in terms of sulphur–sulphur separations. In other words, if the *trans* isomer is kept by the restraints or sulphur–gold bonds and its length is 3/4 Å longer than the equilibrium *cis* length then the probability of *trans*→*cis* conformation change is quite small due to the mechanical (length) restriction. The restriction highly suppresses the internal degrees of freedom of the molecule which are important for the isomer change. In addition, as can be seen from Table 2, the average de-excitation gap is relatively large for equilibrium *trans* length thus making $S_1 \rightarrow S_0$ transition less likely to occur. Hence both mechanical effects and magnitude of the de-excitation gap decrease the probability of the *trans*→*cis* switching for *trans* lengths close to or larger than the equilibrium *trans* length. Contrary, in the compressive regime $\langle E_{\mathrm{deexc}} \rangle$ drops hand-in-hand with reduction of the mechanical hindrance. Hence, the *trans*→*cis* switching may only be possible in compressive regime. This finding may explain the lack of experimental observation of a two-way switching of photochromic molecules anchored to tips [42, 58].

4 Conclusions

We have presented extensive *ab initio* simulations, of three very different types of chemistries applied to organometallic nanojunctions: thermo-, photo-, and mechano-chemistry.

In the first part, results of the first direct comparison of thermal and mechanical breakage of chemical bond was shown for a SAM of thiolates on Cu(111) surface. Two completely *different reaction pathways and product classes* were observed in mechanochemical *versus* thermochemical decomposition. While fluctuations acting in thermochemistry cleave the covalent C–S bond within the thiolate molecule itself, mechanochemistry stabilises selectively and significantly the very same bond. The molecular reason behind this phenomenon is the suppression of fluctuations in concert with a systematic under-coordination of S due to strongly uniaxial stress when a pulling force is applied. This constitutes the novelty of mechanochemistry, which is found to manipulate primarily the molecular coordination and the fluctuations thereof. This externally controllable mechanical manipulation makes the C–S bond one of the *strongest* bonds in stark contrast to the thermochemical scenario where it breaks. This implies immediately that a slight dilating/compressive straining of such molecule/metal junctions and interfaces will increase/decrease their stability.

In the second part, mechanical and opto-mechanical switching of AB molecular junction was studied. Purely mechanical switching from *cis* isomer to *trans* one in the lowest singlet state S_0 was found possible for the tips externally pulled away of each other. The isomerisation in the presence of the gold tips proceeds primarily via the rotation mechanism. Reverse process of mechanically induced *trans*→*cis* switching was also observed. Optical switching via the first excited singlet state S_1 was explored, subject to the mechanical effect of the electrodes modelled by harmonic restraints. The vertical de-excitation gap crucial for non-radiative relaxation was found to significantly vary with the extension/compression applied to the molecule. In line with experiments [42, 43], efficient switching has been shown to be mechanically hindered for stiff junctions under certain conditions due to significantly different isomer lengths, especially in the *trans* → *cis* direction. To avoid the hindrance, the length of the molecule must be mechanically adjusted or the contacts to the tips must not be stiff.

These results indicate the paramount importance mechanochemistry is bound to play in practical applications of these systems.

5 Acknowledgements

Computer resources from Scientific Supercomputing Center Karlsruhe (the HP-XC supercomputer), BoviLab@RUB, Recherverbund–NRW, and CCMS as well as financial support from Volkswagen–Stiftung (StressMol), APVT (20-019202), DFG, and FCI are gratefully acknowledged.

References

1. Ellenbogen, J.C., Love, J.C., Architectures for Molecular Electronic Computers. IEEE, New York, (2000)
2. Joachim, C., Gimzewski, J.K., Aviram, A., Electronics using hybrid-molecular and mono-molecular devices, *Nature* **408**, 541–548 (2000)
3. Park, J., Pasupathy, A.N., Goldsmith, J.I., Chang, C., Yaish, Y., Petta, J.R., Rinkoski, M., Sethna, J.P., Abruña, H.D., McEuen, P.L., Ralph, D.C., Coulomb blockade and the Kondo effect in single-atom transistors, *Nature* **417**, 722–725 (2002)
4. Liang, W., Shores, M.P., Bockrath, M., Long, J.R., Park, H., Kondo resonance in a single-molecule transistor, *Nature* **417**, 725–729 (2002)
5. Carey Lea, M., Disruption of the silver haloid molecule by mechanical force, *Phil. Mag.* **34**, 46–50 (1892)
6. Beyer, M.K., Clausen-Schaumann, H., Mechanochemistry: The mechanical activation of covalent bonds, *Chem. Rev.* **105**, 2921–2948 (2005)
7. Frank, I., Mechanically induced chemistry: New perspective on the nanoscale, *Angew. Chem. Int. Ed.* **45**, 852–854 (2006)
8. Granick, S., Bae, S. C., Physical chemistry – Stressed molecules break down, *Nature* **440**, 160–161 (2006)
9. Rosen, B.M., Percec, V., Mechanochemistry – A reaction to stress, *Nature* **446**, 381–382 (2007)
10. Grandbois, M., et al., How strong is a covalent bond? *Science* **283**, 1727–1730 (1999)
11. Saitta, A.M., et al., Influence of a knot on the strength of a polymer strand, *Nature* **399**, 46–48 (1999)
12. Aktah, D., Frank, I.: Breaking bonds by mechanical stress: When do electrons decide for the other side? *J. Am. Chem. Soc.* **124**, 3402–3406 (2002)
13. Krüger, D., et al., Pulling monatomic gold wires with single molecules: An ab initio simulation, *Phys. Rev. Lett.* **89**, 186402 (2002)
14. Krüger, D., et al., Towards "mechanochemistry": Mechanically induced isomerizations of thiolate-gold clusters, *Angew. Chem. Int. Ed.* **42**, 2251–2253 (2003)
15. Vélez, P., Dassie, S.A., Leiva, E.P.M., First principles calculations of mechanical properties of 4,4(')-bipyridine attached to Au nanowires, *Phys. Rev. Lett.* **95**, 045503 (2005)
16. Muraoka, T., Kinbara, K., Aida, T., Mechanical twisting of a guest photoresponsive host, *Nature* **440**, 512–515 (2006)
17. Novaes, F. D., et al., Oxygen clamps in gold nanowires, *Phys. Rev. Lett.* **96**, 016104 (2006)
18. Sheiko, S. S., et al., Adsorption-induced scission of carbon–carbon bonds, *Nature* **440**, 191–194 (2006)
19. Hickenboth,C. R., et al., Biasing reaction pathways with mechanical force, *Nature* **446**, 423–427 (2007)
20. Gimzewski, J. K. and Joachim, C., Nanoscale science of single molecules using local probes, *Science* **283** 1683–1688 (1999)
21. Moresco, F. and Gourdon, A., Scanning tunneling microscopy experiments on single molecular landers, *PNAS* **102**, 8809–8814 (2005)
22. Otero, R., Rosei, F., Besenbacher, F., Scanning tunneling microscopy manipulation of complex organic molecules on solid surfaces, *Ann. Rev. Phys. Chem.* **57**, 497–525 (2006)

23. Smit R.H.M., Noat, Y., Untiedt, C., Lang, N.D., van Hemert, M.C., van Ruitenbeek, J.M., Measurement of the conductance of a hydrogen molecule, *Nature* **419**, 906–909 (2002)
24. Liu, G. Y., Xu, S., Qian, Y.L., Nanofabrication of self-assembled monolayers using scanning probe lithography, *Acc. Chem. Res.* **33**, 457–466 (2000)
25. James, D.K. and Tour, J.M., Electrical measurements in molecular electronics, *Chem. Mat.* **16**, 4423–4435 (2004)
26. Salaita, K., Wang, Y.H., Mirkin, C.A., Applications of dip–pen nanolithography, *Nat. Nanotech.* **2**, 145–155 (2007)
27. Chen, F., et al., Measurement of single-molecule conductance, *Ann. Rev. Phys. Chem.* **58**, 535–564 (2007)
28. Schreiber, F., Self-assembled monolayers: from "simple" model systems to biofunctionalized interfaces, *J. Phys.: Condens. Matter* **16**, R881–R900 (2004)
29. Love, J.C., et al., Self-assembled monolayers of thiolates on metals as a form of nanotechnology, *Chem. Rev.* **105**, 1103–1169 (2005)
30. Konôpka, M., Turanský, R., Reichert, J., Fuchs, H., Marx, D., Štich, I., Mechanochemistry and thermochemistry are different: Stress-induced strengthening of chemical bonds. *Phys. Rev. Lett.* **100**, 1155031–1–4 (2008)
31. Kondoh, H., Nozoye, H., Molecular process of adsorption and desorption of alkanethiol monolayers on Au(111). *J. Chem. Phys.* **111**, 1175–1184 (1999)
32. Konôpka, M., Rousseau, R., Štich, I., Marx, D., Detaching thiolates from copper and gold clusters: Which bonds to break? *J. Am. Chem. Soc.* **126**, 12103–12111 (2004)
33. Konôpka, M., Rousseau, R., Štich, I., Marx, D., Electronic origin of disorder and diffusion at a molecule-Metal interface: Self-assembled monolayers of CH_3–S on Cu(111). *Phys. Rev. Lett.* **95**, 096102–1–4 (2005)
34. Jackson, G.J., et al., Following local adsorption sites through a surface chemical reaction: CH_3SH on Cu(111). *Phys. Rev. Lett.* **84**, 119 (2000).
35. Lai, Y.-H., et al., Adsorption and thermal decomposition of alkanethiols on Cu(110). *J. Phys. Chem. B* **106**, 5438 (2002).
36. Marx, D., Hutter, J., Ab initio molecular dynamics: Theory and implementation. In: Grotendorst, J. (ed) *Modern Methods and Algorithms of Quantum Chemistry.* NIC, FZ Jülich (2000), pp. 301–449; for downloads see: `www.theochem.ruhr-uni-bochum.de/go/cprev.html`
37. Perdew, J.P., Burke, K., Ernzerhof, M., Generalized gradient approximation made simple, *Phys. Rev. Lett.* **77**, 3865–3868 (1996); *Phys. Rev. Lett.* **78**, 1396–1396 (1997)
38. CPMD, Copyright IBM Corp 1990–2006, Copyright MPI für Festkörperforschung Stuttgart 1997–2001
39. Dürr, H., Bouas-Laurent, H. (ed.), *Photochromism. Molecules and Systems.* Elsevier, Amsterdam (1990)
40. Hartley, G.S., The Cis–form of Azobenzene. *Nature* **140**, 281–281 (1937)
41. Hugel, T., Holland, N.B., Cattani, A., Moroder, L., Seitz, M., Gaub, H.E., Single-molecule optomechanical cycle. *Science* **296**, 1103–1106 (2002)
42. Reichert, J., Klein, S., Konôpka, M., Turanský, R., Marx, D., Štich, I., Fuchs, H., Conductance of an illuminated metal–molecule–metal junction utilizing a near–field probe as counterelectrode. [Submitted to *Rev. Sci. Inst.*(2008)]
43. Dulić, D., van der Molen, S.J., Kudernac, T., Jonkman, H.T., de Jong, J.J.D., Bowden, T.N., van Esch, J., Feringa, B.L., van Wees, B.J., One-way optoelectronic switching of photochromic molecules on gold. *Phys. Rev. Lett.* **91**, 207402–1–4 (2003)

44. Cimelli, C., Granucci, G., Persio, M., Are azobenzenophanes rotation-restricted? *J. Chem. Phys.* **123**, 174317–1–10 (2005)
45. Nonnenberg, C., Gaub, H., Frank, I., First-principles simulation of the photoreaction of a capped azobenzene: The rotational pathway is feasible. *Chem. Phys. Chem.* **7**, 1455–1461 (2006)
46. Turanský, R., Konôpka, M., Reichert, J., Fuchs, H., Marx, D., Štich, I., Mechanical and opto-mechanical switching of azobenzene metal-organic junctions. Submitted (2008)
47. see `http://www.accelrys.com`
48. Jacobsen, K.W., Norskov, J.K., Puska, M.J., Interatomic interactions in the effective-medium theory, *Phys. Rev. B* **35**, 7423–7442 (1987)
49. Frank, I., Hutter, J., Marx, D., Parrinello, M., Molecular dynamics in low-spin excited states, *J. Chem. Phys.* **108**, 4060–4069 (1998)
50. Grimm, S., Nonnenberg, C., Frank, I., Restricted open-shell Kohn–Sham theory for $\pi - \pi^*$ transitions. I. Polyenes, cyanines, and protonated imines, *J. Chem. Phys.* **119**, 11574–11584 (2003)
51. Goedecker, S., Teter, M., Hutter, J., Separable dual–space Gaussian pseudopotentials, *Phys. Rev. B* **54**, 1703–1710 (1996)
52. Hartwigsen, C., Goedecker, S., Hutter, J., Relativistic separable dual-space Gaussian pseudopotentials from H to Rn. *Phys. Rev. B* **58**, 3641–3662 (1998)
53. Andersson, J.-Å., Petterson, R., Tegnér, L., Flash photolysis experiments in the vapour phase at elevated temperatures I: Spectra of azobenzene and the kinetics of its thermal cis–trans isomerization. *J. Photochem.* **20**, 17–32 (1982)
54. Fliegl, H., Köhn, A., Hättig, C., Ahlrichs, R.: Ab Initio Calculations of the Vibrational and Electronic Spectra of *trans*– and *cis*–Azobenzene. *J. Am. Chem. Soc.* **125**, 9821–9827 (2003)
55. Turanský, R., Konôpka, M., Reichert, J., Fuchs, H., Marx, D., Štich, I.: (In preparation)
56. Cembran, A., Bernardi, F., Garavelli, M., Gagliardi, L., Orlandi, G.: On the Mechanism of the cis-trans isomerization in the lowest electronic states of azobenzene: S_0, S_1, and T_1, *J. Am. Chem. Soc.* **126**, 3234–3243 (2004)
57. Gagliardi, L., Orlandi, G., Bernardi, F., Cembran, A., Garavelli, M., A theoretical study of the lowest electronic states of azobenzene: The role of torsion coordinate in the cis-trans photoisomerization, *Theor. Chem. Acc.* **111**, 363–372 (2004)
58. Choi, B.-Y., Kahng, S.-J., Kim, S., Kim, H., Kim, H.W., Song, Y.J., Ihm, J., Kuk, Y., Conformational molecular switch of the azobenzene molecule: A scanning tunneling microscopy Study, *Phys. Rev. Lett.* **96**, 156106–1–4 (2006)

When It Helps to Be Purely Hamiltonian: Acceleration of Rare Events and Enhanced Escape Dynamics

Dirk Hennig[1], Simon Fugmann[1], Lutz Schimansky-Geier[1], and Peter Hänggi[2]

[1] Institut für Physik, Humboldt-Universität zu Berlin, Newtonstr. 15, 12489 Berlin, Germany
hennigd@physik.hu-berlin.de
[2] Institut für Physik, Universität Augsburg, Universitätsstraße 1, 86135 Augsburg, Germany
hanggi@physik.uni-augsburg.de

Abstract. We consider the self-organized escape of a linear chain of coupled units from a metastable state over a barrier in a microcanonical situation. Initially the units of the chain are situated near the bottom of the potential well forming a flat state. In the underlying conservative and deterministic dynamics such a uniform and linear lattice state with comparatively little energy content seems to be restrained to oscillations around the potential bottom preventing escape from the well. It is demonstrated that even small deviations from the flat state entail internal energy redistribution leading to such strong localization that the lattice chain spontaneously adopts a localized pattern resembling a hairpin-like structure. The latter corresponds to a critical equilibrium configuration, that is a transition state, and, being dynamically unstable, constitutes the starting point for the escape process. The collective barrier crossing of the units takes place as kink-antikink motions along the chain. It turns out that this nonlinear barrier crossing in a microcanonical situation is more efficient compared with a thermally activated chain for small ratios between the total energy of the chain and the barrier energy.

1 Introduction

The intensively investigated Kramers problem concerns the escape of a Brownian particle from a metastable state over a barrier (for a review see [1]). There it is implied that the system is in contact with an external heat bath serving as a permanent energy source, causing dissipation and local energy fluctuations which can trigger the escape process. However, the occurrence of the necessary optimal fluctuations enabling the particle to stochastically overcome an energetic barrier can be a rare event. This is particularly the case when the ratio between the thermal energy supplied to the particle by the heat bath and the barrier energy is low and the rate of thermal barrier crossing is exponentially suppressed. The extension of the escape problem to multi-dimensional systems was given in [2]. Recently, in the biophysical context there has been growing interest in thermally activated barrier crossing of a polymer chain occurring e.g. during the transport of long and flexible polymers across membranes and DNA electrophoresis [3–6].

Different to the numerous studies of the thermally induced escape, based on the coupling of the system to an external heat bath, the microcanonical situation, when only the internal energy of a system has to suffice to perform structural transitions, has been studied less. The question is whether under deterministic and conservative conditions a coupled oscillator chain can still overcome a potential barrier when all its units reside initially near the bottom of a metastable potential well? In such a situation the energy is almost equally shared among the units and the system is also far away from the critical equilibrium configuration related with a saddle point in configuration space, referred to as the transition state [7]. Typically, the critical equilibrium configuration is represented by a localized state of the chain.

Concerning the attainment of localized structures it is by now well established that nonlinear systems exhibit localization features giving rise to the formation of coherent structures that emerge even from an initial almost homogeneous state [8]. The concentration of an originally distributed physical quantity to a few degrees of freedom in confined regions of a spatially extended and homogeneous systems proceeds often in a self-organized manner. In recent years the concept of intrinsic localized modes or discrete breathers as time-periodic and spatially localized solutions of nonlinear lattice systems has turned out to present the prototype of excitations describing localization phenomena in numerous physical situations [9–14]. For the creation of localized structures modulational instability leading to a self-induced modulation of an initial linear wave with a subsequent generation of localized pulses has proven to be an effective mechanism. In this way energy localization in a homogeneous system is achievable. For example breathers have been successfully applied to describe localized excitations which reproduce typical features of the thermally induced opening dynamics of DNA duplex molecules such as the magnitude of the amplitude and the time-scale of the oscillating bubble preceding full strand separation (denaturation) [15–21].

In the present study we address the escape problem of a one-dimensional oscillator chain over the barrier of a metastable potential within a conservative and deterministic lattice model [22, 23]. The units of the lattice system are initially near a metastable equilibrium which hinders the system to immediately perform a task that is associated with overall large-amplitude excitations. In more detail, we consider the energy exchange dynamics in a nonlinear oscillator chain. Each oscillator evolves in a local anharmonic potential possessing a barrier that divides regions of bounded motion in the potential well from unbounded ones beyond the barrier. The oscillators are linearly coupled. Concerning the energy redistribution we focus interest on the initial situation of a low-energy mode of the chain. The question which arises now is whether with the low amount of energy supplied to the system the focusing of energy proceeds that effectively that at least one, if not a few units, can gain enough energy to get over the potential barrier? If so, is it possible that the neighbors coupled to the escaping unit(s) also get drawn over the barrier? If more and more units get involved this would initiate a coordinated escape of the entire chain.

2 The oscillator chain model

The linear chain is modeled as a one-dimensional lattice system of harmonically coupled nonlinear oscillators with the Hamiltonian

$$H = \sum_{n=1}^{N} \left\{ \frac{p_n^2}{2} + U(q_n) \right\} + \frac{\kappa}{2} \sum_{n=1}^{N} \left[q_{n+1} - q_n \right]^2 . \tag{1}$$

The coordinates q_n quantify the amplitude of an oscillator at site n. p_n denotes the momentum canonically conjugate to the coordinate q_n. Each oscillator evolves in an anharmonic potential given by

$$U(q) = \frac{\omega_0^2}{2} q^2 - \frac{a}{3} q^3 , \tag{2}$$

where $a > 0$. The metastable equilibrium of the potential is situated at $q = 0$ and the maximum is located at $q = \omega_0^2/a$. There is a potential barrier which particles have to overcome in order to escape from the potential well of depth $\Delta E = \omega_0^6/(6a^2)$. The oscillators, also referred to as units, are coupled harmonically with nearest-neighbor interaction strength κ. The equations of motion derived from the Hamiltonian given in Eq. (1) read

$$\frac{d^2 q_n}{dt^2} + \omega_0^2 q_n - a q_n^2 - \kappa \left[q_{n+1} + q_{n-1} - 2 q_n \right] = 0 . \tag{3}$$

We impose periodic boundary conditions according to $q_{N+1} = q_1$. Note that nonlinearity is solely contained in the local potential term.

3 Spontaneous energy localization

The ability of nonlinear and discrete systems to exhibit spontaneous localization has been demonstrated recently [24–30]. Such formation of localized excitations can be caused by modulational instability leading to the formation of intrinsically localized modes (breathers). This mechanism initiates an instability of an initial linear wave when small perturbations of non-vanishing wavenumbers are imposed. The instability, giving rise to an exponential growth of the perturbation, destroys the initial linear wave at a critical wavenumber so that a localized hump gets formed.

Initially all the oscillators are placed close to the bottom of the potential well. The chain state is expressed as a plane wave solution (phonons) $q_n(t) = (1/2)q_0 \exp[i(k\,n - \omega t)] + c.c.$ obeying the dispersion relation $\omega^2 = \omega_0^2 + 4\kappa \sin^2(k/2)$ with wave number $k \in [0, \pi]$.

Small modulational perturbations on the plane wave solution are imposed taking random initial amplitudes and/or momenta uniformly distributed in small intervals $|q_n(0) - q_0| \leq \Delta q$ and $|p_n(0) - p_0| \leq \Delta p$, respectively. Thus the chain is initialized close to an *almost homogeneous state* and yet such desynchronized ($\Delta q \neq 0$) to have small but nonvanishing initial interaction terms initiating energy exchange between the coupled units.

We recall that an uniform lattice state with amplitude q_0 and wave number k remains stable as long as the nonlinear character related with the aq^3 term of the potential $U(q)$ can be neglected. The chain evolves harmonically and localization of energy does not take place. Otherwise, the nonlinear part of the potential makes a modulational instability of waves possible. That is perturbations with a wave number Q may grow exponentially resulting in accumulation of energy at the expense of energy from the other units.

The exponential growth for the flat state, i.e. $k = 0$, takes place with rate [22, 23, 28]

$$\Gamma = \sqrt{\sin^2\left(\frac{Q}{2}\right) \frac{\kappa}{\omega_0^2} \left(\frac{5a^2}{3\omega_0^2}\, q_0^2 - 4\kappa\, \sin^2\left(\frac{Q}{2}\right)\right)}\,, \tag{4}$$

if the argument of the square root is positive. Thus it must hold that

$$\frac{5a^2}{3\omega_0^2}\, q_0^2 - 4\kappa\, \sin^2\left(\frac{Q}{2}\right) > 0\,, \tag{5}$$

which means that for fixed q_0 the anharmonicity a needs to be large enough or with given a the q_0 has to obtain overcritical values.

The set of coupled equations (3) has been numerically integrated with a fourth-order Runge-Kutta scheme. The accuracy of the calculation was checked by monitoring the conversation of the total energy with precision of at least 10^{-4}. The chain consists of $N = 100$ coupled oscillators.

Starting from an initial lattice state of nearly equipartition the attainment of a nontrivial structure is observed. More precisely, a pattern evolves in the course of time (of the order of $t \sim 2 \times 10^2$) for which at some lattice sites the amplitudes grow considerably whereas they get lowered in the surrounding regions. This localization phenomenon is reflected in the appearance of an array of irregularly spaced breathers on the lattice as seen in Fig. 1 where the spatio-temporal evolution of the energy density

$$E_n = \frac{p_n^2}{2} + U(q_n) + \frac{\kappa}{4}\left[(\,q_{n+1} - q_n\,)^2 + (\,q_{n-1} - q_n\,)^2\right]. \tag{6}$$

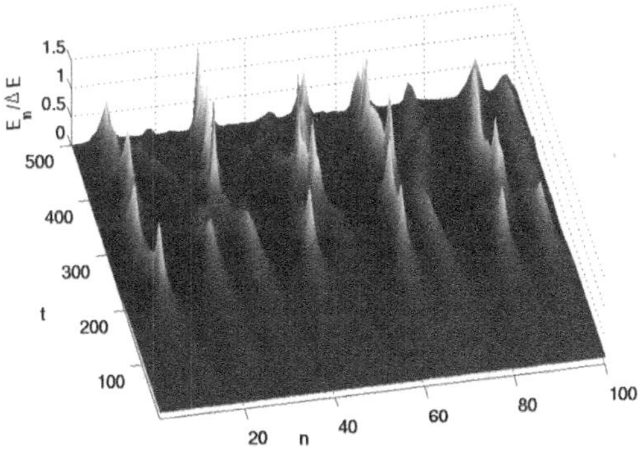

Fig. 1. Spatio-temporal evolution of the energy distribution $E_n(t)$. Initially the coordinates are uniformly distributed in the interval $|q_n(0) - q_0| \leq \Delta q$ with mean $q_0 = 0.45$ and width $\Delta q = 0.01$ and $p_n(0) = \Delta p = 0$. This leads to a total energy $E_{\text{total}} = 17.2 \equiv 12.9 \times \Delta E$. The parameter values are given by $a = 1$, $\omega_0^2 = 2$, $N = 100$ and $\kappa = 0.3$.

is shown. The total energy is $E_{\text{total}} = 17.2$ which is equivalent to $12.9 \times \Delta E$. In the beginning the total energy is virtually evenly shared among all units. The corresponding energy density, i.e. the average amount of energy contained in a single oscillator, lies significantly below the barrier energy. To be precise, the ratio between the energy density and the barrier energy is $E_n(0)/\Delta E = 0.129$. In other words, in order that a unit passes over the energy barrier from the region of bounded into unbounded motion directed flow of energy into this unit is demanded. Actually the energy amount of at least 7 other units has to be transferred into one unit so that this units energy levels that of the barrier.

There exist moving breathers that have the tendency to collide inelastically with other breathers (cf. Fig. 1 at site $n = 30$). In fact, various breathers merge to form larger amplitude breathers proceeding preferably such that the larger amplitude breathers grow on the expense of the smaller ones. As a result energy gets even stronger concentrated into smaller regions of the chain. Such localization scenario has been shown to be characteristic for a number of nonlinear lattice systems [25, 29–33].

For further illustration we depict in Fig. 2 snapshots of the energy density $E_n(t)$ at different instants of time. Starting point is the almost homogeneous state and the first snap shot is taken at $t = 5$ when the pattern is still flat. After a certain time the local energy accumulation is such enhanced that at least one of the involved units possess enough energy to overcome the barrier. As illustrated in Fig. 2 for the snap shot taken at $t = 550$ this happens for the unit at $n = 30$.

The question then is: does an escaped unit continue its flight beyond the barrier or can it even be pulled back into the bound chain formation ($q_n <$

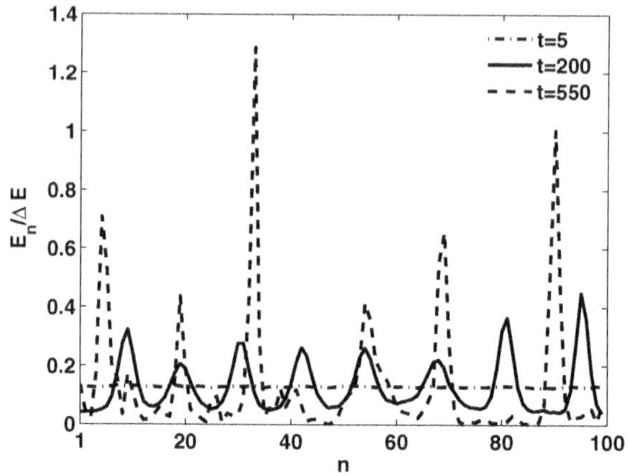

Fig. 2. Snap shots of the energy density at instants of time as indicated in the plot. Initial conditions: $q_0 = 0.45$, $\Delta q = 0.01$, $p_0 = \Delta p = 0$. Parameter values: $\omega_0^2 = 2$, $a = 1$, $N = 100$ and $\kappa = 0.3$.

$q_{\max}$) by the restoring binding forces exercised by its neighbors? On the other hand, the unit that has already escaped from the potential well might drag neighboring ones closer to or in the extreme even over the barrier. Thus, concerted escape of at least parts, if not the whole chain, from the potential valley seems possible.

4 Escape dynamics

Whether a unit of growing amplitude can really escape from the potential well or is held back by the restoring forces of their neighbors depends on the corresponding amplitude ratio as well as on the coupling strength. The critical chain configuration, that is the *transition state*, is determined by $\ddot{q}_n(t) = 0$ resulting in the stationary system

$$-\frac{\partial U}{\partial q_n} + \kappa[q_{n+1} + q_{n-1} - 2q_n] = 0\,. \tag{7}$$

Interpreting n as the 'discrete' time, with $1 \leq n \leq N$, the Eq. (7) describes the motion of a point particle in the inverted potential $-U(q)$. This difference system can be cast in form of a two-dimensional map by defining $x_n = q_n$ and $y_n = q_{n-1}$ [34] which gives:

$$\begin{aligned} x_{n+1} &= (w_o^2 x_n - a x_n^2)/\kappa + 2x_n - y_n \\ y_{n+1} &= x_n\,. \end{aligned} \tag{8}$$

The fixed points of this map are found as $x_0 = y_0 = 0$ and $x_1 = y_1 = \frac{w_0^2}{a}$. A stability analysis reveals that $x_0 = y_0 = 0$ represents an unstable hyperbolic equilibrium while at $x_1 = y_1 = \frac{w_0^2}{a}$ a stable center is located. The map is nonintegrable. The stable and unstable manifold of the hyperbolic point intersect each other yielding homoclinic crossings.

The corresponding homoclinic orbit of the map is on the lattice chain equivalent to a localized hump solution $\tilde{q}_n$ resembling the form of a hairpin. In Fig. 3 the profile of critical equilibrium configurations $\tilde{q}_n$ for several coupling

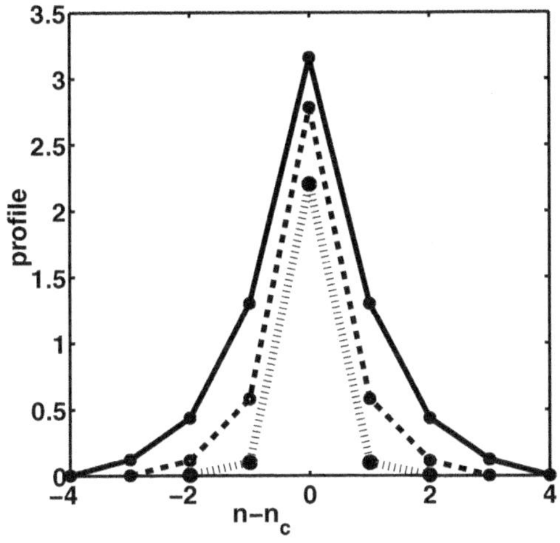

Fig. 3. Amplitude profile of the critical chain configuration for different coupling strengths: $k = 0.1$ (*dashed-dotted line*), $k = 0.5$ (*dashed line*), and $k = 1$ (*solid line*). For better illustration only the part of the lattice chain around the central site n_c with seizable elongations of the bonds is shown.

strengths are displayed. The stronger the coupling is the larger the maximal amplitude of the hump and the wider the width of the latter. Equation (7) can be derived from an energy functional $F = \sum_n \left(U(q_n) + \frac{\kappa}{2}[q_n - q_{n-1}]^2\right)$ as $\partial F/\partial q_n = 0$. Apparently, with increasing coupling more activation energy is needed to get the chain into its critical equilibrium configuration. One obtains $F = 1.33$, $F = 2.77$ and $F = 4.54$ for $\kappa = 0.1$, $\kappa = 0.5$ and $\kappa = 1$, respectively. Most importantly regarding escape, for the elongation of the bond at the central site, i.e. the maximal amplitude $\tilde{q}_{\max}$, it holds that $\partial U(\tilde{q}_{\max})/\partial q < 0$.

In the following we prove the dynamical instability of the critical localized mode. To this end we set $q_n(t) = \tilde{q}_n + w_n(t)$ with $|w_n| \ll 1$ and derive the linearized equations of motion as

$$\ddot{w}_n(t) = -\frac{\partial^2 U(q_n)}{\partial q_n^2}|_{q_n=\tilde{q}_n} w_n(t) + \kappa[w_{n+1}(t) + w_{n-1}(t) - 2w_n(t)] \,. \tag{9}$$

With the ansatz $w_n(t) = \phi_n \exp(\sqrt{\lambda}t)$ for the solution of (9) one arrives at an eigenvalue problem

$$\lambda\phi_n = -V_n\phi_n + \kappa[\phi_{n+1} + \phi_{n-1} - 2\phi_n] \,, \tag{10}$$

with

$$V_n = \frac{\partial^2 U(q_n)}{\partial q_n^2}|_{q_n=\tilde{q}_n} = \omega_0^2 - 2a\tilde{q}_n \,. \tag{11}$$

The second-order difference equation (10) is of the discrete stationary Schrödinger type, with a non-periodic potential, $-V_n$, breaking the translational invariance so that localized solutions exist (so called stop-gap states). The evolution of the two-component vector $(\phi_{n+1}, \phi_n)^T$ is determined by the following Poincaré map:

$$\mathcal{M}: \begin{pmatrix} \phi_{n+1} \\ \phi_n \end{pmatrix} = \begin{bmatrix} E_n & -1 \\ 1 & 0 \end{bmatrix} \begin{pmatrix} \phi_n \\ \phi_{n-1} \end{pmatrix} , \tag{12}$$

with on-site energy $E_n = (\lambda + V_n)/\kappa + 2$. The node-less even-parity ground state of Eq. (10), with its energy under the lower edge of the energy band of the passing band states, corresponds to an orbit of the linear map $\mathcal{M}$ being homoclinic to the hyperbolic equilibrium point at the origin $(0, 0)$ of the map plane. For the presence of a hyperbolic equilibrium the following inequality has to be satisfied:

$$\mathrm{Trace}(\mathcal{M}) = E_n = \frac{\lambda + V_n}{\kappa} + 2 > 2 \,, \tag{13}$$

implying that λ must fulfill the following constraint:

$$\lambda > \max_n(-V_n) = 2a \max_n \tilde{q}_n - \omega_0^2 > 0. \tag{14}$$

With the maximal amplitude of the c.l.m. lying beyond the barrier, viz. $\max_n \tilde{q}_n > \omega_0^2/a$ one finds

$$\lambda > \omega_0^2 > 0 \,. \tag{15}$$

Therefore, the ground state belongs to a positive eigenvalue from which we deduce that perturbations of the corresponding solution in the time domain grow exponentially.

The critical equilibrium solution tell us that for overcritical elongations of the units from their rest positions, $q_n > \tilde{q}_n$, the whole chain performs directed motion over the barrier. Conversely, if the elongations lie below the ones corresponding to $\tilde{q}_n$ then crossing the barrier is excluded.

Since for those states that have passed through the c.l.m. the kinetic energy of the outward motion increases a return backwards over the barrier into the original well is prevented. Figure 4 illustrates the kink-antikink motion in the escape time of the units $T_{\text{esc}}^{(n)}$ (defined as the moment at which a unit passes through a point $q = 20$ far beyond the barrier) versus the position on the lattice. Consecutively all oscillators manages to climb over the barrier one after another in a relatively short time interval.

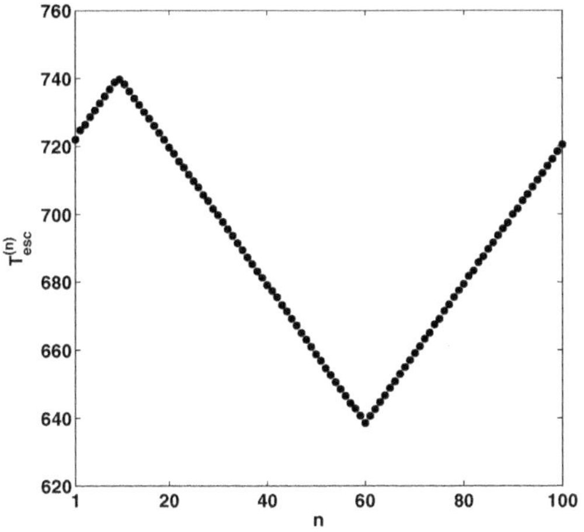

Fig. 4. The escape time of the chain units versus position. Initial conditions: $q_0 = 0.45$, $\Delta q = 0.01$, $p_0 = \Delta p = 0$. Parameter values: $a = 1$, $\omega_0^2 = 2$, $N = 100$ and $\kappa = 0.3$

Finally we compare the microcanonical escape process with a corresponding thermally activated process in the Kramers problem [1]. The Langevin equations read

$$\frac{d^2 q_n}{dt^2} + \gamma \frac{dq_n}{dt} + \frac{dU}{dq_n} - \kappa \left[q_{n+1} + q_{n-1} - 2q_n \right] + \xi_n(t) = 0 \ . \qquad (16)$$

Here γ is the friction parameter and $\xi_n(t)$ is a Gaussian distributed thermal random force with $< \xi_n(t) >= 0$ and $< \xi_n(t)\xi_{n'}(t') >= 2\gamma k_B T \delta_{n,n'} \delta(t - t')$.

Our results are summarized in Fig. 5 showing the mean escape time of the chain. The latter is determined by the mean value of the escape times of its units (see above). We took averages over 500 realizations of random initial conditions in the microcanonical and of noise in the Langevin equations, respectively. The Langevin equations were numerically integrated using a two-order Heun stochastic solver scheme. Results are presented as a function of $E_0/\Delta E$. For the deterministic and conservative system (3) E_0 is given by the initial energy per unit while it corresponds to thermal energy $k_B T$ in case of the Langevin system (16). In both cases there is a rather strong decay of T_{esc}

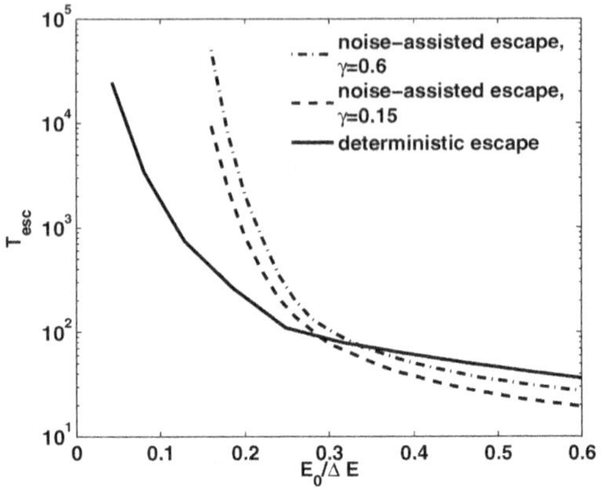

Fig. 5. Mean escape time as a function of the ratio $E_0/\Delta E$ (for details see text) for the microcanonical (*solid line*) and Langevin (*dashed line* and *dashed-dotted line*) dynamics respectively. Parameter values: $N = 100$, $\omega_0^2 = 2$, $a = 1$, $\kappa = 0.3$ and γ as given in the legend.

with growing ratio $E_0/\Delta E$ in the low energy region. This effect weakens gradually for further increasing E_0. Remarkably, for low E_0 the escape proceeds by far faster for the microcanonical system than for the one coupled to a heat bath. For small $k_B T$ the escape time in case of the Langevin system exceeds our simulation time taken as $t = 10^5$ for both depicted values of damping γ.

Concerning the difference between the deterministic and stochastic nature of the formation and stability of the c.l.m. we remark that under microcanonical conditions breather formation proceeds as an inherent and self-organized process. A breather of high enough energy can be created either directly due to a rapidly developing modulational instability or through the subsequent coalescence of smaller-amplitude breathers. In particular, breathers, as coherent structures sustained by the nonlinear chain, are fairly robust, i.e. they are stable with respect to interactions with linear waves. Notably, the deterministic processes take place on a time scale (see above) that is short compared with the time it can take till in the stochastic bath dynamics optimal fluctuations appear that trigger the formation of the c.l.m.. Even if in the stochastic process such a rare event has taken place the formed c.l.m. may readily be destroyed afterwards due to interactions with the heat bath.

5 Conclusions

In conclusion, for the escape of a chain of coupled oscillators from a metastable region over a barrier it is more effective to supply the energy at once and let the system afterwards evolve under microcanonical conditions rather than keeping the system permanently in contact with a heat bath from which energy can be absorbed. In other form, the underlying deterministic chaotic dynamics, which is generated intrinsically through the interaction of the oscillators propels pronounced energy exchange between the units so that the c.l.m. can be formed on the occurrence of which the escape process is conditioned. At least for small initial energies compared to the barrier values we have found faster transition times. More precisely, while at weak thermal noise the rate of thermal escape is exponentially suppressed, a deterministic nonlinear breather dynamics yields a robust critical nucleus configuration, which in turn causes an enhancement of the noise-free escape rate. Thus, the freezing out of noise may prove advantageous for transport in metastable landscapes, whenever the deterministic escape dynamics can be launched in a single shot via an initial energy supply.

We performed also studies with systems which are more complex than the one-dimensional chain model of harmonically interacting units considered here. It turned out that (i) for chain systems with more than one one degree-of-freedom per unit (e.g. taking into account also motions of the units along the direction transversal to the transition direction) and (ii) for interactions going beyond the linear (harmonic) one (e.g. Morse-type interaction potentials) as well as (iii) other on-site potentials (including biased two-well potentials, potentials being periodic in the direction transversal to the transition direction) the escape time in the microcanonical situation can be significantly shorter than for the corresponding system coupled to a heat bath.

Our study demonstrates the enormous capabilities of nonlinear systems to self-promote their functional processes. Particularly, the ability to manage efficiently (coherent escape) despite being initialized in no ideal condition (far too low energy density compared to the barrier height) distinguishes such systems.

6 Acknowledgments

This research has been supported by SFB-555 (L. Sch.-G, S. F.) and, as well, by the joint Volkswagen Foundation projects I/80424 (P. H.) and I/80425 (L. Sch.-G.).

References

1. P. Hänggi, P. Talkner, M. Borkovec: Rev. Mod. Phys. **62**, 251 (1990)
2. J.S. Langer: Ann. Phys. (N.Y.) **54**, 258 (1969)
3. P.J. Park, W. Sung: J. Chem. Phys. **111**, 5259 (1999)
4. K.L. Sebastian, A.K.R. Paul: Phys. Rev. E **62**, 927 (2000)
5. S. Lee, W. Sung: Phys. Rev. E **63**, 021115 (2001)

6. K. Lee, W. Sung: Phys. Rev. E **64**, 041801 (2001)
7. W. Forst: *Theory of Unimolecular Reaction* (Academic Press, New York, 1971)
8. M. Remoissenet: *Waves Called Solitons* (Springer-Verlag, Berlin, 1978)
9. R.S. MacKay, S. Aubry: Nonlinearity **7**, 1623 (1994)
10. S. Aubry: Physica D **103**, 201 (1997)
11. S. Flach, C.R. Willis: Phys. Rep. **295**. 181 (1998)
12. P. Marquié, J.M. Bilbault, M. Remoissenet: Phys. Rev. E **51**, 6127 (1995)
13. H.S. Eisenberg, Y. Silberberg, R. Morandotti, A.R. Boyd, J.S. Aitchison: Phys. Rev. Lett. **81**, 3383 (1998)
14. P. Binder, D. Abraimov, A.V. Ustinov, S. Flach, Y. Zolotaryuk: Phys. Rev. Lett. **84**, 745 (2000)
15. M. Peyrard, A.R. Bishop: Phys. Rev. Lett. **62**, 2755 (1989)
16. T. Dauxois, M. Peyrard, A.R. Bishop: Phys. Rev. E **47**, 684 (1993)
17. L.V. Yakushevich: Quart. Rev. Biophys. **26**, 201 (1993); *Nonlinear Physics of DNA* (Wiley and Sons, 1998)
18. G. Gaeta, C. Reiss, M. Peyrard, T. Dauxois: Riv. Nuovo Cim. **17**, 1 (1994)
19. M. Barbi, S. Cocco, M. Peyrard: Phys. Lett. A **253**, 358 (1999)
20. S. Cocco, R. Monasson: Phys. Rev. Lett. **83**, 5178 (1999)
21. M. Barbi, S. Cocco, M. Peyrard, S. Ruffo: J. Biol. Phys. **24**, 97 (1999)
22. D. Hennig, L. Schimansky-Geier, P. Hänggi: Europhys. Lett. **78**, 20002 (2007)
23. D. Hennig, S. Fugmann, L. Schimansky-Geier, P. Hänggi: Phys. Rev. E **76**, 041110 (2007)
24. Yu.S. Kivshar, M. Peyrard: Phys. Rev. A **46**, 3198 (1992)
25. T. Dauxois, M. Peyard: Phys. Rev. Lett. **70**, 3935 (1993)
26. K.W. Sandusky, J.B. Page: Phys. Rev. B **50**, 866 (1994)
27. Yu.S. Kivshar: Phys. Rev. E **48**, 4132 (1993)
28. I. Daumont, T. Dauxois, M. Peyrard: Nonlinearity **10**, 617 (1997)
29. O. Bang, M. Peyrard: Phys. Rev. E **53**, 4143 (1996)
30. M. Peyrard: Physica D **119**, 184 (1998)
31. J.L. Marin, S. Aubry: Nonlinearity **9**, 1501 (1996)
32. T. Dauxois, S. Ruffo, A. Torcini: Phys. Rev. E **56**, R6229 (1997)
33. T. Cretegny, T. Dauxois, S. Ruffo, A. Torcini: Physica D **121**, 109 (1998)
34. D. Hennig, G.P. Tsironis: Phys. Rep. **307**, 333 (1999)

Liquid Polyamorphism and the Anomalous Behavior of Water

H. E. Stanley[1], S. V. Buldyrev[2], S.-H. Chen[3], G. Franzese[4], S. Han[1], P. Kumar[1], F. Mallamace[5], M. G. Mazza[1], L. Xu[1], and Z. Yan[1]

[1] Center for Polymer Studies and Department of Physics Boston University, Boston, MA 02215 USA
[2] Department of Physics, Yeshiva University, 500 West 185th St., New York, NY 10033 USA
[3] Nuclear Science and Engineering Department, Mass. Inst. of Technology, Cambridge, MA 02139 USA
[4] Departament de Física Fonamental, Univ. de Barcelona, Diagonal 647, Barcelona 08028, SPAIN
[5] Dipartimento di Fisica, Univ. Messina, Vill. S. Agata, C.P. 55, I-98166, Messina ITALY

Abstract. We present evidence from experiments and computer simulations supporting the hypothesis that water displays polyamorphism, i.e., water separates into two distinct liquid phases. This concept of a new liquid–liquid phase transition is finding potential application to other liquids as well as water, such as silicon and silica. Here we review the relation between changes in dynamic and thermodynamic anomalies arising from the presence of the liquid–liquid critical point in (i) Two models of water, TIP5P and ST2, which display a first order liquid–liquid phase transition at low temperatures; (ii) Two spherically symmetric two-scale potentials known to possess a liquid–liquid critical point, in which the competition between two liquid structures is generated by repulsive and attractive ramp interactions; and (iii) A Hamiltonian model of water where the idea of two length/energy scales is built in. This model also displays a first order liquid–liquid phase transition at low temperatures besides the first order liquid-gas phase transition at high temperatures. We find a correlation between the dynamic fragility crossover and the locus of specific heat maxima $C_P^{\max}$ ("Widom line") emanating from the critical point. Our findings are consistent with a possible relation between the previously hypothesized liquid-liquid phase transition and the transition in the dynamics recently observed in neutron scattering experiments on confined water. More generally, we argue that this connection between $C_P^{\max}$ and the dynamic crossover is not limited to the case of water, a hydrogen bonded network liquid, but is a more general feature of crossing the Widom line, an extension of the first-order coexistence line in the supercritical region.

1 Background

One "mysterious" property of liquid water was recognized 300 years ago [1]: although most liquids contract as temperature decreases, liquid bulk water begins to expand when its temperature drops below 4°C. Indeed, a simple kitchen experiment demonstrates that the bottom layer of a glass of unstirred iced water remains at 4°C while colder layers of 0°C water "float" on top (cf., Fig. 1 of Ref. [2]). The mysterious properties of liquid bulk water become more pronounced in the supercooled region below 0°C [3–5]. For example, if the coefficient of thermal expansion α_P, isothermal compressibility K_T, and constant-pressure specific heat C_P are extrapolated below the lowest temperatures measurable they would become infinite at a temperature of $T_s \approx 228$ K [3, 6].

Water is a liquid, but glassy water—also called amorphous ice—can exist when the temperature drops below the glass transition temperature T_g. Although it is a solid, its structure exhibits a disordered molecular liquid-like arrangement. *Low-density* amorphous ice (LDA) has been known for 60 years [7], and a second kind of amorphous ice, *high-density* amorphous ice (HDA) was discovered in 1984 [8, 9]. HDA has a structure similar to that of high-pressure liquid water, suggesting that HDA may be a glassy form of high-pressure water [10, 11], just as LDA may be a glassy form of low-pressure water. Water has at least two different amorphous solid forms, a phenomenon called *polyamorphism* [12], and recently additional forms of glassy water have been the focus of active experimental and computational investigation.

1.1 Current hypotheses

Many classic "explanations" for the mysterious behavior of liquid bulk water have been developed, including a simple two-state model dating back to Röntgen [13] and a clathrate model dating back to Pauling [14]. Two hypotheses are under current discussion:

(i) The *singularity-free hypothesis* [15], considers the possibility that the observed polyamorphic changes in water resemble a genuine transition, but is not. For example, if water is a locally-structured transient gel comprised of molecules held together by hydrogen bonds whose number increases as temperature decreases [16–18], then the local "patches" or bonded subdomains [19, 20] lead to enhanced fluctuations of specific volume and entropy and negative cross-correlations of volume and entropy whose anomalies closely match those observed experimentally.
(ii) The *liquid–liquid (LL) phase transition hypothesis* [21] arose from MD studies on the structure and equation of state of supercooled bulk water and has received considerable support for a variety of model systems [22–30]. Below the hypothesized *second* critical point the liquid

phase separates into two distinct liquid phases: a low-density liquid (LDL) phase at low pressures and a high-density liquid (HDL) at high pressure (Fig.1). Bulk water near the known critical point at 647 K is a fluctuating mixture of molecules whose *local* structures resemble the liquid and gas phases. Similarly, bulk water near the hypothesized LL critical point is a fluctuating mixture of molecules whose local structures resemble the two phases, LDL and HDL. These enhanced fluctuations influence the properties of liquid bulk water, thereby leading to anomalous behavior.

1.2 Selected experimental results

Many precise experiments have been performed to test the various hypotheses discussed in the previous section, but there is as yet no widespread agreement on which physical picture—if any—is correct. The connection between liquid water and the two amorphous ices predicted by the LL phase transition hypothesis is difficult to prove experimentally because supercooled water freezes spontaneously below the homogeneous nucleation temperature T_H, and amorphous ice crystallizes above the crystallization temperature T_X [31–33]. Crystallization makes experimentation on the supercooled liquid state between T_H and T_X almost impossible. However, comparing experimental data on amorphous ice at low temperatures with that of liquid water at higher temperatures allows an indirect discussion of the relationship between the liquid and amorphous states. It is found from neutron diffraction studies [11] and simulations that the structure of liquid water changes toward the LDA structure when the liquid is cooled at low pressures and changes toward the HDA structure when cooled at high pressures, which is consistent with the LL phase transition hypothesis [11]. The amorphous states (LDA and HDA) are presently considered to be "smoothly" connected thermodynamically to the liquid state if the entropies of the amorphous states are small [34, 35], and experimental results suggest that their entropies are indeed small [36].

In principle, it is possible to investigate experimentally the liquid state in the region between T_H and T_X during the extremely short time interval before the liquid freezes to crystalline ice [33, 37, 38]. Because high-temperature liquid bulk water becomes LDA without crystallization when it is cooled rapidly at one bar [39], LDA appears directly related to liquid water. A possible connection between liquid bulk water at high pressure and HDA can be seen when ice crystals are melted using pressure [37]. Other experimental results [33] on the high-pressure ices that might demonstrate a LL first-order transition in the region between T_H and T_X have been obtained.

1.3 Selected results from simulations

Water is challenging to simulate because it is a molecular liquid and there is presently no precise yet tractable intermolecular potential that is universally agreed on. Nevertheless there are some distinct advantages of simulations over experiments. Experiments cannot probe the "No-Man's land" that arises in bulk water from homogeneous nucleation phenomena, but simulations have the advantage that they can probe the structure and dynamics well below T_H since nucleation does not occur on the time scale of computer simulations. Of the two hypotheses above, the LL phase transition hypothesis is best supported by simulations, some using the ST2 potential which *exaggerates* the real properties of bulk water, and others using the SPC/E and TIP4P potentials which *underestimate* them [21, 40–44]. Recently, simulations have begun to appear using the "intermediate" TIP5P potential [45–47]. The precise location of the LL critical point is difficult to obtain since the continuation of the first order line is a locus of maximum compressibility [40, 41, 43].

Further, computer simulations may be used to probe the *local* structure of water. At low temperatures, many water molecules appear to possess one of two principal local structures, one resembling LDA and the other HDA [21, 40, 42, 48]. Experimental data can also be interpreted in terms of two distinct local structures [49–51].

2 Understanding "static heterogeneities"

The systems in which water is confined are diverse—including the rapidly-developing field of artificial "nanofluidic" systems (man-made devices of order of nanometer or less that convey fluids). Among the special reasons for our interest in confined water is that phenomena occurring at a given set of conditions in bulk water occur under perturbed conditions for confined water. For example, the coordinates of the hypothesized LL critical point lie in the experimentally-inaccessible No-Man's land of the bulk water phase diagram, but appear to lie in an accessible region of the phase diagrams of both two-dimensionally and one-dimensionally confined water [52, 53]. Simulations have been carried out to understand the effect of purely geometrical confinement [54–59] and of the interaction with hydrophilic [60–64] or hydrophobic [65–67, 132–134] surfaces. It is interesting also to study the effects that confinement may have on the phase transition properties of supercooled water [59], in order to clarify the possible presence of a LL phase transition in the water. Recent work on the phase behavior of confined water suggests a sensitive dependence on the interaction with the surfaces [67], as a LL phase transition appears to be consistent with simulations of water confined between two parallel flat hydrophobic walls [57]. Works are in progress to extend this work to hydrophilic pores, such as those in Vycor glasses or biological situations, and to hydrophobic hydrogels, systems of current experimental interest.

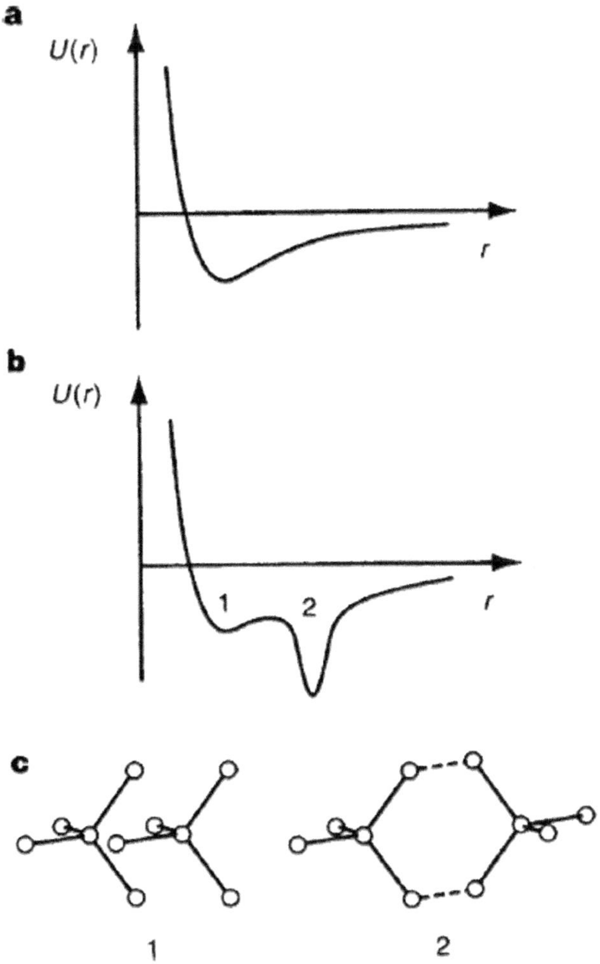

Fig. 1. (**a**) Idealized system characterized by a pair interaction potential with a single attractive well. At low enough T ($T < T_C$) and high enough P ($P > P_C$), the system condenses into the "liquid" well shown. (**b**) Idealized system characterized by a pair interaction potential whose attractive well has two sub-wells, the outer of which is deeper and narrower. For low enough T ($T < T_{c'}$) and low enough P ($P < P_{c'}$), the one-phase liquid can "condense" into the narrow outer "LDL" sub-well, thereby giving rise to a LDL phase, and leaving behind the high-density liquid phase occupying predominantly the inner subwell. (**c**) Two idealized interaction clusters of water molecules in configurations that may correspond to the two sub-wells of (**b**).

2.1 Potentials with two characteristic length scales: physical arguments

A critical point appears if the pair potential between two particles of the system exhibits a minimum, and Fig. 1a sketches the potential of such an idealized system. At high temperature, the system's kinetic energy is so large that the potential well does not have an effect, and the system is in a single "fluid" (or gas) phase. At low enough temperature ($T < T_C$) and large enough pressure ($P > P_C$) the fluid is sufficiently influenced by the minimum in the pair potential that it can condense into the low specific volume liquid phase. At lower pressure ($P < P_C$), the system explores the full range of distances—the large specific volume gas phase.

If the potential well has the form shown in Fig. 1b—the attractive potential well of Fig. 1a has now bifurcated into a deeper outer sub-well and a more shallow inner sub-well. Such a two-minimum ("two length scale") potential can give rise to the occurrence at low temperatures of a LL critical point at $(T_{C'}, P_{C'})$ [68]. At high temperature, the system's kinetic energy is so large that the two sub-wells have no appreciable effect on the thermodynamics and the liquid phase can sample both sub-wells. However, at low enough temperature ($T < T_{C'}$) and not too high a pressure ($P < P_{C'}$) the system must respect the depth of the outer sub-well so the liquid phase "condenses" into the outer sub-well (the LDL phase). At higher pressure it is forced into the shallower inner sub-well (the HDL phase).

The above arguments concern the average or "thermodynamic" properties, but they may also be useful in anticipating the local properties in the neighborhood of individual molecules [69]. Consider, again, an idealized fluid with a potential of the form of Fig. 1a and suppose that T is, say, 1.2 T_C so that the macroscopic liquid phase has not yet condensed out. Although the system is not entirely in the liquid state, small clusters of molecules begin to coalesce into the potential well, thereby changing their characteristic interparticle spacing (and hence their local specific volume) and their local entropy, so the fluid system will experience spatial fluctuations characteristic of the liquid phase even though this phase has not yet condensed out of the fluid at $T = 1.2\, T_C$. Specific volume fluctuations are measured by the isothermal compressibility and entropy fluctuations by the constant-pressure specific heat, so these two functions should start to increase from the values they would have if there were no potential well at all. As T decreases toward T_C, the magnitude of the fluctuations (and hence of the compressibility and the specific heat) increases monotonically and in fact diverges to infinity as $T \to T_C$. The cross-fluctuations of specific volume and entropy are proportional to the coefficient of thermal expansion, and this (positive) function should increase without limit as $T \to T_C$.

Consider an idealized fluid with a potential of the form of Fig. 1b, and suppose that T is now *below* T_C but is 20 percent *above* $T_{C'}$, so that the LDL phase has not yet condensed out. The liquid can nonetheless begin to sample the two sub-wells and clusters of molecules will begin to coalesce in each well, with the result that the liquid will experience spatial fluctuations characteristic of the LDL and HDL phase even though the liquid has not yet phase separated. The specific volume fluctuations and entropy fluctuations will increase, and so the isothermal compressibility K_T and constant-pressure specific heat C_P begin to diverge. Moreover, if the outer well is narrow, then when a cluster of neighboring particles samples the outer well it has a larger specific volume and a smaller entropy, so the anticorrelated cross-fluctuations of specific volume (the isothermal expansion coefficient α_P) is now *negative*, and approaching $-\infty$ as T decreases toward $T_{C'}$.

Now if by chance the value of $T_{C'}$ is lower than the value of T_H, then the phase separation discussed above would occur only at temperatures so low that the liquid would have frozen! In this case, the "hint" of the LL critical point C' is the presence of these local fluctuations whose magnitude would grow as T decreases, but which would never actually diverge if the point C' is never actually reached. Functions would be observed experimentally to increase as if they would diverge to ∞ or $-\infty$ but at a temperature below the range of experimental accessibility.

Now consider not the above simplified potential, but rather the complex (and unknown) potential between nonlinear water molecules. The tetrahedrality of water dictates that the outermost well corresponds to the ordered configuration with lower entropy. Thus although we do not know the actual form of the intermolecular potential in bulk water, it is not implausible that the same considerations apply as those discussed for the simplified potential of Fig. 1b. Indeed, extensive studies of such pair potentials have been carried out recently and the existence of the LL critical point has been demonstrated in such models.

To make more concrete how plausible it is to obtain a bifurcated potential well of the form of Fig. 1b, consider that one can crudely approximate water as a collection of 5-molecule groups called Walrafen pentamers (Fig. 1c) [50]. The interaction strengths of two adjacent Walrafen pentamers depends on their relative orientations. The first and the second energy minima of Fig. 1b correspond to the two configurations of adjacent Walrafen pentamers with different mutual orientations (Fig. 1c).

The two local configurations—#1 and #2 in Fig. 1c—are (i) a high-energy, low specific volume, high-entropy, non-bonded #1-state, or (ii) a low-energy, high specific volume, low-entropy, bonded #2-state. The difference in local structure is also the difference in the local structure between a high-pressure crystalline ice (such as ice VI or ice VII) and a low-pressure crystalline ice (such as ice I_h) (Fig. 1c).

The region of the P-T plane along the line continuing from the LDL-HDL coexistence line extrapolated to higher temperatures above the second critical point is the locus of points where the LDL on the low-pressure side and the HDL on the high-pressure side are continuously transforming—it is called the Widom line, defined to be the locus of points where the correlation length is maximum. Near this line, two different kinds of local structures, having either LDL or HDL properties, "coexist" [48, 70, 71]. The entropy fluctuations are largest near the Widom line, so C_P increases to a maximum, displaying a λ-like appearance [72]. The increase in C_P [35] resembles the signature of a glass transition as suggested by mode-coupling theory [73–75]. Careful measurements and simulations of static and dynamic correlation functions [70, 76–79] may be useful in determining the exact nature of the apparent singular behavior near 220 K.

2.2 Potentials with two characteristic length scales: tractable models

The above discussion is consistent with the possible existence of two well-defined classes of liquids: simple and water-like. The former interact via spherically-symmetric non-softened potentials, do not exhibit thermodynamic nor dynamic anomalies. We can calculate translational and orientational order parameters (t and q), and project equilibrium state points onto the (t, q) plane thereby generating what is termed the Errington-Debenedetti (ED) order map [20, 80]. In water-like liquids, interactions are orientation-dependent; these liquids exhibit dynamic and thermodynamic anomalies, and their ED "order map" is in general two-dimensional but becomes linear (or quasi-linear) when the liquid exhibits structural, dynamic or thermodynamic anomalies.

Hemmer and Stell [81] showed that in fluids interacting via pairwise-additive, spherically-symmetric potentials consisting of a hard core plus an attractive tail, softening of the repulsive core can produce additional phase transitions. This pioneering study elicited a considerable body of work on so-called core-softened potentials which can generate water-like density and diffusion anomalies [81–97]. This important finding implies that strong orientational interactions, such as those that exist in water and silica, are not a necessary condition for a liquid to have thermodynamic and dynamic anomalies.

Scattering experiments for materials such as Cs and Ce give rise to effective "core-softened" pair potentials [4, 68, 84, 98, 99]. Theoretical works in 1D and 2D suggests a LL phase transition and anomalous behavior [84, 86]. In 3D we showed that a squared potential with a repulsive shoulder and an attractive well displays a phase diagram with a LL critical point and no density anomaly [87–90]. The continuous version of the same shouldered attractive potential shows not only the LL critical point, but also water like anomalies [91, 97]. Soft-core potentials show a relationship between configurational entropy S_{conf} and diffusion coefficient D similar to what has been found in SPC/E water potential [100], suggesting that the maximum of S_{conf} tracks the density maxima line.

Two questions arise naturally from this emerging taxonomy of liquid behavior. First, is structural order in core-softened fluids hard-sphere or water-like? Second, is it possible to seamlessly connect the range of liquid behavior from hard spheres to water-like by a simple and common potential, simply by changing a physical parameter?

Yan et al. [101–103] used a simple spherically-symmetric "hard-core plus ramp" potential to address the first question. They found that this core-softened potential with two characteristic length scales not only gives rise to water-like diffusive and density anomalies, but also to an ED water-like order map, implying that orientational interactions are not necessary in order for a liquid to have structural anomalies. The ED order map evolves from water-like to hard-sphere-like upon varying the ratio λ of hard to soft-core diameters between 4/7 and 6/7, traversing the range of liquid behavior encompassed by hard spheres ($\lambda = 1$) and water-like ($\lambda \sim 4/7$).

3 Understanding "dynamic heterogeneities"

3.1 Recent experiments on confined water

Simulations and experiments both are consistent with the possibility that the LL critical point, if it exists at all, lies in the experimentally inaccessible No-Man's land. If this statement is valid, then at least two reactions are possible:

(i) If something is not experimentally accessible, then it does not deserve discussion.
(ii) If something is not experimentally accessible, but its influence *is* experimentally accessible, then discussion is warranted.

Option (ii) has guided most research thus far, since the manifestations of a critical point extend far away from the actual coordinates of that point. Indeed, accepting option (i) means there is nothing more to discuss. However if we confine water, the homogeneous nucleation temperature decreases and it becomes possible to enter the No-Man's land and, hence, search for the LL critical point. In fact, recent experiments at MIT and Messina by the Chen and Mallamace groups demonstrate that for nanopores of typically 1.5 nm diameter, the No-Man's land actually ceases to exist—one can supercool the liquid state all the way down to the glass temperature. Hence studying confined water offers the opportunity of directly testing, for the first time, the LL phase transition hypothesis.

In fact, using two independent techniques, neutron scattering and NMR, the MIT and Messina groups found a sharp kink in the dynamic properties (a "dynamic crossover") at the same temperature $T_L \approx 225$K [53, 104]. Our calculations on *bulk* models [105] are not inconsistent with one tentative interpretation of this dynamic crossover as resulting from the system passing from the high-temperature high-pressure "HDL" side of the Widom line (where the liquid might display fragile behavior) to the low-temperature low-pressure "LDL" side of the Widom line (where the liquid might display strong behavior).

The interpretation of the dynamic crossover could have implications for nanofluidics and perhaps even for natural confined water systems, e.g., some proteins appear to undergo a change in their flexibility at approximately the same temperature T_L that the MIT-Messina experiments identify for the dynamic crossover in pure confined water.

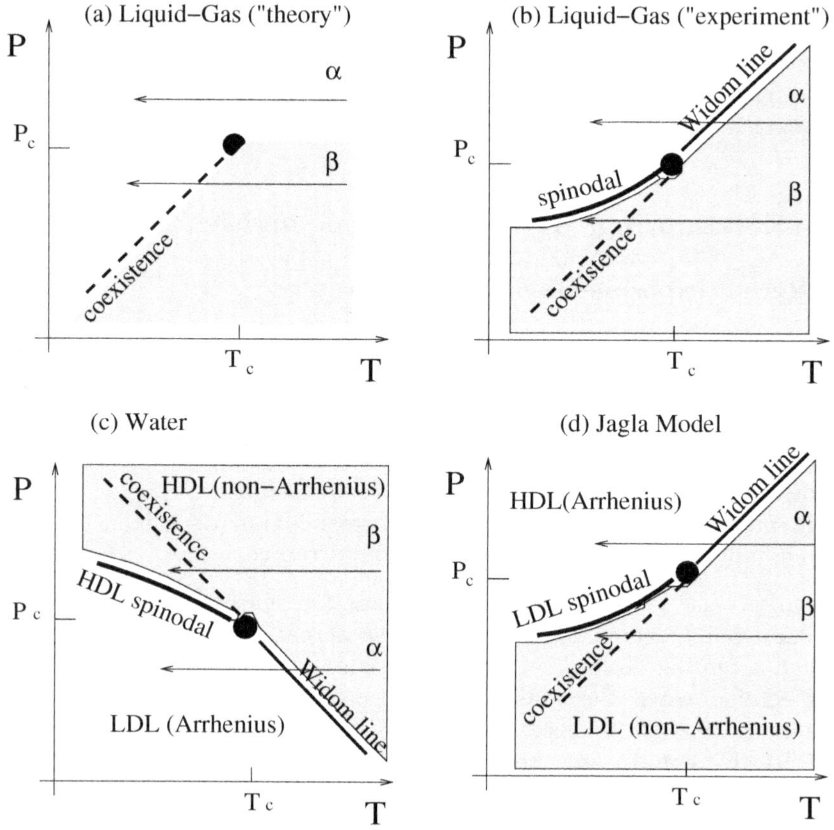

Fig. 2. (**a**) Schematic phase diagram for the critical region associated with a liquid-gas critical point. Two features display mathematical singularities: the critical point and the liquid-gas coexistence. (**b**) Same, with the addition of the gas-liquid spinodal and the Widom line. Along the Widom line, thermodynamic response functions have extrema in their T dependence. (**c**) A hypothetical phase diagram for water of possible relevance to recent confined water neutron scattering experiments [53, 104, 106, 107]. (**d**) A sketch of the $P-T$ phase diagram for the two-scale Jagla model.

3.2 Possible significance of the Widom line

The conjectured interpretation of the MIT-Messina experiments relies on the concept of the Widom line, a concept not widely appreciated even though it has been known by experimentalists dating back to the 1958 Ph.D. thesis of J. M. H. Levelt (now Levelt-Sengers). Since a Widom line arises only from a critical point, if the MIT-Messina experiments can be rationalized by a Widom line then they are consistent with the existence of a LL critical point in confined water.

By definition, in a first order phase transition, thermodynamic functions discontinuously change as we cool the system along a path crossing the equilibrium coexistence line [Fig. 2(a), path β]. However in a *real* experiment, this discontinuous change may not occur at the coexistence line since a substance can remain in a supercooled metastable phase until a limit of stability (a spinodal) is reached [4] [Fig. 2(b), path β].

If the system is cooled isobarically along a path above the critical pressure P_c [Fig. 2(b), path α], the state functions continuously change from the values characteristic of a high temperature phase (gas) to those characteristic of a low temperature phase (liquid). The thermodynamic response functions which are the derivatives of the state functions with respect to temperature [e.g., C_P] have maxima at temperatures denoted $T_{\max}(P)$. Remarkably these maxima are still prominent far above the critical pressure [108], and the values of the response functions at $T_{\max}(P)$ (e.g., $C_P^{\max}$) diverge as the critical point is approached. The lines of the maxima for different response functions asymptotically approach one another as the critical point is approached, since all response functions become expressible in terms of the correlation length. This asymptotic line is sometimes called the Widom line, and is often regarded as an extension of the coexistence line into the "one-phase regime."

Suppose now that the system is cooled at constant pressure P_0. (i) If $P_0 > P_C$ ("path α"), experimentally-measured quantities will change dramatically but continuously in the vicinity of the Widom line (with huge fluctuations as measured by, e.g., C_P). (ii) If $P_0 < P_C$ ("path β"), experimentally-measured quantities will change discontinuously if the coexistence line is actually seen. However the coexistence line can be difficult to detect in a pure system due to metastability, and changes will occur only when the spinodal is approached where the gas phase is no longer stable.

In the case of water—the most important solvent for biological function [109]—a significant change in dynamical properties has been suggested to take place in deeply supercooled states [110–113]. Unlike other network forming materials [114], water behaves as a fragile liquid in the experimentally accessible window [111, 115, 116]. Based on analogies with other network forming liquids and with the thermodynamic properties of the amorphous forms of water, it has been suggested that, at ambient pressure, liquid water should show a crossover between fragile behavior at high T to strong behavior at low T [82, 83, 112, 117, 118] in the deep supercooled region of the phase diagram below the homogeneous nucleation line. This region may contain the hypothesized LL critical point [21], the terminal point of a line of first order LL phase transitions. Recently, dynamic crossovers in confined water were studied experimentally [53, 59, 106, 119] since nucleation can be avoided in confined geometries. Also, a dynamic crossover has been associated with the LL phase transition in both silicon and silica [120, 121]. In the following, we offer a very tentative interpretation of the observed fragility transition in water as arising from crossing the Widom line emanating from the hypothesized LL critical point [121] [Fig. 2, path α].

4 Hamiltonian model of water

In Ref. [122], we investigated the generality of the dynamic crossover in a Hamiltonian model of water which displays a liquid–liquid phase transition at low temperatures. We consider a cell model that reproduces the fluid phase diagram of water and other tetrahedral network forming liquids [27–30]. The model includes a van der Waals attractive interaction with a characteristic energy ϵ, a hydrogen bond formation term with a characteristic energy $J < \epsilon$, a intramolecular interaction driving three nearest neighbor molecules toward a energetically favorable tetrahedral configuration with characteristic energy $J_\sigma < J$. The formation of a hydrogen bond induces a local increase of the volume occupied by the molecule. For $J_\sigma > 0$ the model reproduces the LL phase transition scenariol; for $J_\sigma = 0$ it reproduces the singularity-free scenario.

We find that different response functions such as C_P, α_p show maxima and these maxima increase and seem to diverge as the critical pressure is approached, consistent with the picture of Widom line. Moreover we find that the temperature derivative of the number of hydrogen bonds dN_{HB}/dT displays a maximum in the same region where the other thermodynamic response functions have maxima; suggesting that the fluctuations in the number of hydrogen bonds is maximum at the Widom line temperature T_W.

To further test if this model system also displays a dynamic crossover as found in the other models of water, we study the total spin relaxation time of the system as a function of T for different pressures. We find that for $J_\sigma/\epsilon = 0.05$ (*liquid–liquid critical point scenario*) the crossover occurs at the Widom line $T_W(P)$ for $P < P_{C'}$. For completeness we study the system also in the case of *singularity free scenario*, corresponding to $J_\sigma = 0$. For $J_\sigma = 0$ the crossover is at $T(C_P^{\max})$, the temperature of $C_P^{\max}$.

We next calculate the Arrhenius activation energy $E_{\rm A}(P)$ from the low-T slope of $\log \tau$ vs. $1/T$. We extrapolate the temperature $T_{\rm A}(P)$ at which τ reaches a fixed macroscopic time $\tau_{\rm A} \geq \tau_{\rm C}$. We choose $\tau_{\rm A} = 10^{14}$ MC steps > 100 sec [75] We find that $E_{\rm A}(P)$ and $T_{\rm A}(P)$ decrease upon increasing P in both scenarios, providing no distinction between the two interpretations. Instead, we find a dramatic difference in the P dependence of the quantity

$E_A/(k_B T_A)$ in the two scenarios, increasing for the LL critical point scenario and approximately constant for the singularity free scenario.

5 Outlook

It is possible that other phenomena that appear to occur on crossing the Widom line are in fact not coincidences, but are related to the changes in local structure that occur when the system changes from the "HDL-like" side to the "LDL-like" side. In this work we concentrated on reviewing the evidence for changes in dynamic transport properties, such as diffusion constant and relaxation time. Additional examples include: (1) a breakdown of the Stokes-Einstein relation for $T < T_W(P)$ [123–128], (2) systematic changes in the static structure factor $S(q)$ and the corresponding pair correlation function $g(r)$ revealing that for $T < T_W(P)$ the system more resembles the structure of LDL than HDL, (3) appearance for $T < T_W(P)$ of a shoulder in the dynamic structure factor $S(q, \omega)$ at a frequency $\omega \approx 60$ cm^{-1} ≈ 2 THz [129–131], (4) rapid increase in hydrogen bonding degree for $T < T_W(P)$ [122], (5) a *minimum* in the density at low temperature [135], and (6) a scaled equation of state near the critical point [136]. It is important to know how general a given phenomenon is, such as crossing the Widom line which by definition is present whenever there is a critical point. Using data on other liquids which have local tetrahedral symmetry, such as silicon and silica, which appear to also display a liquid-liquid critical point and hence must possess a Widom line emanating from this point into the one-phase region. For example, we learned of interesting new work on silicon, which also interprets structural changes as arising from crossing the Widom line of silicon [137]. It might be interesting to test the effect of the Widom line on simple model systems that display a liquid-liquid critical point, such as two-scale symmetric potentials of the sort recently studied by Xu and her collaborators [138] or by Franzese [91] and Barros de Oliveira and coworkers [97].

Very recently, Mallamace and his collaborators succeeded in locating the Widom line by finding a clearcut maximum in the coefficient of thermal expansion, at $T_W \approx 225$ K [139–141], which remarkably is the same temperature as the specific heat maximum [142]. Also, private discussions with Jacob Klein reveal a possible reason for why confined water does not freeze at –38°C, the bulk homogeneous nucleation temperature: Klein and co-workers [143] noted that confined water behaves differently than typical liquids in that water does not experience the huge increase in viscosity characteristic of other strongly confined liquids. They interpret this experimental finding as arising from the fact that strong confinement hampers the formation of a hydrogen bonded network, and we know from classic work of Linus Pauling that without the extensive hydrogen bonded network, water's freezing temperature will be depressed by more than 100°. Thus confinement reduces the extent of the hydrogen bonded network and hence lowers the freezing temperature, but leaves the key tetrahedral local geometry of the water molecule itself unchanged.

References

1. R. Waller, trans., *Essayes of Natural Experiments* [original in Italian by the Secretary of the Academie del Cimento]. Facsimile of 1684 English translation (Johnson Reprint Corporation, New York, 1964).
2. H. E. Stanley, [edited transcript of Turnbull Prize lecture], *Materials Research Bulletin* **24**(5), 22–30 (May 1999).
3. C. A. Angell, M. Oguni, and W. J. Sichina, *J. Phys. Chem.* **86**, 998–1002 (1982).
4. P. G. Debenedetti and H. E. Stanley, *Phys. Today* **56**(6), 40–46 (2003).
5. O. Mishima and H. E. Stanley, *Nature* **396**, 329–335 (1998).
6. R. J. Speedy and C. A. Angell, *J. Chem. Phys.* **65**, 851–858 (1976)
7. E. F. Burton and W. F. Oliver, "The Crystal Structure of Ice at Low Temperatures," *Proc. Roy. Soc. London Ser. A* **153**, 166–172 (1936).
8. O. Mishima, L. D. Calvert, and E. Whalley, *Nature* **310**, 393–395 (1984).
9. O. Mishima, L. D. Calvert, and E. Whalley, *Nature* **314**, 76–78 (1985).
10. M. C. Bellissent-Funel, L. Bosio, A. Hallbrucker, E. Mayer, and R. Sridi-Dorbez, *J. Chem. Phys.* **97**, 1282–1286 (1992).
11. M. C. Bellissent-Funel and L. Bosio, *J. Chem. Phys.* **102**, 3727–3735 (1995).
12. P. H. Poole, T. Grande, F. Sciortino, H. E. Stanley, and C. A. Angell, *J. Comp. Mat. Sci.* **4**, 373–382 (1995).
13. W. C. Röntgen, *Ann. d. Phys. u. Chem.* **45**, 91–97 (1892).
14. L. Pauling, in *Hydrogen Bonding*, edited by D. Hadzi (Pergamon Press, New York, 1959), pp. 1–5.
15. S. Sastry, P. Debenedetti, F. Sciortino, and H. E. Stanley, *Phys. Rev. E* **53**, 6144–6154 (1996).
16. H. E. Stanley, J. Teixeira, A. Geiger, and R. L. Blumberg, *Physica A* **106**, 260–277 (1981).
17. H. E. Stanley, *J. Phys. A* **12**, L329–L337 (1979).
18. H. E. Stanley and J. Teixeira, *J. Chem. Phys.* **73**, 3404–3422 (1980).
19. A. Geiger and H. E. Stanley, *Phys. Rev. Lett.* **49**, 1749–1752 (1982).
20. J. R. Errington, P. G. Debenedetti, and S. Torquato, *Phys. Rev. Lett.* **89**, 215503 (2002).
21. P. H. Poole, F. Sciortino, U. Essmann, and H. E. Stanley, *Nature* **360**, 324–328 (1992).
22. E. G. Ponyatovskii, V. V. Sinitsyn, and T. A. Pozdnyakova, *JETP Lett.* **60**, 360–364 (1994).
23. C. T. Moynihan, *Mat. Res. Soc. Symp. Proc.* **455**, 411–425 (1997).
24. P. H. Poole, F. Sciortino, T. Grande, H. E. Stanley, and C. A. Angell, *Phys. Rev. Lett.* **73**, 1632–1635 (1994).
25. S. S. Borick, P. G. Debenedetti, and S. Sastry, *J. Phys. Chem.* **99**, 3781–3793 (1995).
26. C. F. Tejero and M. Baus, *Phys. Rev. E* **57**, 4821–4823 (1998).
27. G. Franzese and H. E. Stanley, *Physica A* **314**, 508 (2002).
28. G. Franzese and H. E. Stanley, *J. Phys.: Cond. Mat.* **14**, 2193 (2002).

29. G. Franzese, M. I. Marqués, and H. E. Stanley, *Phys. Rev. E.* **67**, 011103 (2003).
30. G. Franzese and H. E. Stanley, *J. Phys.: Cond. Mat.* **19**, 205126 (2007).
31. H. Kanno, R. Speedy, and C. A. Angell, *Science* **189**, 880–881 (1975).
32. O. Mishima, *J. Chem. Phys.* **100**, 5910–5912 (1994).
33. O. Mishima and H. E. Stanley, *Nature* **392**, 164–168 (1998).
34. E. Whalley, D. D. Klug, and Y. P. Handa, *Nature* **342**, 782–783 (1989).
35. G. P. Johari, G. Fleissner, A. Hallbrucker, and E. Mayer, *J. Phys. Chem.* **98**, 4719–4725 (1994).
36. R. J. Speedy, P. G. Debenedetti, R. S. Smith, C. Huang, and B. D. Kay, *J. Chem. Phys.* **105**, 240–244 (1996).
37. O. Mishima, *Nature* **384**, 546–549 (1996).
38. L. S. Bartell and J. Huang, *J. Phys. Chem.* **98**, 7455–7457 (1994).
39. P. Brüggeller and E. Mayer, *Nature* **288**, 569–571 (1980).
40. P. H. Poole, U. Essmann, F. Sciortino, and H. E. Stanley, *Phys. Rev. E* **48**, 4605–4610 (1993).
41. H. Tanaka, *J. Chem. Phys.* **105**, 5099–5111 (1996).
42. S. Harrington, R. Zhang, P. H. Poole, F. Sciortino, and H. E. Stanley, *Phys. Rev. Lett.* **78**, 2409–2412 (1997).
43. F. Sciortino, P. H. Poole, U. Essmann, and H. E. Stanley, *Phys. Rev. E* **55**, 727–737 (1997).
44. S. Harrington, P. H. Poole, F. Sciortino, and H. E. Stanley, *J. Chem. Phys.* **107**, 7443–7450 (1997).
45. W. L. Jorgensen, J. Chandrasekhar, J. Madura, R. W. Impey, and M. Klein, *J. Chem. Phys.* **79**, 926 (1983).
46. M. Yamada, S. Mossa, H. E. Stanley, F. Sciortino, *Phys. Rev. Lett.* **88**, 195701 (2002).
47. D. Paschek, *Phys. Rev. Lett.* **94**, 217802 (2005).
48. E. Shiratani and M. Sasai, *J. Chem. Phys.* **108**, 3264–3276 (1998).
49. M.-C. Bellissent-Funel, *Europhys. Lett.* **42**, 161–166 (1998).
50. H. E. Stanley, S. V. Buldyrev, M. Canpolat, O. Mishima, M. R. Sadr-Lahijany, A. Scala, and F. W. Starr, *Phys. Chem. Chem. Phys. (PCCP)* **2**, 1551–1558 (2000).
51. A. K. Soper and M. A. Ricci, *Phys. Rev. Lett.* **84**, 2881 (2000) and references cited therein.
52. J.-M. Zanotti, M.-C. Bellissent-Funel, and S.-H. Chen, *Europhys. Lett.* **71**, 91–97 (2005).
53. L. Liu, S.-H. Chen, A. Faraone, C.-W. Yen, and C. Y. Mou, *Phys. Rev. Lett.* **95**, 117802 (2005).
54. M. E. Green and J. Lu, *J. Coll. Int. Sci.* **171**, 117–126 (1995).
55. K. Koga, X. C. Zeng, and H. Tanaka, *Phys. Rev. Lett.* **79**, 5262–5265 (1997).
56. J. Slovak, K. Koga, H. Tanaka, and X. C. Zeng, *Phys. Rev. E* **60**, 5833–5840 (1999).
57. K. Koga, X. C. Zeng, and H. Tanaka, *Chem. Phys. Lett.* **285**, 278–283 (1998).
58. K. Koga, H. Tanaka, and X. C. Zeng, *Nature.* **408**, 564 (2000).
59. R. Bergman and J. Swenson, *Nature* **403**, 283–286 (2000).
60. J. Teixeira, J. M. Zanotti, M.-C. Bellissent-Funel, and S. H. Chen, *Physica B* **234**, 370–374 (1997).
61. P. Gallo, *Phys. Chem. Phys.* **2**, 1607–1611 (2000).

62. P. Gallo, M. Rovere, M. A. Ricci, C. Hartnig, and E. Spohr, *Europhys. Lett.* **49**, 183–188 (2000).
63. P. Gallo, M. Rovere, M. A. Ricci, C. Hartnig, and E. Spohr, *Philos. Mag. B* **79**, 1923–1930 (1999).
64. M. Rovere, M. A. Ricci, D. Vellati, and F. Bruni, *J. Chem. Phys.* **108**, 9859–9867 (1998).
65. M.-C. Bellissent-Funel, R. Sridi-Dorbez, and L. Bosio, *J. Chem. Phys.* **104**, 10023–10029 (1996).
66. J. Forsman, B. Jonsson, and C. E. Woodward, *J. Phys. Chem.-US* **100**, 15005–15010 (1996).
67. M. Meyer and H. E. Stanley, *J. Phys. Chem. B* **103**, 9728–9730 (1999).
68. P. C. Hemmer and G. Stell, *Phys. Rev. Lett.* **24**, 1284–1287 (1970).
69. M. Canpolat, F. W. Starr, M. R. Sadr-Lahijany, A. Scala, O. Mishima, S. Havlin, and H. E. Stanley, *Chem. Phys. Lett.* **294**, 9–12 (1998).
70. E. Shiratani and M. Sasai, *J. Chem. Phys.* **104**, 7671–7680 (1996).
71. H. Tanaka, *Phys. Rev. Lett.* **80**, 113–116 (1998).
72. C. A. Angell, J. Shuppert, and J. C. Tucker, *J. Phys. Chem.* **77**, 3092–3099 (1973).
73. F. Sciortino, L. Fabbian, S.-H. Chen, and P. Tartaglia, *Phys. Rev. E* **56**, 5397–5404 (1997).
74. F. Sciortino, P. Gallo, P. Tartaglia, and S.-H. Chen, *Phys. Rev. E* **54**, 6331–6343 (1996).
75. P. Kumar, G. Franzese, S. V. Buldyrev, H. E. Stanley, *Phys. Rev. E* **73**, 041505 (2006).
76. Y. Xie, K. F. Ludwig, G. Morales, D. E. Hare, and C. M. Sorensen, *Phys. Rev. Lett.* **71**, 2051–2053 (1993).
77. F. Sciortino, P. Poole, H. E. Stanley, and S. Havlin, *Phys. Rev. Lett.* **64**, 1686–1689 (1990).
78. A. Luzar and D. Chandler, *Nature* **379**, 55–57 (1996); *Phys. Rev. Lett.* **76**, 928–931 (1996).
79. F. W. Starr, J. K. Nielsen, and H. E. Stanley, *Phys. Rev. Lett.* **82**, 2294–2297 (1999); *Phys. Rev. E* **62**, 579–587 (2000).
80. J. R. Errington and P. G. Debenedetti, *Nature* **409**, 318–321 (2001).
81. J. M. Kincaid, G. Stell, and C. K. Hall, *J. Chem. Phys.* **65**, 2161 (1976).
82. E. A. Jagla, *J. Phys. Cond. Mat.* **11**, 10251–10258 (1999).
83. E. A. Jagla, *Phys. Rev. E* **63**, 061509 (2001).
84. M. R. Sadr-Lahijany, A. Scala, S. V. Buldyrev, and H. E. Stanley, *Phys. Rev. Lett.* **81**, 4895–4898 (1998).
85. A. Scala, M. R. Sadr-Lahijany, N. Giovambattista, S. V. Buldyrev, and H. E. Stanley, *Phys. Rev. E* **63** 041202 (2001).
86. A. Scala, M. Reza Sadr-Lahijany, N. Giovambattista, S. V. Buldyrev, and H. E. Stanley, *J. Stat. Phys.* **100**, 97–106 (2000).
87. G. Franzese, G. Malescio, A. Skibinsky, S. V. Buldyrev, and H. E. Stanley, *Nature* **409**, 692–695 (2001).
88. G. Franzese, G. Malescio, A. Skibinsky, S. V. Buldyrev, and H. E. Stanley, *Phys. Rev. E* **66**, 051206 (2002).
89. A. Skibinsky, S. V. Buldyrev, G. Franzese, G. Malescio, and H. E. Stanley, *Phys. Rev. E* **69**, 061206 (2004).
90. G. Malescio, G. Franzese, A. Skibinsky, S. V. Buldyrev, and H. E. Stanley, *Phys. Rev. E* **71**, 061504 (2005).

91. G. Franzese, *J. Mol. Liq.* **136**, 267 (2007).
92. P. Kumar, S. V. Buldyrev, F. Sciortino, E. Zaccarelli, and H. E. Stanley, *Phys. Rev. E* **72**, 021501 (2005).
93. P. G. Debenedetti, V. S. Raghavan, and S. S. Borick, *J. Phys. Chem.* **95**, 4540–4551 (1991).
94. V. B. Henriques and M. C. Barbosa, *Phys. Rev. E* **71**, 031504 (2005).
95. B. Guillot and Y. Guissani, *J. Chem. Phys.* **119**, 11740–11752 (2003).
96. V. B. Henriques, N. Guisoni, M. A. Barbosa, M. Thielo, and M. C. Barbosa, *Molec. Phys.* **103**, 3001–3007 (2005).
97. A. Barros de Oliveira, G. Franzese, P. A. Netz, and M. C. Barbosa, *J. Chem. Phys.* **128**, 064901 (2008).
98. E. A. Jagla, *J. Chem. Phys.* **111**, 8980–8986 (1999).
99. T. H. Hall, L. Merril, and J. D. Barnett, *Science* **146**, 1297–1299 (1964).
100. A. Scala, F. W. Starr, E. La Nave, F. Sciortino, and H. E. Stanley, *Nature* **406**, 166–169 (2000).
101. Z. Yan, S. V. Buldyrev, N. Giovambattista, and H. E. Stanley, *Phys. Rev. Lett.* **95**, 130604 (2005).
102. Z. Yan, S. V. Buldyrev, N. Giovambattista, P. G. Debenedetti, and H. E. Stanley, *Phys. Rev. E* **73**, 051204 (2006).
103. Z. Yan, S. V. Buldyrev, P. Kumar, N. Giovambattista, P. G. Debenedetti, and H. E. Stanley, *Phys. Rev. E* **76**, 051201 (2007).
104. F. Mallamace, M. Broccio, C. Corsaro, A. Faraone, U. Wanderlingh, L. Liu, C.-Y. Mou, and S.-H. Chen, *J. Chem. Phys.* **124**, 161102 (2006).
105. L. Xu, P. Kumar, S. V. Buldyrev, S.-H. Chen, P. H. Poole, F. Sciortino, and H. E. Stanley, *Proc. Natl. Acad. Sci.* **102**, 16558–16562 (2005).
106. A. Faraone, L. Liu, C.-Y. Mou, C.-W. Yen, and S.-H. Chen, *J. Chem. Phys.* **121**, 10843–10846 (2004).
107. L. Liu, Ph.D. thesis, M.I.T., September 2005.
108. M. A. Anisimov, J. V. Sengers, and J. M. H. Levelt-Sengers, in *Aqueous System at Elevated Temperatures and Pressures: Physical Chemistry in Water, Stream and Hydrothermal Solutions*, edited by D. A. Palmer, R. Fernandez-Prini, and A. H. Harvey (Elsevier, Amsterdam, 2004).
109. M.-C. Bellissent-Funel, ed., *Hydration Processes in Biology: Theoretical and Experimental Approaches* [Proc. NATO Advanced Study Institutes, Vol. 305] (IOS Press, Amsterdam, 1999).
110. C. A. Angell, R. D. Bressel, M. Hemmatti, E. J. Sare, and J. C. Tucker, *Phys. Chem. Chem. Phys. (JCCP)* **2**, 1559–1566 (2000).
111. P. G. Debenedetti, *J. Phys.: Condens. Matter* **15**, R1669–R1726 (2003).
112. C. A. Angell, *J. Phys. Chem.* **97**, 6339–6341 (1993).
113. F. W. Starr, C. A. Angell, and H. E. Stanley, *Physica A* **323**, 51–66 (2003).
114. J. Horbach and W. Kob, *Phys. Rev. B* **60**, 3169–3181 (1999).
115. E. W. Lang and H. D. Lüdemann, *Angew Chem. Intl. Ed. Engl.* **21**, 315–329 (1982).
116. F. X. Prielmeier, E. W. Lang, R. J. Speedy, H. D. Lüdemann, *Phys. Rev. Lett.* **59**, 1128–1131 (1987) .
117. K. Ito, C. T. Moynihan, and C. A. Angell, *Nature* **398**, 492–495 (1999).
118. H. Tanaka, *J. Phys.: Condens. Matter* **15**, L703–L711 (2003).
119. J. Swenson, H. Jansson, W. S. Howells, and S. Longeville, *J. Chem. Phys.* **122**, 084505 (2005).

120. S. Sastry and C. A. Angell, *Nat. Mater.* **2**, 739–743 (2003).
121. I. Saika-Voivod, P. H. Poole, and F. Sciortino, *Nature* **412**, 514–517 (2001).
122. P. Kumar, G. Franzese, and H. E. Stanley, *Phys. Rev. Lett.* **100**, 084102 (2008).
123. L. Xu, F. Mallamace, Z. Yan, F. W. Starr, S. V. Buldyrev, and H. E. Stanley (preprint).
124. P. Kumar, S. V. Buldyrev, S. L. Becker, P. H. Poole, F. W. Starr, and H. E. Stanley, *Proc. Natl. Acad. Sci. USA* **104**, 9575–9579 (2007).
125. P. Kumar, S. V. Buldyrev, and H. E. Stanley (submitted).
126. S.-H. Chen, F. Mallamace, C.-Y. Mou, M. Broccio, C. Corsaro, and A. Faraone, *Proc. Nat. Acad. Sciences USA* **103**, 12974–12978 (2006)
127. M. G. Mazza, N. Giovambattista, F. W. Starr, and H. E. Stanley, *Phys. Rev. Lett.* **96**, 057803 (2006).
128. M. G. Mazza, N. Giovambattista, H. E. Stanley, and F. W. Starr, *Phys. Rev. E* **76**, 031202 (2007).
129. S.-H. Chen, L. Liu, E. Fratini, P. Baglioni, A. Faraone, and E. Mamontov, *Proc. Natl. Acad. Sci. USA* **103**, 9012 (2006).
130. F. Mallamace, S.-H. Chen, M. Broccio, C. Corsaro, V. Crupi, D. Majolino, V. Venuti, P. Baglioni, E. Fratini, C. Vannucci, and H. E. Stanley, *J. Chem. Phys.* **127**, 045104 (2007).
131. F. Mallamace, C. Branca, M. Broccio, C. Corsaro, N. Gonzalez-Segredo, H. E. Stanley, and S.-H. Chen, *Euro. Phys. J.* **161**, 19 (2008).
132. P. Kumar, S. V. Buldyrev, F. W. Starr, N. Giovambattista, and H. E. Stanley, *Phys. Rev. E* **72**, 051503 (2005).
133. P. Kumar, F. W. Starr, S. V. Buldyrev, and H. E. Stanley, *Phys. Rev. E* **75**, 011202 (2007).
134. S. Han, P. Kumar, and H. E. Stanley, *Phys. Rev. E* **77**, 030201(R) (2008).
135. D. Liu, Y. Zhang, C.-C. Chen, C.-Y. Mou, P. H. Poole, and S.-H. Chen, *Proc. Natl. Acad. Sci. USA* **104**, 9570–9574 (2007).
136. D. A. Fuentevilla and M. A. Anisimov, *Phys. Rev. Lett.* **97**, 195702 (2006).
137. T. Morishita, *Phys. Rev. Lett.* **97**, 165502 (2006).
138. L. Xu, S. V. Buldyrev, C. A. Angell, and H. E. Stanley, *Phys. Rev. E* **74**, 031108 (2006).
139. F. Mallamace, C. Branca, M. Broccio, C. Corsaro, C.-Y. Mou, and S.-H. Chen, *Proc. Natl. Acad. Sci. USA* **104**, 18387–18391 (2007).
140. F. Mallamace, C. Corsaro, M. Broccio, C. Branca, N. González-Segredo, J. Spooren, S.-H. Chen, and H. E. Stanley PNAS **104**, 18387 (2008).
141. S.-H. Chen, F. Mallamace, L. Liu, D. Z. Liu, X. Q. Chu, Y. Zhang, C. Kim, A. Faraone, C.-Y. Mou, E. Fratini, P. Baglioni, A. I. Kolesnikov, and V. Garcia-Sakai, in *Fifth International Workshop on Complex Systems, Sendai, Japan*, edited by M. Tokuyama, I. Oppenheim, and H. Nishiyama, *AIP Conf. Proc.* **982**, (2008).
142. S. Maruyama, K. Wakabayashi, and M. Oguni, *AIP Conf. Proc.* **708**, 675–676 (2004).
143. U. Raviv, P. Laurat, and J. Klein, *Nature* **413**, 51–54 (2001).

Part V

Dynamical Effects, Rectification and Nonlinearities

Terahertz Detection of Many-Body Signatures in Semiconductor Heterostructures

Sangam Chatterjee[1], Torben Grunwald[1], Stephan W. Koch[1], Galina Khitrova[2], Hyatt M. Gibbs[2], and Rudolf Hey[3]

[1] Faculty of Physics, Philipps-Universität Marburg, Renthof 5, D-35032 Marburg, Germany
sangam.chatterjee@physik.uni-marburg.de

[2] College of Optical Sciences, The University of Arizona, 1630 E. University Blvd., Tucson, AZ, 85721, USA

[3] Paul-Drude-Institut für Festkörperelektronik, Hausvogteiplatz 5-7, D-10117 Berlin, Germany

Abstract. This article reviews our THz-spectroscopy results on the intra-excitonic transitions of different III–V multi quantum well structures. Following a brief introduction to THz Time-Domain Spectroscopy a detailed description to the data analysis and extraction of refractive index and absorption is given. The behavior of the induced excitonic THz absorption in GaAs/(AlGa)As and (GaIn)As/GaAs multi quantum well structures is compared. Good agreement with previous experiments on the GaAs/(AlGa)As system [1, 2] is obtained. A distinctly different behaviour is observed for the (GaIn)As/GaAs structures.

1 Introduction

The fundamental many-body interactions of elementary excitations in semiconductors lead to a wide range of complex correlation effects. In particular, the Coulomb-interaction between electrical charges is responsible for characteristic quasi-particle states in matter such as excitons or plasmons [3, 4]. To investigate the many-body interactions and their dynamics, semiconductors feature ideal premises. There, electrons (e) can be excited electrically or optically from the highest filled valence band to the conduction band. The depleted states in the valence band are called defect electrons or holes (h). The oppositely charged electrons and holes can form bound hydrogen-like pairs (excitons) as a result of Coulomb interaction [5]. As a consequence of their small reduced electron and hole masses and the high dielectric constants of semiconductors, the excitonic binding energy as well as the energetic differences between different excitation states are on the order of several to tens of meV, which is several orders of magnitude below the semiconductor band gap energy.

In the past, near-visible optical experiments were conducted to obtain information about excitons or emission from the plasma. These experiments are mostly based on photoluminescence spectroscopy. The temporal build-up of an excitonic resonance was generally interpreted as creation [6–11] and its decay as annihilation [12, 13] processes of excitons. However, only excitons with very small center-of-mass momentum directly contribute to the luminiscence, due to the photon's small momentum. Further, the results are, in general, distorted by emission from the electron-hole plasma via the excitonic recombination channel [14]. To clearly identify excitonic populations some experiments use a combination of photoluminescence and absorption spectroscopy in connection with a detailed analysis based on a systematic microscopic theory [15, 16]. An even more direct approach may be gained by looking at atomic spectroscopy, where populations are measured by observing the absorption by means of intra-atomic transitions from one atomic energy state to another. Transferring this idea to the exciton problem, one has to measure intra-excitonic transitions, e.g. from the 1s- to the 2p-state, located in the THz frequency regime [1, 17, 18]. Those experiments provide information about excitons integrated over all $\boldsymbol{k}$-states, since they do not depend on the exciton's center-of-mass momentum.

2 Experimental detail

A powerful tool to study transitions between different many-body states is a combination of THz time-domain spectroscopy with the optical-pump THz-probe technique [19]. In our case, an optical pump pulse excites carriers some of which may exist in the form of excitons. The excited states are probed by a subsequent THz pulse [1]. A sketch of the experiment is shown in Fig. 1

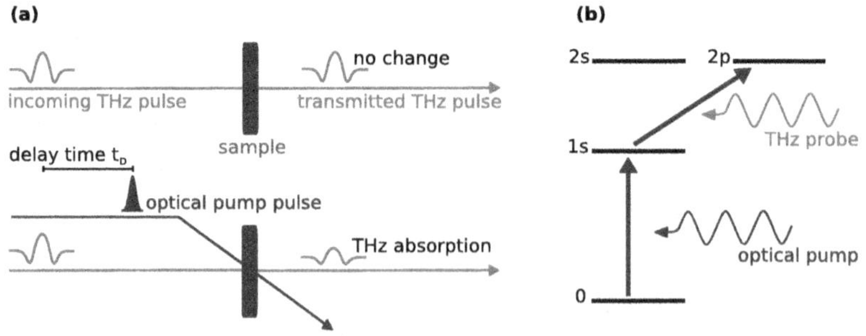

Fig. 1. (**a**) Illustration of optical-pump THz-probe technique. (**b**) Schematic energy diagram of optical-pump THz-probe for excitons.

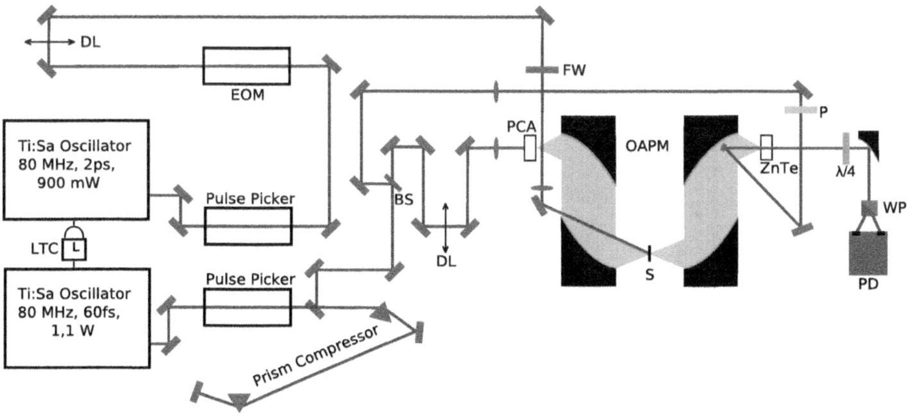

Fig. 2. Schematic drawing of the THz time-domain spectrometer used for this study. Abbreviations: "BS" = beam splitter, "DL" = delay line, "EOM" = electro optical modulator, "FW" = filter wheel, "LTC" = "Lok-To-Clock" repetition rate locking unit, "$\lambda/4$" = quarter wave plate, "OAPM" = off-axis parabolic mirrors, "P" = polarizer, "PCA" = photoconductive THz antenna, "PD" = photo diodes, "S" = sample in cryostat, "WP" = Wollaston prism, "ZnTe" = ZnTe detection crystal.

Tn adjustment of the variable time delay between optical-pump and THz-probe, t_D, allows for time-resolved measurements to investigate the exciton dynamics. A typical THz time-domain spectrometer, capable of using the mentioned pump-probe technique, is shown in Fig. 2.

A Ti:Sa-oscillator provides 60 fs pulses used for the generation of THz radiation via a photoconductive THz emitter and detection by using electro optical effects in a 1 mm thick (110)-oriented ZnTe crystal. A pair of off-axis parabolic mirrors focuses the generated THz beam onto the sample, located in a cryostat. Another pair of off-axis parabolic mirrors images the THz radiation onto the detection crystal. The signal is detected by electro-optical sampling. For that purpose parts of the fs oscillator laser beam are split-off and focused on the detection crystal. There, the incoming THz field induces a change of polarization of the optical laser beam, which, using a so-called optical bridge, is transformed into two beams of different intensity by a combination of a quarter-wave plate and Wollaston prism. These two beams are detected by a balanced pair of biased photo diodes. In this manner, information about the current THz field amplitude is transformed into a difference of currents. The respective voltage drop-off on a high-pass filter is measured by a lock-in amplifier. The modulation on the photoconductive THz emitter is used as a reference. To obtain the THz field amplitude for all times, a delay line is used to gradually change the delay time between THz pulse and optical detection pulse at the ZnTe detection crystal. This technique allows us to measure the THz field amplitude with a signal-to-noise ratio of three to four orders of

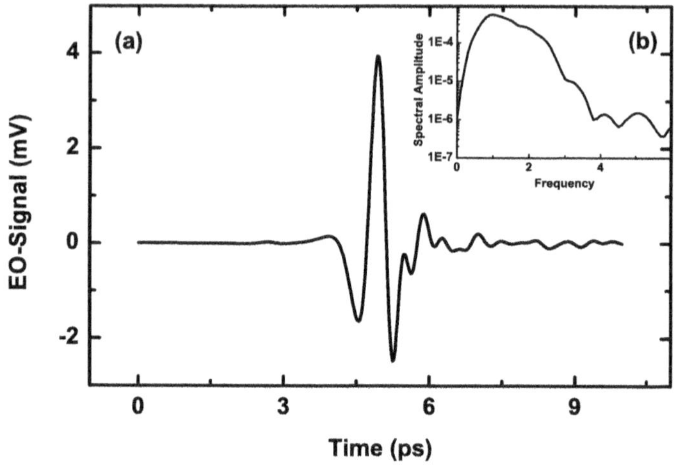

Fig. 3. Typical THz time trace (**a**) in time domain and its Fourier transformed spectral amplitude (**b**).

magnitude in time domain. The spectral information is obtained by Fourier transforming the time-domain data (Fig. 3 a,b).

The area around the off-axis parabolic mirrors, THz emitter and detector crystal is encapsulated and purged with dry nitrogen gas in order to eliminate spectral absorption lines caused by water vapor. A second Ti:Sa-oscillator provides short picosecond pulses to optically pump the sample and excite the carriers. The time delay between optical pump and THz probe pulse is adjusted by a second mechanical translation stage in the pump beam pathway. This line allows for 6 ns time delay corresponding to about 1 m travel distance. To ensure proper overlap between optical pump and THz probe beam, the pump beam is locked in space using piezoelectric actuators and a position sensitive detector. Both lasers are phase-locked by commercially available electronics. The phase jitter of the Spectra-Physics "Lok-To-Clock" unit is below 2 ps.

The complex refractive index or, alternatively, the real and imaginary part of the dielectric function can be calculated directly from a measured transient THz signal $E(t)$ and the pump-induced differential THz signal $\Delta E(t)$. A more detailed description of the data analysis is given in the following section.

3 Data analysis

Starting point for our data analysis are the measured signals $E(t)$ and $\Delta E(t)$ in the time domain. These signals are Fourier transformed in order to obtain the spectral information $E(\omega)$ and $\Delta E(\omega)$, respectively. Choosing the appropriate sampling step size Δt and time window for the measured data is crucial for good spectral results. The time window $n \cdot \Delta t$ should at least have a length corresponding to the smallest frequency that is to be measured, e.g., 10 ps to measure frequencies down to 0.1 THz. Most problems arise from the fact that in general the measured signal is not periodic in time, since only a certain time window is taken into account. To account for these features in the discrete Fourier transform algorithm one has to correct linear offsets and use window functions in time domain to smoothly fade in and out the measured information, prior to Fourier transforming. Additionally, the sampling step size Δt should be chosen small enough, so that the Nyquist-Frequency $\pi/\Delta t$ is much higher than the highest frequencies of the measured signal. To increase the density of data points in the spectral domain, one can apply zero-padding in the time domain. However, this does not increase the actual spectral resolution itself. Using zero-padding, one can only retrieve information that was not visible before, due to a too large spacing of data points in the spectral domain. True spectral resolution is only influenced by the length of the original measured time window $n \cdot \Delta t$ [20].

To calculate, e.g., the change in the complex refractive index $\Delta \widetilde{n}$, the following ansatz, similar to [21] is made: The transient signal, without pump, is described by

$$E(\omega) = E_0(\omega) t t' e^{i \widetilde{n} \omega d/c} ,$$

where $E_0(\omega)$ denotes the incident THz field on the sample, t,t' are the transmission coefficients, c is the speed of light, $\widetilde{n}$ the complex refractive index, and d the sample thickness. The THz transient $E_{\mathrm{P}}(\omega) = E(\omega) + \Delta E(\omega)$ through the pumped sample is then given by

$$E_{\mathrm{P}}(\omega) = E_0(\omega) t t' e^{i(\widetilde{n} + \Delta \widetilde{n}) \omega d/c} .$$

Dividing and taking the logarithm leads to

$$ln \left(\frac{E_{\mathrm{P}}(\omega)}{E(\omega)} \right) = i \, \Delta \widetilde{n} \frac{\omega d}{c} .$$

With $\Delta \widetilde{n} = \Delta n + i \, \Delta \kappa$ one obtains

$$\Delta n = \frac{c}{\omega \, d} (\Phi_{\mathrm{P}} - \Phi),$$

$$\Delta \kappa = - \frac{c}{\omega d} \, ln \left(\frac{|E_{\mathrm{P}}|}{|E|} \right) .$$

In the equations above, Φ_P and Φ are the spectral phases of $E_P = |E_P|e^{i\Phi_P}$ and $E = |E|e^{i\Phi}$, respectively. The differential absorption $\Delta\alpha L$ follows from $\Delta\kappa$ to

$$\Delta\alpha L = -2\, ln\left(\frac{|E_P|}{|E|}\right).$$

4 Intra-excitonic 1s-2p transition in GaAs/(AlGa)As quantum wells

The investigated sample is a $GaAs/Al_{0.35}Ga_{0.65}As$ multi quantum well structure. It consists of 20 GaAs layers of 15 nm thickness between 15 nm thick $Al_{0.35}Ga_{0.65}As$ barriers. The whole structure is sandwiched in between 100 nm GaAs layers. An AlAs etch stop is grown on top of the GaAs buffer and substrate. The structure is chemically etched and attached to a sapphire substrate using UV-cured cement. The polymer is transparent both in the visible and THz spectral regime. A typical absorption spectrum of the etched sample for a temperature of 5 K is shown in Fig. 4.

The sample is held at 5 K in a helium flow cryostat and is excited resonantly at the 1s-hh position, corresponding to a pump wavelength of 810.5 nm. We measure the differential absorption $\Delta\alpha L$ for different excitation densities

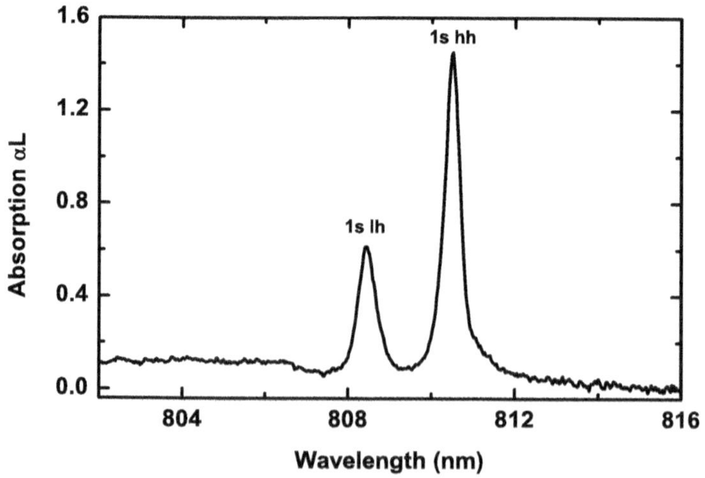

Fig. 4. Absorption spectrum of the 20-times $GaAs/Al_{0.35}Ga_{0.65}As$ multi quantum well structure.

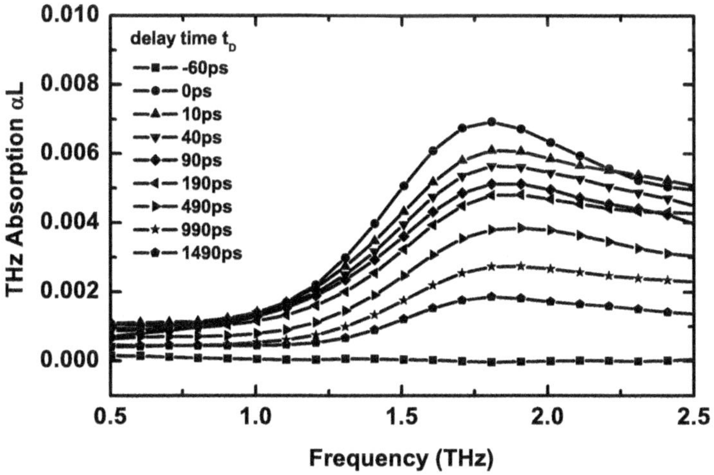

Fig. 5. Induced THz-absorption spectra of the $GaAs/Al_{0.35}Ga_{0.65}As$ quantum well sample for an optical excitation of $2 \cdot 10^{12}$ photons/cm^2 at the 1s resonance. The induced absorption signal subsequently decays over about 2 ns.

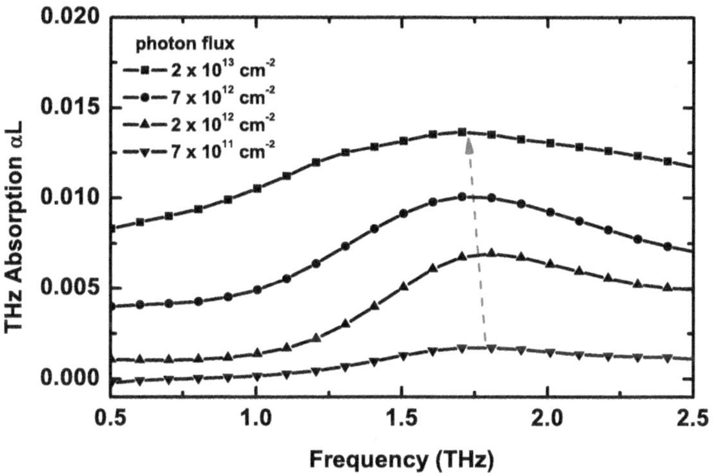

Fig. 6. Induced THz-absorption spectra of the $GaAs/Al_{0.35}Ga_{0.65}As$ quantum well sample for optical excitation at the 1s resonance. The spectra are taken shortly after the excitation pulse. The pump fluence is varied from ca $7 \cdot 10^{11}$ photons/cm^2 to $7 \cdot 10^{13}$ photons/cm^2. The *dashed arrow* illustrates the peak energy shifting.

and different pump-probe delay times. A typical result for a given excitation photon flux of $2 \cdot 10^{12}$ photons/cm^2 is given in Fig. 5.

One can nicely observe the rapid buildup of a intra-excitonic absorption during the excitation pulse width and its slow decay over an 800 ps time period. The asymmetric shape is due to contributions of transitions to higher bound as well as continuum states.

The density dependence is shown in Fig. 6. It can be seen, that the peaked resonance begins to bleach, while the whole absorption shows a strong broadening as the excitation density increases. Additionally, the absorption peak shifts to lower energies. This is in good agreement with previous experiments [1, 2].

5 Intra-excitonic 1s-2p transition in (GaIn)As/GaAs quantum wells

In order to test the material dependent aspects of the GaAs/(AlGa)As results, we now investigate a high quality (GaIn)As/GaAs multi quantum well structure. It consists of twenty quantum wells with an In-content of 6 % and a thickness of 8 nm each. The wells are separated by 130 nm thick GaAs barriers, the topmost one acting as cap layer. The whole structure is grown on a 0.5 mm thick GaAs substrate selected for high optical quality. The GaAs

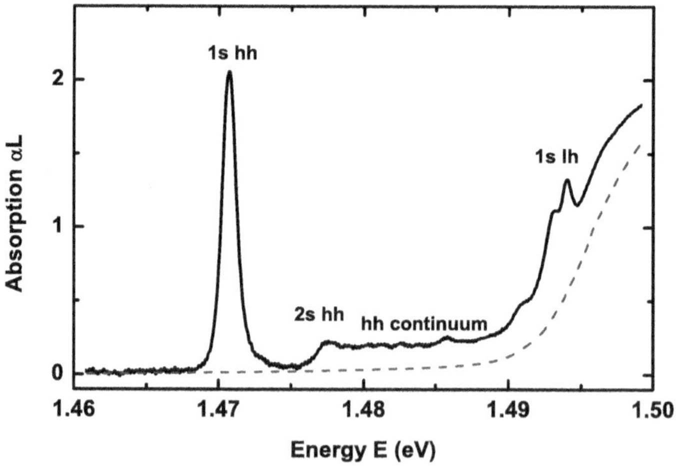

Fig. 7. Absorption spectrum of (GaIn)As/GaAs quantum well. The *dashed curve* shows the substrate absorption taken on the same wafer. The light hole resonance is shifted to higher energies due to strain in the quantum well.

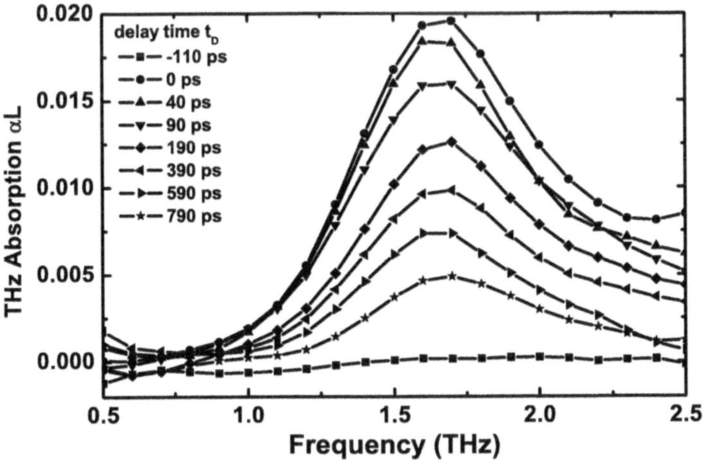

Fig. 8. Induced THz-absorption spectra of the (GaIn)As/GaAs quantum well sample for an optical excitation of $7 \cdot 10^{11}$ photons/cm^2 at the 1s resonance. The induced absorption signal subsequently decays over about 1 ns

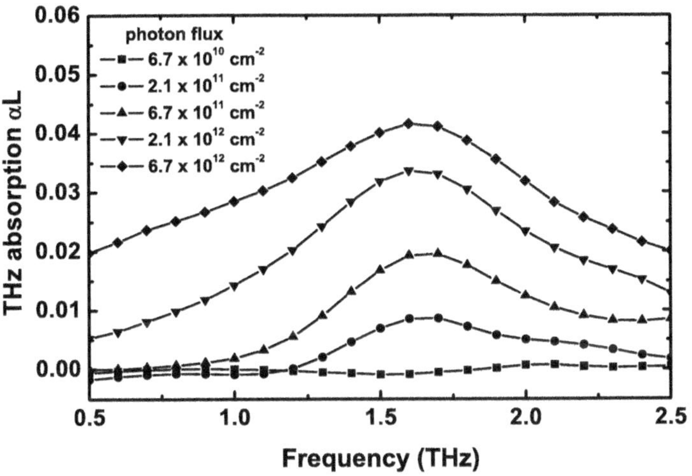

Fig. 9. Induced THz-absorption spectra of the (GaIn)As/GaAs quantum well sample for optical excitation at the 1s resonance. The spectra are taken shortly after the excitation pulse. The pump fluence is varied from ca $7 \cdot 10^{10}$ photons/cm^2 to $7 \cdot 10^{12}$ photons/cm^2.

substrate is transparent both in the visible to near infrared and in the investigated THz spectral regime and thus does not need to be removed. A typical absorption spectrum for low temperatures is shown in Fig. 7.

The experiments were done under low temperature conditions (5 K), as the GaAs/(AlGa)As sample described above. The sample is excited at the 1s-hh resonance at 1.471 eV, corresponding to a pump wavelength of 843 nm. The differential absorption $\Delta\alpha L$ is measured for different excitation densities and different pump-probe delay times. Figure 8 shows a typical result for a excitation photon flux of $7 \cdot 10^{11}$ photons/cm^2.

As for the investigated GaAs/(AlGa)As sample we observe an asymmetric absorption peak. However, the measured decay time in this case is considerably shorter than for the GaAs/(AlGa)As sample. The density dependence of the absorption signal is illustrated in Fig. 9.

It can be seen that the resonance mainly broadens. However, no peak shift is observed. Thus, the observed behavior is significantly different from the dynamics in GaAs/(AlGa)As quantum wells as described in the previous section and in [1, 2]. Further experiments, including the incorporation of optical pump and probe experiments under rigorously identical conditions are necessary to clarify the exact nature and origin of the differences in behavior.

6 Acknowledgements

The authors want to thank the German ministry of education and research (Bundesministerium für Bildnug und Forschung, BMBF) under contract number 13N9405 and the Optodynamics research center for financial support. Technical discussions with A. Leitensdorfer, M. Koch and E. Bründermann are gratefully acknowledged. T.G. and S.C. thank Wolfgang W. Rühle for continuous experimental support. The Tucson group thanks NSF for support.

References

1. R. A. Kaindl, M. A. Carnahan, D. Hägele, R. Lävenich, D. S. Chemla: Ultrafast terahertz probes of transient conducting and insulating phases in an electron-hole gas, Nature **423**, 734–738 (2003) doi:10.1038/nature01676
2. R. Huber, R. A. Kaindl, B. A. Schmid, D. S. Chemla: Broadband terahertz study of excitonic resonance in the high-density regime in GaAs/$Al_xGa_{1-x}As$ quantum wells, Phys. Rev. B **72**, R161314–R161318 (2005) doi:10.1103/PhysRevB.72.R161314
3. D. S. Chemla, J. Shah: Many-body and correlation effects in semiconductors, Nature **411**, 549–557 (2001) doi:dx.doi.org/10.1038/35079000
4. H. Haug, S. W. Koch: *Quantum Theory of the Optical and Electronic Properties of Semiconductors* (World Scientific Publishing, London, 2004)
5. R. J. Elliott: Intensity of optical absorption by excitons, Phys. Rev. **108**, 1384–1389 (1957) doi:10.1103/PhysRev.108.1384

6. J.-i. Kusano, Y. Segawa, Y. Aoyagi, S. Namba, H. Okamoto: Extremely slow energy relaxation of a two-dimensional exciton in a gaAs superlattice structure, Phys. Rev. B **40**, 1685–1691 (1989) doi:10.1103/PhysRevB.40.1685
7. T. C. Damen, J. Shah, D. Y. Oberli, D. S. Chemla, J. E. Cunningham, J. M. Kuo: Dynamics of exciton formation and relaxation in gaas quantum wells, Phys. Rev. B **42**, 7434–7438 (1990) doi:10.1103/PhysRevB.42.7434
8. R. Eccleston, R. Strobel, W. W. Rühle, J. Kuhl, B. F. Feuerbacher, K. Ploog: Exciton dynamics in a gaas quantum well, Phys. Rev. B **44**, 1395–1398 (1991) doi:10.1103/PhysRevB.44.1395
9. P. W. M. Blom, P. J. van Hall, C. Smit, J. P. Cuypers, J. H. Wolter: Selective exciton formation in thin GaAs/$Al_xGa_{1-x}As$ quantum wells, Phys. Rev. Lett. **71**, 3878–3881 (1993) doi:10.1103/PhysRevLett.71.3878
10. R. Kumar, A. S. Vengurlekar, S. S. Prabhu, J. Shah, L. N. Pfeiffer: Picosecond time evolution of free electron-hole pairs into excitons in gaas quantum wells, Phys. Rev. B **54**, 4891–4897 (1996) doi:10.1103/PhysRevB.54.4891
11. M. Gulia, F. Rossi, E. Molinari, P. E. Selbmann, P. Lugli: Phonon-assisted exciton formation and relaxation in GaAs/$Al_xGa_{1-x}As$ quantum wells, Phys. Rev. B **55**, R16049–R16052 (1997) doi:10.1103/PhysRevB.55.R16049
12. B. Deveaud, F. Clérot, N. Roy, K. Satzke, B. Sermage, D. S. Katzer: Enhanced radiative recombination of free excitons in gaas quantum wells, Phys. Rev. Lett. **67**, 2355–2358 (1991) doi:10.1103/PhysRevLett.67.2355
13. J. Feldmann, G. Peter, E. O. Göbel, P. Dawson, K. Moore, C. Foxon, R. J. Elliott: Linewidth dependence of radiative exciton lifetimes in quantum wells, Phys. Rev. Lett. **59**, 2337–2340 (1987) doi:10.1103/PhysRevLett.59.2337
14. M. Kira, F. Jahnke, S. W. Koch: Microscopic theory of excitonic signatures in semiconductor photoluminescence, Phys. Rev. Lett. **81**, 3263–3266 (1998) doi:10.1103/PhysRevLett.81.3263
15. S. Chatterjee, C. Ell, S. Mosor, G. Khitrova, H. M. Gibbs, W. Hoyer, M. Kira, S. W. Koch, J. P. Prineas, H. Stolz: Excitonic photoluminescence in semiconductor quantum wells: Plasma versus excitons, Phys. Rev. Lett. **92**, 067402 (2004) doi:10.1103/PhysRevLett.92.067402
16. W. Hoyer, C. Ell, M. Kira, S. W. Koch, S. Chatterjee, S. Mosor, G. Khitrova, H. M. Gibbs, H. Stolz: Many-body dynamics and exciton formation studied by time-resolved photoluminescence, Phys. Rev. B (Condensed Matter and Materials Physics) **72**, 075324 (2005) doi:10.1103/PhysRevB.72.075324 URL `http://link.aps.org/abstract/PRB/v72/e075324`
17. J. Černe, J. Kono, M. S. Sherwin, M. Sundaram, A. C. Gossard, G. E. W. Bauer: Terahertz dynamics of excitons in GaAs/AlGaAs quantum wells, Phys. Rev. Lett. **77**, 1131–1134 (1996) doi:10.1103/PhysRevLett.77.1131
18. I. Galbraith, R. Chari, S. Pellegrini, P. J. Phillips, C. J. Dent, A. F. G. van der Meer, D. G. Clarke, A. K. Kar, G. S. Buller, C. R. Pidgeon, B. N. Murdin, J. Allam, G. Strasser: Excitonic signatures in the photoluminescence and terahertz absorption of a GaAs/$Al_xGa_{1-x}As$ multiple quantum well, Phys. Rev. B (Condensed Matter and Materials Physics) **71**, 073302 (2005) doi: 10.1103/PhysRevB.71.073302 URL `http://link.aps.org/abstract/PRB/v71/e073302`
19. R. Huber, F. Tauser, A. Brodschelm, M. Bichler, G. Abstreiter, A. Leitenstorfer: Ultrafast terahertz probes of transient conducting and insulating phases in an electron-hole gas, Nature **414**, 286–289 (2001) doi:dx.doi.org/10.1038/35104522

20. T. Butz: *Fouriertransformation für Fußgänger* (B.G. Teubner, Stuttgart · Leipzig, 1998)
21. M. C. Nuss, J. Orenstein: *Millimeter-Wave Spectroscopy of Solids*, vol. 74, Springer Topics in Applied Physics (Springer, Berlin, 1998)

Theory of Ultrafast Dynamics of Electron-Phonon Interactions in Two Dimensional Electron Gases: Semiconductor Quantum Wells, Surfaces and Graphene

Marten Richter, Stefan Butscher, Norbert Bücking, Frank Milde, Carsten Weber, Peter Kratzer, Matthias Scheffler, and Andreas Knorr

[1] Institut für Theoretische Physik, Technische Universität Berlin, Hardenbergstr. 36, 10623 Berlin, Germany
marten.richter@tu-berlin.de
[2] Mathematical Physics, Lund University, Box 118, 22100 Lund, Sweden
[3] Fachbereich Physik, Universität Duisburg-Essen, Lotharstr. 1, 47048 Duisburg, Germany
[4] Fritz-Haber-Institut der Max-Planck-Gesellschaft, Faradayweg 4-6, 14195 Berlin, Germany

Abstract. Two-dimensional semiconductors are ideal model systems for investigating the dynamics of the electron-phonon interaction under spatial confinement of electronic excitations. In this contribution, the simultaneous quantum dynamics of electrons and phonons on ultrafast timescales is theoretically addressed. Typical examples include the ultrafast electron transfer at silicon surfaces, optical intersubband transitions in doped quantum wells, and non-equilibrium phonon generation in graphene.

1 Introduction

Two-dimensional electron gases appear in different semiconductor and semimetal nano-structures like quantum wells [1–3], graphene [4–6], and at surfaces [7, 8]. Recent experiments focus on the ultrafast dynamics of these 2D electron gases. A theoretical description of relaxation processes in such systems at low electronic densities and high temperature requires an understanding of how electrons interact with the vibrations of the lattice (phonons) [9, 10].

The main goal of this paper is to give a brief introduction to the theory of optical response for different electron gases and an overview of corresponding experimental investigations. The paper is organized as follows: First, the Hamilton operator is introduced and calculation techniques are presented. Second, the phonon-induced relaxation dynamics of electronic excitations at the silicon (001) 2×1 surface is described. Third and fourth, electron-phonon scattering processes in quantum cascade lasers and non-equilibrium phonons in graphene are described. Finally, the scattering-induced quantum emission of an equilibrium 2D electron gas is addressed.

2 Theoretical framework

2.1 Hamilton operator

In order to investigate the dynamics of a 2D electron gas, including many-particle effects, the Hamilton operator of the system has to be specified. Here, we focus on the interaction of electrons with phonons. The Hamilton operator for the free phonons and electron dynamics reads:

$$H_0 = \sum_{l\boldsymbol{k}} \varepsilon_{l\boldsymbol{k}} a^\dagger_{l\boldsymbol{k}} a_{\boldsymbol{k}} + \sum_{\boldsymbol{q}} \hbar\omega_{LO} b^\dagger_{\boldsymbol{q}} b_{\boldsymbol{q}}. \tag{1}$$

The operators $a^\dagger_{l\boldsymbol{k}}$, $a_{l\boldsymbol{k}}$ and $b^\dagger_{\boldsymbol{q}}$, $b_{\boldsymbol{q}}$ are the creation and annihilation operators for electrons and phonons. In most examples below, optical phonons with a phonon frequency ω_{LO} and quasi momentum $\boldsymbol{q}$ are considered, apart from Sect. 3, where acoustic phonons have been included as well. For the electronic states, the 2D momentum $\boldsymbol{k}$ and a (sub-)band index l are introduced as quantum numbers. The electron phonon interaction reads [10, 11]:

$$H_{\mathrm{EPI}} = \sum_{n\boldsymbol{k},n'\boldsymbol{k}',\boldsymbol{q}} g^{n'\boldsymbol{k}'}_{n\boldsymbol{k},\boldsymbol{q}} a^\dagger_{n\boldsymbol{k}} a_{n'\boldsymbol{k}'} b_{i\boldsymbol{q}} + h.a. \quad . \tag{2}$$

Here, g is the electron-phonon coupling element describing on the phonon-induced electronic transitions. To describe optical excitation, the interaction with the electrical field is included as well. As long as spontaneous emission can be neglected, a classical description of the electrical field through the vector potential $\boldsymbol{A}(t)$ within the dipole approximation is chosen [12]:

$$H_{\mathrm{ext-field}} = \sum_{\boldsymbol{k}n\neq n'} \mathbf{A}(t)\mathbf{p}^{n'\boldsymbol{k}}_{n\boldsymbol{k}}\, a^\dagger_{n\boldsymbol{k}} a_{n'\boldsymbol{k}}. \tag{3}$$

If spontaneous emission is considered, the interaction with the electromagnetic field has to be formulated using photon operators $c^\dagger, c$ [13]:

$$H_{qfield} = \sum_{\boldsymbol{k}\boldsymbol{k}'n\neq n'} \left(\mathbf{F}^{n'\boldsymbol{k}'}_{n\boldsymbol{k}Ks} c^\dagger_{Ks} + (\mathbf{F}^{n\boldsymbol{k}}_{n'\boldsymbol{k}'Ks})^* c_{Ks} \right) a^\dagger_{n\boldsymbol{k}} a_{n'\boldsymbol{k}}, \tag{4}$$

$$H_{0,phtn} = \hbar \sum_{Ks} \omega_{Ks} c^\dagger_{Ks} c_{Ks}. \tag{5}$$

Here, photon modes are characterized by their momentum indices K, polarization s and their frequency ω_{Ks}.

2.2 Dynamics

The electron dynamics of macroscopic quantities, such as the real space density, can be described via the electronic transitions $p_{n\boldsymbol{k}}^{m\mathbf{l}} := \langle a_{n\boldsymbol{k}}^{\dagger} a_{m\mathbf{l}} \rangle$, electronic (sub-)band densities $f_{n\boldsymbol{k}} := \langle a_{n\boldsymbol{k}}^{\dagger} a_{n\boldsymbol{k}} \rangle$, where $\langle \ldots \rangle$ denotes the quantum mechanical expectation value of the operator combination. The equations for these quantities are derived using the Heisenberg equation. As a typical example the equation for the polarization is presented [11]:

$$
\begin{aligned}
i\hbar \frac{d}{dt} p_{m\mathbf{l}}^{m'\mathbf{l}'} &= (\varepsilon_{m\mathbf{l}} - \varepsilon_{m'\mathbf{l}'}) p_{m\mathbf{l}}^{m'\mathbf{l}'} + \mathbf{A}(t) \sum_{n\mathbf{k}} \left(\mathbf{p}_{m\mathbf{l}}^{n\mathbf{k}} p_{n\mathbf{k}}^{m'\mathbf{l}'} - \mathbf{p}_{n\mathbf{k}}^{m'\mathbf{l}'} p_{m\mathbf{l}}^{n\mathbf{k}} \right) \\
&\quad + \sum_{n\mathbf{k}, i\mathbf{q}} \Bigg(g_{m\mathbf{l},i\mathbf{q}}^{n\mathbf{k}} \left(s_{m'\mathbf{l}',i\mathbf{q}}^{n\mathbf{k}} \right)^* + g_{m\mathbf{l},i\mathbf{q}}^{n\mathbf{k}} s_{a_{n\mathbf{k}},i\mathbf{q}}^{m'\mathbf{l}'} - g_{n\mathbf{k},i\mathbf{q}}^{m'\mathbf{l}'} \left(s_{n\mathbf{k},i\mathbf{q}}^{m\mathbf{l}} \right)^* \\
&\quad - g_{n\mathbf{k},i\mathbf{q}}^{m'\mathbf{l}'} s_{m\mathbf{l},i\mathbf{q}}^{n\mathbf{k}} \Bigg),
\end{aligned} \tag{6}
$$

where phonon assisted quantities $s_{m\mathbf{l},i\mathbf{q}}^{m'\mathbf{l}'} = \langle a_{m\mathbf{l}}^{\dagger} a_{m'\mathbf{l}'} b_{i\mathbf{q}} \rangle$ appear. It is obvious that Eq. (6) is not directly solvable, since it couples to phonon-assisted densities, whose dynamics have to be derived as well:

$$
\begin{aligned}
i\hbar \frac{d}{dt} \left(s_{m'\mathbf{l}' i\boldsymbol{q}}^{m\mathbf{l}} \right)^* &= \left(\varepsilon_{m\mathbf{l}} - \varepsilon_{m'\mathbf{l}'} + \hbar\omega_{i\mathbf{q}} \right) \left(s_{m'\mathbf{l}',i\boldsymbol{q}}^{m\mathbf{l}} \right)^* \\
&\quad + \mathbf{A}(t) \sum_{n\boldsymbol{k}} \left(\mathbf{p}_{m\mathbf{l}}^{n\mathbf{k}} \left(s_{m'\mathbf{l}',i\boldsymbol{q}}^{n\mathbf{k}} \right)^* - \mathbf{p}_{n\mathbf{k}}^{m'\mathbf{l}'} \left(s_{n\mathbf{k},i\boldsymbol{q}}^{m\mathbf{l}} \right)^* \right) \\
&\quad + \sum_{n\mathbf{k}, i'\mathbf{q}'} \Bigg(g_{m\mathbf{l},i'\mathbf{q}'}^{n\mathbf{k}} \left(R_{n\boldsymbol{k},i\boldsymbol{q}}^{m'\mathbf{l}',i'\boldsymbol{q}'} \right)^* + g_{m\mathbf{l},i'\mathbf{q}'}^{n\mathbf{k}} T_{n\boldsymbol{k},i\boldsymbol{q}}^{m'\mathbf{l}',i'\boldsymbol{q}'} \\
&\quad - g_{n\mathbf{k},i'\mathbf{q}'}^{m'\mathbf{l}'} \left(R_{m\mathbf{l},i\boldsymbol{q}}^{n\mathbf{k},i'\boldsymbol{q}'} \right)^* - g_{n\mathbf{k},i'\mathbf{q}'}^{m'\mathbf{l}'} T_{m\mathbf{l},i\boldsymbol{q}}^{n\mathbf{k}',i'\boldsymbol{q}'} \Bigg) \\
&\quad + \sum_{n\mathbf{k}, n'\mathbf{k}'} g_{n\mathbf{k},i\mathbf{q}}^{n'\mathbf{k}'} \langle a_{m\mathbf{l}}^{\dagger} a_{n\mathbf{k}}^{\dagger} a_{m'\mathbf{l}'} a_{n'\mathbf{k}'} \rangle.
\end{aligned} \tag{7}
$$

Within the spirit of a correlation expansion, $R_{n\boldsymbol{k},i\boldsymbol{q}}^{m\mathbf{l},i'\boldsymbol{q}'} = \langle a_{n\boldsymbol{k}}^{\dagger} a_{m\mathbf{l}} b_{i\boldsymbol{q}} b_{i'\boldsymbol{q}'} \rangle$ and $T_{n\boldsymbol{k},i\boldsymbol{q}}^{m\mathbf{l},i'\boldsymbol{q}'} = \langle a_{n\boldsymbol{k}}^{\dagger} a_{m\mathbf{l}} b^{\dagger}{}_{i\boldsymbol{q}} b_{i'\boldsymbol{q}'} \rangle$ are factorized at the second order Born level $(g_{n\mathbf{k},i\mathbf{q}}^{n'\mathbf{k}'})^2$ or coupled to higher-order correlations within a self-consistent Born approximation [14, 15]. In second-order Born approximation, we obtain e.g.: $\langle a_{n\boldsymbol{k}}^{\dagger} a_{m\mathbf{l}} b^{\dagger}{}_{i\boldsymbol{q}} b_{i'\boldsymbol{q}'} \rangle \approx \langle a_{n\boldsymbol{k}}^{\dagger} a_{m\mathbf{l}} \rangle \langle b^{\dagger}{}_{i\boldsymbol{q}} b_{i'\boldsymbol{q}'} \rangle$. Depending on the situation, occurring phonon occupations $n_{i\boldsymbol{q}} = \langle b_{i\boldsymbol{q}}^{\dagger} b_{i\boldsymbol{q}} \rangle$ might be treated within a bath approximation or calculated dynamically within the same second-order Markov Born approximation to consider non-equilibrium phonon distributions. Here, as example, the dynamic phonon density equation, later discussed for calculations non equilibrium calculations in graphene, is given [16]:

$$\dot{n}_{\mathbf{q}} = \frac{2\pi}{\hbar} \sum_{i,\mathbf{k}} |g_{\mathbf{k},\mathbf{q}}|^2 \delta(-\varepsilon_{i\mathbf{k}} + \varepsilon_{i\mathbf{k}+\mathbf{q}} - \hbar\omega_{\mathbf{q}}) \times [(n_{\mathbf{q}} + 1) f^{+}_{i\mathbf{k}+\mathbf{q}} f^{-}_{i\mathbf{k}} - n_{\mathbf{q}} f^{+}_{i\mathbf{k}} f^{-}_{i\mathbf{k}+\mathbf{q}}]. \tag{8}$$

We have introduced $f^{+}_{i\mathbf{k}} := f_{i\mathbf{k}}$ and $f^{-}_{i\mathbf{k}} := (1 - f_{i\mathbf{k}})$ and kept indices relevant for the graphene case. Within the same level of the Born-Markov approximation, the equations for the electron distributions read [10, 11]:

$$\frac{d}{dt} f_{n\boldsymbol{k}} = 2 \sum_{n'} \left(\Gamma^{n'\,in}_{n\boldsymbol{k}} (1 - f_{n\boldsymbol{k}}) - \Gamma^{n'\,out}_{n\boldsymbol{k}} f_{n\boldsymbol{k}} \right), \tag{9}$$

$$\Gamma^{n'\,in}_{n\boldsymbol{k}} = \frac{\pi}{\hbar} \sum_{\boldsymbol{k}' q \pm} \left| g^{n'\boldsymbol{k}'}_{n\boldsymbol{k},iq} \right|^2 \delta(\epsilon_{n\boldsymbol{k}} - \epsilon_{n'\boldsymbol{k}'} \pm \hbar\omega_{iq})(n_q + \frac{1}{2} \pm \frac{1}{2}) f_{n'\boldsymbol{k}'},$$

$$\Gamma^{n'\,out}_{n\boldsymbol{k}} = \frac{\pi}{\hbar} \sum_{\boldsymbol{k}' q \pm} \left| g^{n'\boldsymbol{k}'}_{n\boldsymbol{k},iq} \right|^2 \delta(\epsilon_{n\boldsymbol{k}} - \epsilon_{n'\boldsymbol{k}'} \pm \hbar\omega_{iq})(n_q + \frac{1}{2} \mp \frac{1}{2})(1 - f_{n'\boldsymbol{k}'}).$$

Equations (8–9) form a self-consistent set of equations. To describe the system accurately enough in a weak coupling regime, the used approximations might already be sufficient. For stronger coupling regimes, the influence of higher electron-phonon couplings have to be included at least approximatively (cf. [14, 15]).

3 Phonon-induced relaxation dynamics at the silicon (001) 2×1 surface

Due to the interest in device miniaturisation, electron relaxation effects at semiconductor surfaces play an increasing role in recent research. In contrast to embedded low-dimensional quantum systems (quantum dots, quantum wells), where the band structure and the matrix elements $g^{n'\boldsymbol{k}'}_{n\boldsymbol{k};iq}$ may be described a few parameters (effective masses, dielectric constant, etc. [17]), ab initio calculations are typically necessary to calculate the dynamics of a surface structure [18]: In our approach, density-functional theory (DFT) is used to obtain Kohn-Sham-orbitals to calculate the matrix elements, e.g. Eqs. (6), (2). The evaluation of the dynamics is performed in two steps: first, the DFT calculations are processed for a silicon (001) 2×1 surface and all interaction matrix elements $g^{n'\boldsymbol{k}'}_{n\boldsymbol{k};iq}$ and the relevant electronic bandstructure $\varepsilon_{l\boldsymbol{k}}$ are determined from the Kohn-Sham wavefunctions, and, in a second step, are inserted into the dynamical equations (9). Equation (9) describes the electron relaxation dynamics for fixed phonon distributions (bath approximation), after the injection of a non-equilibrium electron occupation via optical excitation [11].

The resulting dynamics of the phonon-induced relaxation of hot electrons within the conduction bands for the silicon (001) 2×1 surface [7, 8] is dominated by two timescales, cf. Fig. 1, which shows the partial relaxation from surface states to bulk states on short times and then back to surface states after longer times. The two timescales can be rationalized to different relaxation processes due to acoustic and optical phonons and specific features of the bandstructure of the Si (001) surface. In addition to the bulk bands, a surface-related conduction band (D^{down}-band) is partly located in the semiconductor band gap (and thus constitutes the conduction band minimum at the $\overline{\Gamma}$-point).

While the relaxation within the bulk and within the D^{down}-band is fast (1 ps), the interband relaxation from the bulk to the D^{down}-band is much slower [8, 11, 19]. Starting from an optical excitation process with a 1.69 eV pulse of 50 fs duration, the relaxation of optically injected electrons is rather complex: Initially, almost the whole population in the conduction bands is found in the D^{down}-band at an energy about 0.4 eV above the bulk conduction minimum. In Fig. 1, this can be seen by a real space population located near the surface after injection at $t = 0$ ps (lhs). At a later timestep (2 ps), the conduction band population distributes into the bulk: by phonon emission, a part of the population is transfered from the D^{down}-band to the bulk bands [15]. At the end of the relaxation, we find the entire population near the surface again. The final quasi equilibrium state is the total conduction band minimum (a D^{down}-state at the $\overline{\Gamma}$-point), and all relaxation channels lead to this final energetically lowest state, spatially located at the surface.

By combining the density matrix formalism with density-functional theory and applying it to a silicon 2×1 (001) surface structure, the timescale for the

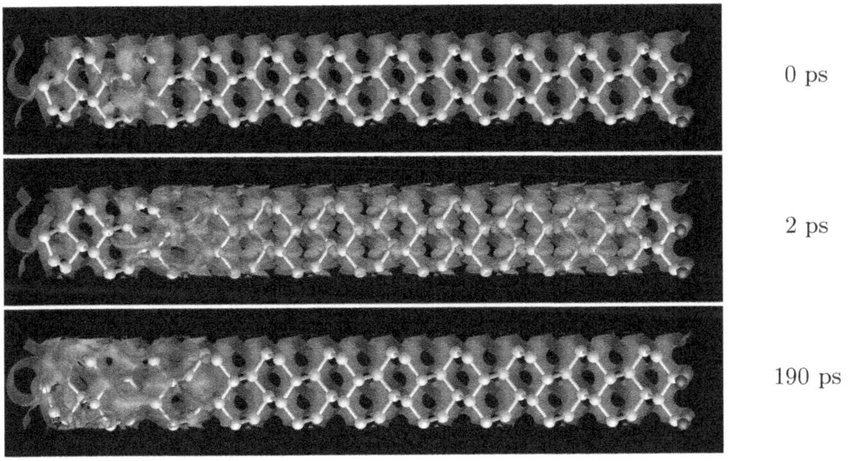

Fig. 1. Projection of the conduction band population into real space for timesteps of 0, 2 and 190 ps. While the initial population is strongly localized at the surface, the population partly shifts to the bulk at 2 ps, and finally returns into a surface state (minimum of D^{down}-band).

fast relaxation process in the D^{down} band of about 1 ps has been obtained in good agreement with experimental findings [7, 8, 19].

4 Scattering response and spatiotemporal wavepackets in quantum cascade lasers

In this section, ultrafast dynamics of the subband population and spatial electronic density evolution in quantum cascade lasers is considered. The quantum cascade laser (QCL) is an electrically pumped intersubband laser consisting of up to several hundreds of quantum wells [20, 21]. Via scattering and tunneling processes, the electrons move from the injector into the active region of one period where the laser transition takes place. The electrons are then collected in the injector of the next period, leading to a cascade of emitted photons. Through engineering of the quantum well widths and potential offsets, the lifetimes and energies of the quantum states can be designed to obtain optimal lasing conditions. For instance, in many THz devices, an inversion is obtained by depleting the lower laser subband on a short time scale via emission of longitudinal optical (LO) phonons [cf. Fig. 2(a)].

In our approach, the interaction of the electronic system with LO phonons and ionized doping centers is described within a second order Born-Markov and bath approximation [cp. Eq. (9)]. The complex multi-period structure is modelled by considering only interaction of next-neighbor elementary cells of the superlattice and applying periodic boundary conditions to the density matrix and the coupling elements [22].

In Fig. 2(b), the dynamics of the subband populations $f_n = 2/A \sum_{\boldsymbol{k}} f_{n\boldsymbol{k}}$ of the THz QCL structure, discussed in Ref. [23], is shown at $T = 10$ K. Here, a strong 170 fs pump pulse saturates the gain inversion at the transition $3 \to 2$, leading to Rabi flopping of the involved subbands [22]. After the passage of the pulse, non-radiative relaxation from the lower laser state into subband 1 via phonon emission depletes the population in subband 2, while

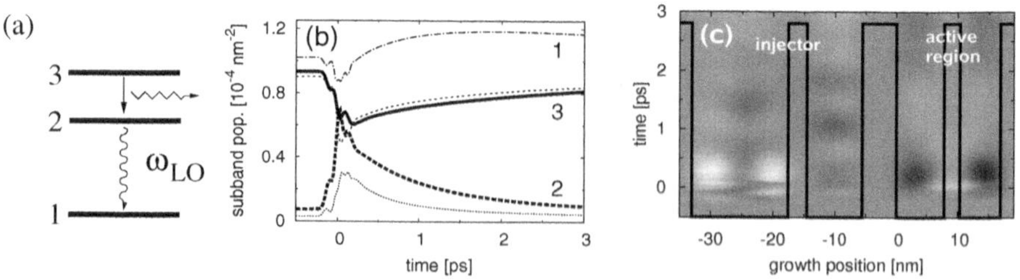

Fig. 2. (**a**) Resonant phonon design of THz QCL with the laser states 2, 3. The transition $2 \to 1$ depletes the lower laser level through phonon emission. (**b**) Subband population dynamics for an excitation with a strong ultrafast pulse. (**c**) Corresponding spatiotemporal evolution of the electronic density (barriers are shown).

the upper laser state 3 is refilled due to elastic scattering from the injector states of the prior period. The population dynamics is dominated by incoherent evolution after the passage of the pulse i.e. by the cooling of the electron distributions within the spatially extended structure. Figure 2(c) shows the corresponding spatiotemporally resolved evolution of the electronic density $n(z,t) = 2/A \sum_{n,m} \xi_n^*(z)\, \xi_m(z) \sum_{\boldsymbol{k}} p_{n\boldsymbol{k}}^{m\boldsymbol{k}}(t)$, where $\xi_n(z)$ denotes the envelope of the quantum confined state (white denotes excess, black decreased density compared to the stationary values). Shortly after the pulse, the electronic density in level 3 in the active region is reduced due to the gain saturation of the laser transitions. After this, the system slowly returns to the stationary state on a picosecond time scale. In addition, coherent effects such as density (gain) oscillations between the injector and the active region can be observed. They result from a coherent charge transfer through the main tunneling barrier connecting the two regions [24, 25].

As can be seen, optically induced electron-phonon scattering dynamics in quantum cascade laser reveals the relaxation channels of the structure as well as coherent effects due to resonant tunneling through the barriers [22].

5 Non-equilibrium phonon dynamics in graphene

When peeling off the graphite constituting slices of honeycomb-arranged carbon atoms to the thinnest possible form, mono-layered '2D-graphite' – or graphene – can be produced. Due to its unique electronic properties, graphene has recently drawn a lot of attention: For example, theoretical investigations focused on the quasiparticle properties and dynamics of so-called Dirac-Fermions [26] or electron-phonon interaction [27]. Recently also ultrafast relaxation processes of photo-excited electrons have been studied in experiments [6].

In this section, the relaxation of an optically excited conduction band carrier population into thermal equilibrium by energy dissipation through phonon emission is discussed [16]. It turns out that, for a proper understanding, the dynamics of photo-excited electrons and heated phonons in graphene has to be treated simultaneous. Given optical phonon energies of almost 200 meV and an intermediately strong electron-phonon interaction, the latter process provides an efficient cooling mechanism in this two dimensional system. The relevant quantities for the dynamics are the interband coherence $p_{\boldsymbol{k}}$ at wave vector $\boldsymbol{k}$, the valence (v) and conduction (c) band population $f_{i\boldsymbol{k}}$ ($i = c, v$), as well as the phonon occupation number $n_{\boldsymbol{q}}$ at wave vector $\boldsymbol{q}$, all defined around Eq. (6) in the introduction. We treat the electron-phonon interaction within second-Born Markov equations, known as Bloch-Boltzmann-Peierls equations, (8), (9). To determine the electron-phonon coupling (EPC) matrix elements $g_{\boldsymbol{k},\boldsymbol{q}}$ Kohn anomalies in the phonon dispersions are used [27]. The EPC matrix elements describe intravalley (Γ-phonons) as well as intervalley ($\boldsymbol{K}$-phonons) scattering processes. Due to the anisotropic optical matrix elements [28] anisotropic optical excitation and relaxation processes occur.

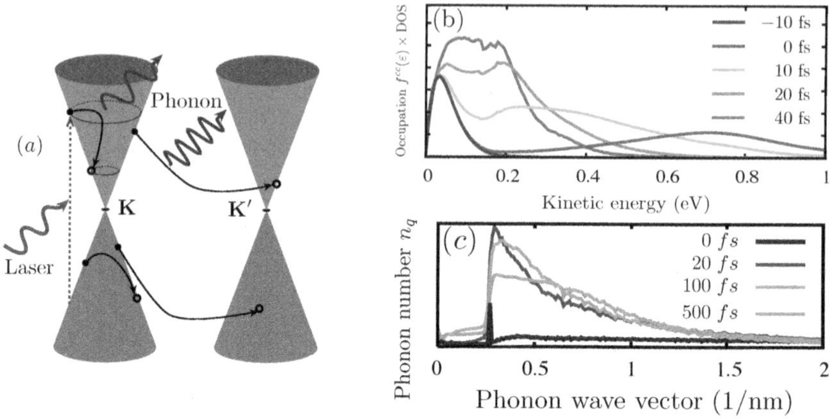

Fig. 3. (**a**) After laser excitation into the conduction band electrons in the graphene sample relax via intra- and inter-valley scattering with optical phonons.(**b**) Shows the temporal evolution of the conduction band occupation f_{ck} times the DOS. Here f_{ck} has been integrated with respect to the angle. (c) Shows the occupation of the $\Gamma - E_{2g,LO}$ phonon mode of graphene for different times.

Results of the coupled dynamics are shown in Fig. 3. Electron relaxation of a photo-excited non-equilibrium distribution takes place on a femtosecond timescale, but is dramatically slowed down after 20 fs. The early drastic electron cooling (b) results in formation of a finite population of the initially quasi unoccupied phonon modes (c). Hot phonons (c) clearly reduce the possibility of energy dissipation by the electronic system after 20 fs.

Our findings provide insight into the ultrafast dynamics of the first 500 fs after excitation not yet accessible by experiment. We show that generation of non-equilibrium (hot) phonons has a noticeable impact on the relaxation dynamics of the excited carriers.

6 Terahertz light emission

In this section we focus on the spontaneous terahertz (THz) quantum emission from a 2D equilibrium electron gas in a doped semiconductor quantum well [13]. Only one subband in effective mass approximation inside the quantum well is considered. The interaction of the electron gas with the light field is treated using a quantized light field Eq. (5) considering a spatially inhomogeneous and frequency dependent dielectric function of the well environment [13, 29–31]. This allows one to incorporate also the sample geometry and the influence of transversal optical (TO)-phonons of the barrier material, having similar resonance frequency as the electron gas THz emission. Emission perpendicular to the quantum well is observed, cf. Fig. 4. This geometry selects intraband emission for the electron gas (not intersubband processes),

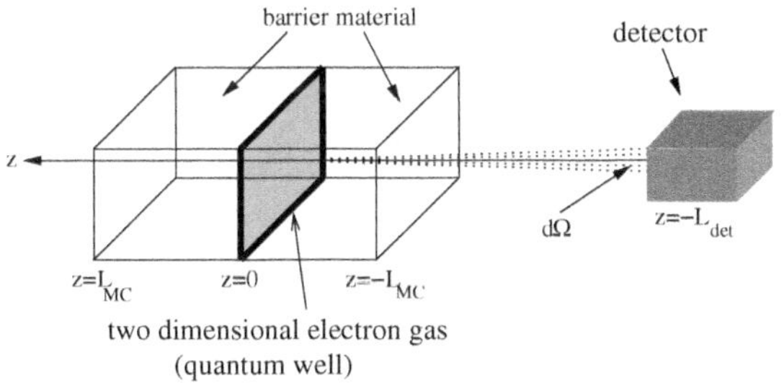

Fig. 4. Geometric setup of the terahertz quantum light emission.

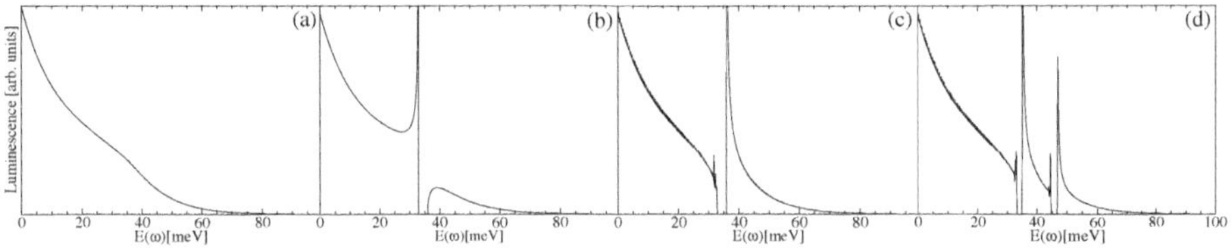

Fig. 5. Luminescence of a Fermi distribution in a GaAs quantum well in AlGaAs for parameters $T = 300\,\mathrm{K}$, $m = 0.07\,m_e$, well width 10 nm and a electron density of $n = 1.2 \cdot 10^{13} m^{-2}$ (**a**) for a free standing quantum well, (**b**) in GaAs barrier phonons (**c**) including finite sample boundaries and (**d**) additionally with AlAs phonons.

since the dipole moment is in-plane with the quantum well and leads to a mostly perpendicular emission. The quantum light emission is calculated using a correlation expansion approach in second-order Born approximation and describes a momentum and energy conservation ensuring joint process of electron-phonon and electron-photon interaction. We find the following formula for the stationary light emission of a specific mode $\boldsymbol{k}, \sigma$ [13, 32]:

$$\partial_t m_{\boldsymbol{k}\sigma} = \sum_{\boldsymbol{c}\boldsymbol{q}} 2\pi |g_{\boldsymbol{q}}|^2 |\zeta(\omega_{\boldsymbol{k}\sigma})|^2 |F^{\boldsymbol{q}}_{\boldsymbol{q}\boldsymbol{k}\sigma}|^2 f_{\boldsymbol{c}+\boldsymbol{q}_\parallel}(1 - f_{\boldsymbol{c}}) \cdot \\ \cdot \left[(1 + n_{\boldsymbol{q}})\, \delta(\omega_{\boldsymbol{k}\sigma} + \varepsilon_{\boldsymbol{c}} - \varepsilon_{\boldsymbol{c}+\boldsymbol{q}_\parallel} + \omega_{LO}) \right. \\ \left. + n_{\boldsymbol{q}}\, \delta(\omega_{\boldsymbol{k}\sigma} + \varepsilon_{\boldsymbol{c}} - \varepsilon_{\boldsymbol{c}+\boldsymbol{q}_\parallel} - \omega_{LO})\right], \quad (10)$$

where ζ is the Heitler-Zeta function and $m_{\boldsymbol{k}\sigma}$ is the photon number, which is directly connected to the observed stationary emission spectra $S(\omega_{\boldsymbol{k}\sigma}) \propto \hbar\omega_{\boldsymbol{k}\sigma}\partial_t m_{\boldsymbol{k}\sigma}$. Inspecting the spectra given by Eq. (10) shows, that the photons are generated through the stimulated/spontaneous emission $(1+n_{\boldsymbol{q}})$ or the induced absorption $(n_{\boldsymbol{q}})$ of a phonons , where $n_{\boldsymbol{q}}$ is the phonon occupation. The emission is propertional to a Pauli blocking term $f_{\boldsymbol{c}+\boldsymbol{q}_\parallel}(1-f_{\boldsymbol{c}})$ and the phonon occupation, jointly participating in the emission process. The argument of the delta function ensures energy conservation. It is possible to evaluate Eq. (10) with different quantization schemes for the photons, cf. Fig. 5. In Fig. 5(a) the electron gas of a free standing quantum well in vacuum leads to a broad THz emission spectrum with an enhancement at the TO-phonon-frequency.

After including also barrier TO-phonons inside an infinitely extended sample [cf. Fig. 5(b)], the emission at the TO-phonon frequency is furthermore intensified and a longitudinal-transversal splitting between LO- and TO-phonon frequency appears. Considering additionally the geometric constrains due to a finite sample in the quantization, the emission at the LO-frequency is intensified while it is suppressed at the TO-frequency (cf. Fig. 5(c)). This occurs because the photon modes in the finite sample geometry at LO-frequency are stronger at the quantum well position while the TO-frequency modes are reduced at this position. Furthermore, in Fig. 5(d) also AlAs-like phonons of the barrier material are included.

Obviously the quantum light THz emission from intrasubband processes inside a quantum well is assisted by electron-phonon relaxation and is strongly influenced by the TO-phonons of the sample.

7 Summary

We have applied a non-equilibrium density matrix approach to a broad variety of 2D electron gases. Based on a unified approach for the electron-phonon interaction, the combined electron-phonon dynamics is shown to determine the features of electron cooling and their spatiotemporal propagation in systems of different character: silicon surfaces, THz quantum cascade lasers, graphene, and GaAs quantum wells.

8 Acknowledgement

We acknowledge support from the Deutsche Forschungsgemeinschaft through KN 427, Sfb 787 and CoE UniCAT.

References

1. R. Huber, R. A. Kaindl, B. A. Schmid, D. S. Chemla: Broadband terahertz study of excitonic resonances in the high-density regime in GaAs/$Al_xGa_{1-x}As$ quantum wells, Phys. Rev. B **72**, 161314 (2005)
2. T. Shih, K. Reimann, M. Woerner, T. Elsaesser, I. Waldmüller, A. Knorr, R. Hey, K. H. Ploog: Nonlinear response of radiatively coupled intersubband transitions of quasi-two-dimensional electrons, Phys. Rev. B. **72**, 195338 (2005)
3. Z. Wang, K. Reimann, M. Woerner, T. Elsaesser, D. Hofstetter, J. Hwang, W. J. Schaff, L. F. Eastman: Optical phonon sidebands of electronic intersubband absorption in strongly polar semiconductor heterostructures, Phys. Rev. Lett. **94**, 037403 (2005)
4. P. R. Wallace: The band theory of graphite, Phys. Rev. **71**, 622–634 (1947)
5. K. S. Novoselov, A. K. Geim, S. V. Morozov, D. Jiang, Y. Zhang, S. V. Dubonos, I. V. Grigorieva, A. A. Firsov: Electric field effect in atomically thin carbon films, Science **306**, 666–669 (2004)

6. T. Kampfrath, L. Perfetti, F. Schapper, C. Frischkorn, M. Wolf: Strongly coupled optical phonons in the ultrafast dynamics of the electronic energy and current relaxation in graphite, Phys. Rev. Lett. **95**, 187403 (2005)
7. S. Tanaka, K. Tanimura: Time-resolved two-photon photoelectron spectroscopy of the Si(001)-(2×1) surface, Surf. Sci. **529**, 251 (2003)
8. M. Weinelt, M. Kutschera, T. Fauster, M. Rohlfing: Dynamics of exciton formation at the Si(100) $c(4 \times 2)$ surface, Phys. Rev. Lett. **92**, 126801 (2004)
9. A. Thränhardt, S. Kuckenburg, A. Knorr, T. Meier, S. W. Koch: Quantum theory of phonon-assisted exciton formation and luminescence in semiconductor quantum wells, Phys. Rev. B **62**, 2706–2720 (2000)
10. F. Rossi, T. Kuhn: Theory of ultrafast phenomena in photoexcited semiconductors, Rev. Mod. Phys. **74**, 895–950 (2002)
11. N. Buecking, P. Kratzer, M. Scheffler, A. Knorr: Theory of optical excitation and relaxation phenomena at semiconductor surfaces: Linking density functional and density matrix theory, Appl. Phys. A **88**, 505 (2007)
12. M. Kira, W. Hoyer, S. W. Koch: Microscopic theory of the semiconductor terahertz response, Phys. Status Solidi B **238**, 443–450 (2003)
13. M. Richter, S. Butscher, M. Schaarschmidt, A. Knorr: Model of thermal terahertz light emission of a two-dimensional electron gas, Phys. Rev. B **75**, 115331 (2007)
14. S. Butscher, A. Knorr: Theory of strong electron-phonon coupling for ultrafast intersubband excitations, Phys. Status Solidi B **243**, 2423–2427 (2006)
15. N. Buecking, S. Butscher, M. Richter, C. Weber, S. Declair, M. Woerner, Kreimann, P. Kratzer, M. Scheffler, A. Knorr: Theory of electron-phonon interactions on nanoscales: semiconductor surfaces and two dimesional electron gases, *Proc. SPIE* **6892**, 689209 (2008)
16. S. Butscher, F. Milde, M. Hirtschulz, E. Malic, A. Knorr: Hot electron relaxation and phonon dynamics in graphene, Appl. Phys. Lett. **91**, 203103 (2007)
17. A. Thränhardt, I. Kuznetsova, C. Schlichenmaier, S. W. Koch, L. Shterengas, G. Belenky, J.-Y. Yeh, L. J. Mawst, N. Tansu, J. Hader, J. V. Moloney, W. W. Chow: Nitrogen incorporation effects on gain properties of gainnas lasers: Experiment and theory, App. Phys. Lett. **86**, 201117 (2005)
18. J. Dabrowski, M. Scheffler: Self-consistent study of the electronic and structural properties of the clean Si(001)(2x1) surface, Appl. Surf. Sci. **56**, 15 (1992)
19. N. Buecking, P. Kratzer, M. Scheffler, A. Knorr: Linking density-functional and density-matrix theory: Picosecond electron relaxation at the Si(100) surface, Phys. Rev. B **77**, 233305 (2008)
20. C. Gmachl, F. Capasso, D. L. Sivco, A. Y. Cho: Recent progress in quantum cascade lasers and applications, Rep. Prog. Phys. **64**, 1533–1601 (2001)
21. B. S. Williams: Terahertz quantum-cascade lasers, Nat. Photon. **1**, 517 (2007)
22. C. Weber, F. Banit, S. Butscher, A. Knorr, A. Wacker: Theory of the ultrafast nonlinear response of terahertz quantum cascade laser structures, Appl. Phys. Lett. **89**, 091112 (2006)
23. S. Kumar, B. S. Williams, S. Kohen, Q. Hu, J. L. Reno: Continuous-wave operation of terahertz quantum-cascade lasers above liquid-nitrogen temperature, Appl. Phys. Lett. **84**, 2494–2496 (2004)

24. F. Eickemeyer, K. Reimann, M. Woerner, T. Elsaesser, S. Barbieri, C. Sirtori, G. Strasser, T. Müller, R. Bratschitsch, K. Unterrainer: Ultrafast coherent electron transport in semiconductor quantum cascade structures, Phys. Rev. Lett. **89**, 047402 (2002)
25. M. Woerner, K. Reimann, T. Elsaesser: Coherent charge transport in semiconductor quantum cascade structures, J. Phys.: Condens. Matter **16**, R25–R48 (2004)
26. A. Bostwick, T. Ohta, T. Seyller, K. Horn, E. Rotenberg: Quasiparticle dynamics in graphene, Nat. Phys. **3**, 36–40 (2007)
27. S. Piscanec, M. Lazzeri, F. M. A. C. Ferrari, J. Robertson: Kohn anomalies and electron-phonon interactions in graphite, Phys. Rev. Lett. **93**, 185503 (2004)
28. A. Grueneis, R. Saito, G. G. Samsonidze, T. Kimura, M. A. Pimenta, A. Jorio, A. G. S. Fihho, G. Dresselhaus, M. S. Dresselhaus: Inhomogeneous optical absorption around the k point in graphite and carbon nanotubes, Phys. Rev. B **67**, 165402 (2003)
29. D. J. Santos, R. Loudon: Electromagnetic-field quantization in inhomogeneous and dispersive one-dimensional systems, Phys. Rev. A **52**, 1538 (1995)
30. Z. Lenac: Quantum optics of dispersive dielectric media, Phys. Rev. A **68**, 063815 (2003)
31. W. Hoyer, M. Kira, S. W. Koch, J. V. Moloney, E. M. Wright: Light-matter interaction in finite-size plasma systems, Phys. Status Solidi B **244**, 3540–3557 (2007)
32. M. Richter, M. Schaarschmidt, A. Knorr, W. Hoyer, J. V. Moloney, E. M. Wright, M. Kira, S. W. Koch: Quantum theory of incoherent THz emission of an interacting electron-ion plasma, Phys. Rev. A **71**, 053819 (2005)

Optical Microcavities as Quantum-Chaotic Model Systems: Openness Makes the Difference!

Martina Hentschel

Max-Planck-Institut für Physik komplexer Systeme, Nöthnitzer Str. 38, 01187 Dresden, Germany
martina@pks.mpg.de

Abstract. Optical microcavities are open billiards for light in which electromagnetic waves can, however, be confined by total internal reflection at dielectric boundaries. These resonators enrich the class of model systems in the field of quantum chaos and are an ideal testing ground for the correspondence of ray and wave dynamics that, typically, is taken for granted. Using phase-space methods we show that this assumption has to be corrected towards the long-wavelength limit. Generalizing the concept of Husimi functions to dielectric interfaces, we find that curved interfaces require a semiclassical correction of Fresnel's law due to an interference effect called Goos-Hänchen shift. It is accompanied by the so-called Fresnel filtering which, in turn, corrects Snell's law. These two contributions are especially important near the critical angle. They are of similar magnitude and correspond to ray displacements in independent phase-space directions that can be incorporated in an adjusted reflection law. We show that deviations from ray-wave correspondence can be straightforwardly understood with the resulting adjusted reflection law and discuss its consequences for the phase-space dynamics in optical billiards.

1 Introduction

Mesoscopic and nanoscopic systems [1], ranging from quantum dots and nanoparticles to carbon nanotubes and most recently to graphene [2, 3], still receive growing attention. These systems are characterized by a phase-coherence length that is larger than the system size such that interference effects play a crucial role and a quantum mechanical description is in order. Often, however, a semiclassical description is sufficient in order to explain the observations: mesoscopic systems, with typical sizes on the micrometer scale, are truly *in the middle* (greek *méssi*) between the microscopic and the macroscopic world. The motivation for their investigation comes from both the application-oriented as well as from the basic-research side. Examples are the quest for the miniaturisation of electronic and optical devices or the challenge to observe many-body effects known from bulk metals in finite systems, a prime example here is the observation of the Kondo effect in quantum dots [4, 5]. Another inspiration comes from the field of quantum chaos[1] where the focus lies on the dependence of physical observables on the system geometry. More precisely, the crucial property is whether the dynamics of the underlying classical system is chaotic or integrable, corresponding to a chaotic or integrable (or, in the most generic case, a mixed) phase space. The first observation of such a sensitivity was in the magnetoconductance through circular (integrable) and stadium-shaped (chaotic) structures that shows a triangular and Lorentzian coherent-backscattering signature, respectively, in the magnetoconductance [6].

The objective of the present paper is to give an overview over some of the recent work on quantum chaos and semiclassical aspects in *optical* mesoscopic systems which we introduce in the following section. We will then address a number of deviations from ray-wave correspondence and show how they can be explained when correcting the usual specular reflection law for light by semiclassical effects important on the mesoscopic scale. We end the paper with a discussion of the implications of such an adjusted reflection law.

1.1 Quantum dots and optical microcavities: billiards for electrons and light

Quantum dots realized in semiconductor heterostructures are often the first systems that come into mind in the mesoscopic context. Electrons are conveniently trapped and manipulated by various gate electrodes. Especially many-electron ballistic quantum dots are often described as billiards with hard walls and used as model systems in the field of quantum chaos [6].

Another class of experimental and theoretical qunatum-chaotic model systems are optical microcavities – billiards for light instead of electrons. This is possible because of the analogy between the Schrödinger and Helmholtz equations for electrons and light, respectively, that holds in two dimensions and for the so-called TM (transverse magnetic) polarization direction (where the magnetic field lies in the resonator plane). One fundamental difference is, however, the confinement mechanism: For light, there is no charge to manipulate, and gate-voltage based confinement has to be replaced. The concept used instead is that of total internal reflection at optical interfaces with different refractive indices n_1 and n_2. Total internal reflection occurs at the optically thinner medium (e.g., when going from glass with refractive index $n_1 = n \approx 1.5$ to air with $n_2 = 1$) for angles of incidence χ (measured to the boundary normal) larger than the critical angle $\chi_c = \arcsin(1/n)$. The limit $n \to \infty$ corresponds to that of a closed system with hard walls. This implies in turn that generic optical systems are open systems where the openness is

[1] For a review of quantum chaos in mesoscopic systems, see, e.g., H.-J. Stöckmann: *Quantum Chaos: An Introduction* (Cambridge University Press, Cambridge, England, 1999), and references therein.

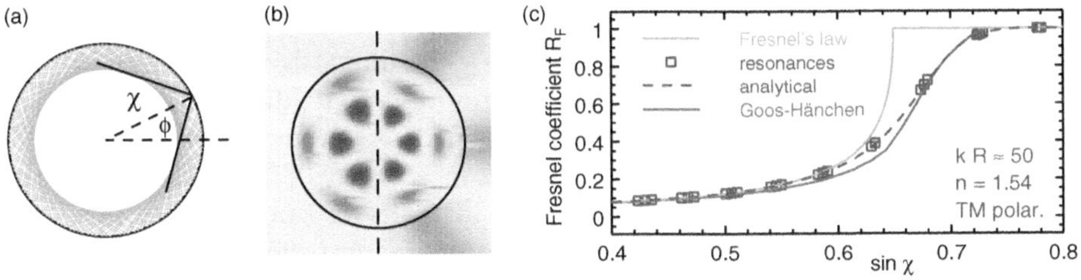

Fig. 1. Ray and wave description of the dielectric disk. (**a**) Whispering-gallery type ray trajectory. (**b**) Resonances of the closed (*left half*, corresponding to a wavenumber $kR = 9.761$) and dielectric (*right half*, $n = 1.54$, $nkR = 11.428 - 0.254i$) disk. The openness of optical systems considerably changes the wave intensity outside the cavity (evanescent wave, or refractive escape). (**c**) Fresnel's law with deviations from the naive ray-wave correspondence clearly visible in the region of critical incidence. See text for details.

related to, and defined by, the possibility of refractive escape. In other words: Optical microcavities are ideally suited to theoretically as well as experimentally study quantum (wave) chaos in *open* systems. Note that the openness is not induced by leads and that we are not interested in transport through the cavity. Rather, the openness exists everywhere along the boundary which is reflected in mixed boundary conditions with nonvanishing wavefunction *and* nonvanishing derivative for the Helmholtz equation.

1.2 Ray-wave correspondence!?

One paradigm intimately related to the field of quantum chaos is the quantum-classical, in the case of optical microcavities, the wave-ray correspondence. It is usually taken for granted and one of the footings of our present understanding of the relation between the classical and the quantum world. We shall see below that, although useful and extremely simple to implement numerically, the ray picture is not able to comprehensively explain *all* the results found for optical microresonators. A detailed list of observed deviations will be given in the following section. Note that these deviations are related to the fact that the wavelength λ of the electromagnetic field is smaller, but still comparable to the cavity size R (and, therefore, the true ray limit $\lambda \to 0$ is not fulfilled): In typical experimental setups based on semiconductor heterostructures ($n = 3.3$), the cavity size is of the order $R \sim 50\,\mu\text{m}$ and infrared light with $\lambda \sim 850$ nm is used. In numerical *wave* simulations, values nkR exceeding, say, 250, are very hard to reach at present.

1.3 Description of optical microcavities

At this point, a few words concerning the description of optical microcavities in the ray and wave picture are in order.

Ray picture: The ray optics description of optical microcavities is based on ray tracing simulations. The trajectory is determined by assuming specular reflections at the resonator walls. In addition, one introduces a variable that monitors the (decaying) intensity of the light ray given by Fresnel's law. It is even possible to account for the interference of rays [7]. Predictions made for the experimentally accessible far-field intensity are now mostly based on the refractive escape of rays taken from the steady-probability distribution [8, 9]. The Poincaré surface of section summarizes the information about the reflection points at the (outer) resonator boundary in reduced phase space given by a spatial variable parametrizing the position along the boundary (such as the polar angle ϕ or the arclength s) and the angular momentum $\sin\chi$, see Figs. 1(a) and 2(c). The condition for total internal reflection, $|\sin\chi| > 1/n$ is violated in the so-called leaky (forbidden) region $-1/n < \sin\chi < 1/n$ [marked by dashed-dotted lines in Fig. 2(c)]. Trajectories hitting this phase-space region will (more or less) easily escape the cavity by refraction. Periodic orbits with stable islands in the leaky region are considered to be not populated by cavity modes.

Wave Picture: The objective is to compute the resonances (or quasibound states) of the cavity. The by far most popular approach and a versatile tool is the boundary element method [10] that gives the resonances directly in the complex plane. The imaginary part of the dimensionless complex wavenumber kR (with $k = 2\pi/\lambda$ and R the radius of curvature) in free space defines the life time and the Q-factor of the resonance, $Q = 1/[2\,\mathrm{Im}(kR)]$. An example of a resonant wave pattern is shown in Fig. 1(b) where that of a closed system (left half) is compared with that of an open, optical system with refractive index $n = 1.54$ (right half). Another approach is to use an S-matrix method [11, 12] that describes the resonator as an open system being "probed" from outside with (test) plane waves. The position of resonances can be read-off from the Wigner delay time. This approach is straightforward by implemented for rotationally invariant systems such as the annular billiard, see Fig. 2(b).

Husimi functions at dielectric interfaces: The mapping of the resonance wave pattern to phase space is realized by means of the Husimi function [13]: Simply speaking, this function measures the overlap of the resonance wave function with a minimal-uncertainty wavepacket centered around a certain position ϕ_0, χ_0 in phase space. As illustrated for the example of the annular billiard in Fig. 2, Husimi functions are a particularly useful tool to study ray-wave correspondence. We point out that in order to do so, the concept of Husimi functions has first to be generalized to dielectric interfaces [13]. In hard-wall systems with Dirichlet boundary conditions the wave function vanishes at the system's boundary, and the Husimi function can be defined based on the wave function derivative alone. At optical interfaces, however, both the wave function and its derivative are non-zero and it is not clear at all which of the two should be the used to define the Husimi function. Moreover, there are now four rays (incident and outgoing on either side of the interface), and the existence of four corresponding Husimi functions would certainly be

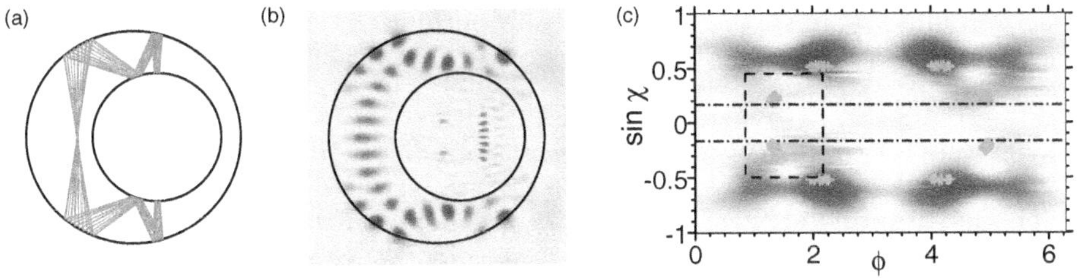

Fig. 2. Ray-wave correspondence in the dielectric annular billiard (with refractive indices $n_a = 3$ and $n_i = 6$ of the annular region and the inner disk). (**a**) Stable periodic orbit typical for the geometry shown. (**b**) (One) corresponding resonance wave pattern. (**c**) Phase-space representation in terms of the Poincaré surface of section (*green/light crosses*) and the (inside incident) Husimi function (maxima in *red/dark regions*) that reveal deviations from ray-wave correspondence on the quantitative level. The *dashed rectangle* marks a region where violation of forward-backward symmetry ($\sin\chi \to -\sin\chi$) is especially evident. The leaky region characterized by the refractive escape of rays is located in between the *dashed-dotted lines*.

desirable and support the concept of ray-wave correspondence. As a matter of fact, both issues can (solely) be solved simultaneously as shown in Ref. [13] to where we refer the reader for details.

Active (lasing) micrcavities: From an application-oriented point of view, the most interesting application of optical microcavities is that as microlasers. This requires the presence of an active material that allows for lasing operation. Then, in addition to the nonlinearities originating from the (chaotic) resonator geometry those from the lasing operation are important and determine the behavior. Although a description of the active material can, to a certain degree, be achieved using a complex refractive index n [14], application of the Schrödinger-Bloch model [15] is in order. We will not further consider the wave description of active cavities in the present paper. We point out that, interestingly, the ray picture (where no active medium can be accounted for) may provide a very reasonable description of the far-field characteristics even for lasing microcavities [16–19]. Given the goal of building microlasers with directional emission (which comes close to being the holy grail in this field at present), ray simulations have proven to be a valuable tool even away from the ray limit for $\lambda \leq R$ [20].

2 Deviations from ray-wave correspondence

In the following, we list a number of deviations from ray-wave correspondence that have accumulated over the past, say 10, years. Though each individual fact might be considered just as a slight mismatch, in its entirety these observations suggest that the ray description needs to be corrected away from the exact limit. The necessary corrections will be discussed in the following section, and we finish the paper with an outlook about further consequences of an adjusted ray model.

2.1 Delayed onset of total internal reflection in Fresnel's law

One nice example illustrating the interplay between the ray and the wave description is the dielectric disk. On the ray side, the rotational invariance of the system conserves the angle of incidence χ. If χ is larger than the critical angle, light is confined in a so-called whispering-gallery (WG) orbit, cf. Fig. 1(a). On the wave side, this is reflected in an azimuthal quantum number m (e.g., $m = 3$ for the WG mode in Fig. 1(b); here, the radial quantum number $\rho = 2$) that further characterizes resonances with complex wavenumbers kR. There is a one-to-one relation between the ray and wave picture quantities (which are χ and the Fresnel reflection coefficient $R_F(\chi)$ on the ray, and m and kR on the wave side), namely [11, 12]

$$\sin\chi = \frac{m}{n\mathrm{Re}(kR)}\,, \tag{1}$$

$$R_F = \exp(4n\mathrm{Im}(kR)\cos\chi)\,. \tag{2}$$

This translation works well, except near the critical angle where the onset of total internal reflection is significantly delayed in the wave picture, see Fig. 1(c). The green (light solid) curve shows the Fresnel law (for TM polarization) with $R_F = 1$ for $\chi \geq \chi_c \approx \arcsin 0.65$. The squares correspond to the Fresnel reflection coefficient, Eq. (2), for resonances with $\mathrm{Re}(kR) \approx 50$. Note that a closed analytical expression for R_F (dashed line) was derived and is given in Ref. [21].

The origin of the deviations near χ_c must be, and is, the curvature of the dielectric interface. The question arises (especially when taking a quantum-chaos inspired point of view) whether a semiclassical explanation of this deviation is possible, i.e., one that is based on the ray picture but takes corrections originating in the wave character of light into account. That this is indeed possible can be seen by the purple (dark solid) line for a ray picture completed by such a wave (interference) correction known as the Goos-Hänchen shift [22, 23] that then closely follows the squares [21] . We will discuss this effect in detail in **Sect. 3**.

2.2 Correspondence of orbits and resonances in configuration and phase space: qualitatively, not quantitatively

In Fig. 2 ray-wave correspondence is illustrated for the example of the dielectric annular billiard. For the geometry chosen, the trajectory/resonance shown in Fig. 2(a) in the ray, and in Fig. 2(b) in the wave description belongs to one typical family of stable orbits (or resonances). The similarity between the two patterns is evident and certainly supports the concept of ray-wave correspondence. When going from configuration to phase space, however, the agreement becomes somewhat less convincing, cf. Fig. 2(c). The larger red (dark) maxima of the (incoming) Husimi function coincide reasonably well with the Poincaré signature of the orbit (green crosses), but the smaller maxima (corresponding to the right reflection points closer to the constriction) clearly deviate from the ray signature which is at least partially related to the Goos-Hänchen shift mentioned above. Such a behavior of qualitative, but not quantitative ray-wave correspondence is the typical case. We will investigate the reasons and mechanisms leading to these deviations below.

2.3 Husimi functions reveal violation of forward-backward symmetry

One intrinsic property of the ray picture is the principle of reversibility of ray trajectories – in other words, forward-backward or time reversal symmetry. The corresponding symmetry operation, $\sin\chi \to -\sin\chi$, is strictly obeyed in the ray picture. This is, however, not the case in the wave description as can be seen in the Husimi function, Fig. 2(c), especially in the area marked by the dashed rectangle. There are numerous examples of such a behavior, e.g. [24], that was recognized in a number of cases but could not be understood. Usually, one was content that time-reversal and spatial (mirror axis of the billiard, e.g., x-axis in the annular billiard with the symmetry operation $\phi \to 2\pi - \phi$) symmetries *together* were obeyed. We shall see in the next section that another semiclassical effect, the so-called Fresnel filtering, is responsible for the violation of time-reversal symmetry and implies, when taken into account in an corrected ray picture, non-Hamiltonian dynamics in optical microcavities.

2.4 Existence of regular modes in chaotic systems

Optical microresonators studied recently in quite some detail both in theory [25] and in experiment [26–29] are cavities with spiral shape, $r(\phi) = R(1 + \epsilon\phi/2\pi)$, where ϵ measures the radial offset. In terms of classical dynamics, the spiral billiard is (for all purposes of a physicist) chaotic. The more surprising was the observation of predominantly regular orbits of triangular and star-like shape reported in Ref. [25]. There are strictly no periodic ray orbits corresponding to these patterns, neither stable (those are missing in chaotic systems) nor unstable (that could become visible as so-called scarred resonances). The observed regular orbits were named quasiscars, and vicinity of the angle of incidence to the critical angle was suspected to be crucial (indeed, triangular orbits were found in cavities with $n = 2$, the star-like type for $n = 3$). Note that the existence of these regular orbits suggests that the system apparently possesses rotational symmetry. Clearly, this cannot be the case in a classical description of spiral billiards. The question arises what mechanism re-establishes the rotational invariance in the wave picture. Again, it is the Fresnel filtering effect that we explain in Sect. 3.

2.5 Violation of Fresnel's and Snell's law at curved interfaces

Eventually, we consider the (single) reflection of a Gaussian test ray at a curved interface. For convenience we choose as a cavity an air hole of radius R in a glass matrix, or equivalently, consider a hole with refractive index $n = 0.66 < 1$ in air. (Note that this implies concave instead of the convex curvature considered so far.) In Fig. 3 the ray and wave result are shown for the specific case of angle of incidence $\chi_0 = 45°$ (controlled via the impact parameter s). Accordingly, the reflected light *ray* will leave the cavity vertically. This is, however, not the case in the wave description: Here, the reflected beam leaves the cavity under a larger angle, see Fig. 3(b).

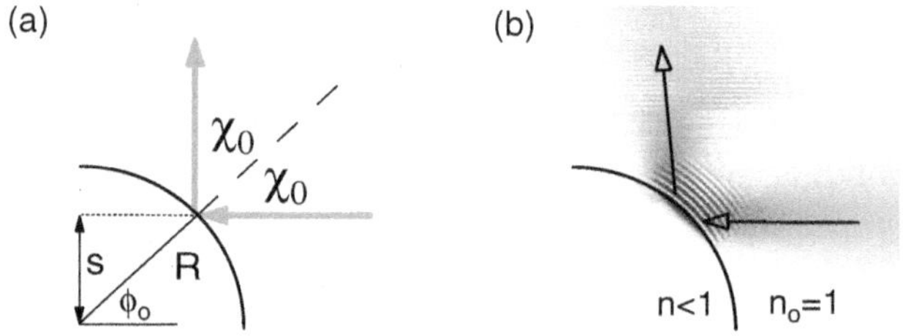

Fig. 3. Reflection of (**a**) a light ray and (**b**) a Gaussian beam at a curved interface. Clearly, the ray picture prediction fails towards the wave regime $\lambda \leq R$. The reason are Goos-Hänchen shift and Fresnel filtering, see also Fig. 5, that vanish only in the pure ray limit $\lambda/R \to 0$.

3 Correcting ray optics by wave effects: Goos-Hänchen shift and Fresnel filtering

At this point a systematic study of the effects causing the deviations from ray-wave correspondence is in order. Inspired by the doubtless advantages of the ray model, such as its easy implementation and its conceptual success, our objective will be to identify semiclassical corrections such that the resulting (adjusted) ray model can quantitatively better capture the wave properties of the system. Following the last example in the previous section where deviations between the ray and wave behavior are clearly visible in a single, near-critical reflection, we analyze this situation in some more detail [30]. Our means of choice are Husimi functions at dielectric interfaces. Incident and outgoing Husimi functions for the reflection of a Gaussian beam at a curved interface, cf. Fig. 3, are shown in Fig. 4. The intersection of the green/dashed lines marks the position of the maxima as expected from the ray picture.

In the near field, the incident Husimi function exactly coincides with the ray position: This choice defines our initial conditions for the incident Gaussian beam (in Fig. 4, the angle of incidence is $\chi_0 = 40°$). The coincidence holds, qualitatively, also in the far field. Deviations become visible when looking

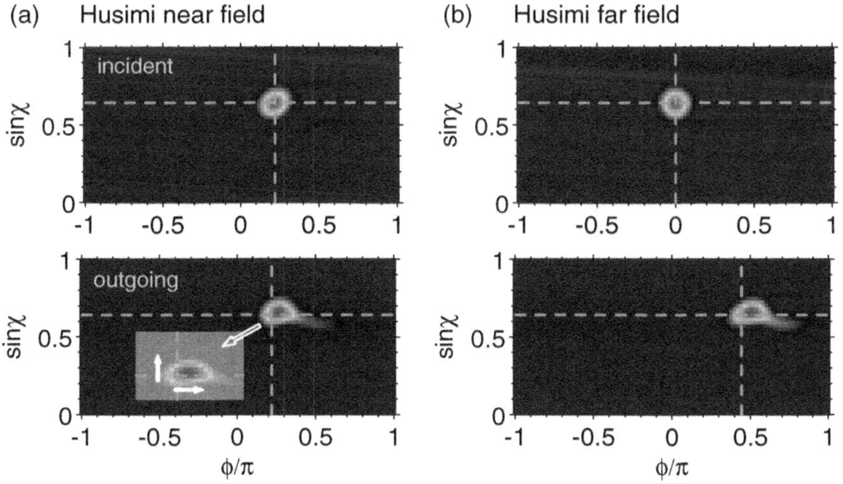

Fig. 4. Husimi functions for the reflection of a Gaussian light beam at an circular inclusion with smaller refractive index as shown in Fig. 3. The parameters of the beam correspond to a light ray incident under $\chi_0 = 40°$, the critical angle is $\chi_c = 41.75°$. Shown are the Husimi functions of the incident and reflected beam in the near and in the far field, see text and [30] for details. The *intersection points of the dashed green/light lines* indicate the ray model expectations. Clearly, the signature of the outgoing Husimi function deviates in both phase-space directions from it due to Goos-Hänchen shift $\Delta\phi_{\mathrm{GH}}$ and Fresnel filtering $\Delta\chi_F$.

at the outgoing Husimi function: The maximum deviates from the ray model prediction *in two independent directions in phase space*, marked by the arrows in the inset of Fig. 4: The shift $\Delta\phi_{\mathrm{GH}}$ in ϕ-direction is known as the Goos-Hänchen shift [22, 23], and the shift $\Delta\chi_F$ in χ-direction has been termed Fresnel filtering [31]. Both effects are schematically illustrated in Fig. 5(b). Their magnitude depends on the wavenumber and is typically several degrees [30], with the Goos-Hänchen correction being the larger.

Crucial for the understanding of both effects is to realize that in optical microcavities each light ray is actually a light *beam*, i.e., composed of light rays with similar but not exactly equal angles of incidence. This becomes immediately clear when recalling that in the mesoscopic regime a light ray assumes a transversal extension of the order λ. At a reflection point, the corresponding angles of incidence will acquire a certain distribution because of the interface curvature [21]. In addition, an electromagnetic wave, e.g., a Gaussian beam, always contains a range of angles of incidence.

Goos and Hänchen showed in a nice experiment in 1947 [22] that, for angles of incidence larger than the critical angle, this leads to an interference effect that results in the lateral shift (of the order λ) of the beam along a *planar* interface before it is reflected. This is illustrated in Fig. 5(a). The reflection can be thought of as to take place at an effective interface somewhat inside

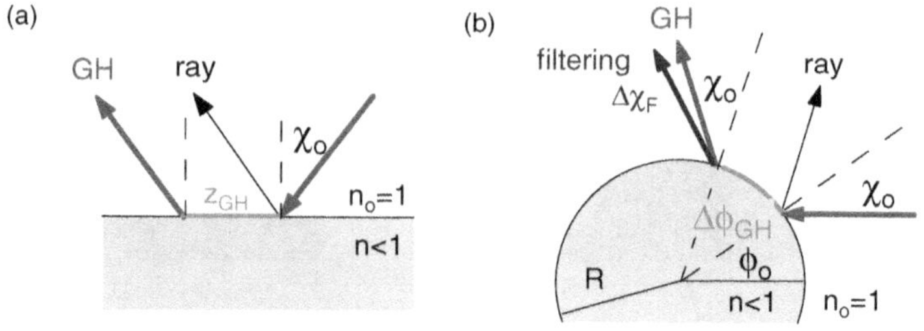

Fig. 5. Goos-Hänchen shift (GH) and Fresnel filtering at (**a**) a planar and (**b**) a curved interface. At planar interfaces, a slight variation in the angles of incidence around χ_0 gives rise to the GH shift z_{GH}, a lateral shift of the reflected beam in the regime of total internal reflection. At curved interfaces, the GH shift $\Delta\phi_{\mathrm{GH}}$ is directly observable as a change in the far field emission direction. In addition, Fresnel filtering $\Delta\chi_F$ leads to non-specular reflection and a deviation from Snell's law that further corrects the far-field emission characteristics.

the optically thinner material. Whereas the angle of incidence is the *same* at the real and the effective interface for *planar* interfaces, this is not true at the curved optical boundary of microcavities [21]: The angle of incidence is smaller (larger) at convex (concave) boundaries. Applying Fresnel's law to this effective angle of incidence can quantitatively explain the deviations in the Fresnel reflections coefficient discussed above, cf. the purple line in Fig. 1(c) and Ref. [21] for details.

Fresnel filtering is even more classically explained than the Goos-Hänchen shift as there is no underlying interference effect. In a collection of rays with angles of incidence around the critical angle, the rays with the largest angles will already be totally reflected whereas subcritical rays are still refracted. This can be seen in the signatures of the outgoing Husimi functions in Fig. 4: The faint signature below the horizontal dashed green/light line corresponds to the refracted beam (that leaves the cavity when hitting the boundary the next time). It is indeed composed of angles $\chi < \chi_c \approx 41.76°$. In turn, the Husimi signature of the reflected beam is shifted above the dashed line that defines the position in the ray model. (Note that the opposite shifts of the reflected and refracted beam ensure conservation of angular momentum.) In other words, the maxima of the incident (Gaussian) beam and the reflected (asymmetrically perturbed) beam do not coincide: Around critical incidence, the angle of the reflected beam is always *larger* by $\Delta\chi_F$, cf. also Fig. 5(b), and the law of specular reflection, and consequently forward-backward symmetry, is violated.

Goos-Hänchen shift and Fresnel filtering have a common origin – the beam rather than a pure ray nature of electromagnetic *waves* – but act in orthogonal directions in phase space and, therefore, cannot be comprised in one and the same correction to the ray model. On the other hand, they also exhaust the number of possible corrections because there are no more independent directions in phase space. We shall see in the last section that the different nature of Goos-Hänchen shift and Fresnel filtering manifests itself in strikingly different effects on the dynamics in optical billiards described with an adjusted ray model that takes the above-discussed non-specular reflection near critical incidence into account [32].

4 Outlook: non-Hamiltonian dynamics in quantum-chaotic model systems

Given the increasing activity in the field of optical microcavities and quantum chaos over the past years, the question arises whether effects of such a non-specular reflection law have not been observed before. We already mentioned that the deviations in Fresnel's law, cf. Fig. 1(c), can be fully understood with the Goos-Hänchen shift. It can also qualitatively explain the differences between ray orbits and resonance patterns, Fig. 2, via an adjustment at the reflection points with the steeper (near critical) angle of incidence. In fact, all deviations from ray-wave correspondence discussed in Sect. 2 can be addressed based on Goos-Hänchen shift (that is important for all angles of incidence $\chi > \chi_c$) and Fresnel filtering (that is important especially for $\chi \approx \chi_c$). The two remaining examples, the existence of regular orbits in spiral microcavities and the breaking of time reversal symmetry in the Husimi functions, can be explained by Fresnel filtering [32].

In the spiral cavity, the filtering correction $\Delta\chi_F$ in the outgoing beam re-establishes the conservation of angular momentum as it counteracts the change in curvature that decreases χ. Therefore, regular orbits (similar to those in the disk) may exist again [32]. They are unstable and, consequently, host *true* scars. In general, a finite $\Delta\chi_F$ destroys the principle of ray path reversibility. This is easiest seen when considering a ray with $\chi_0 \approx \chi_c$ that leaves then under an angle $\chi_0 + \Delta\chi_F > \chi_c$. Tracing its trajectory in opposite direction will yield a (nearly) zero filtering correction (such a situation can easily be constructed), and the reflected ray does not coincide with the original ray. It is precisely this type of mechanism that causes the observed loss of time-reversal symmetry in the Husimi functions.

Most remarkably, a billiard dynamics based on the adjusted reflection law leads to non-Hamiltonian dynamics [32]. Responsible is the filtering correction that causes deviations of the Jacobian matrix determinant from unity, see [32] for details. The origin is the openness of the billiard, effectively described by the adjusted reflection law. Note that dissipative as well as attractive dynamics with the formation of repellors and attractors, respectively, is possible [32]. We are optimistic that more signatures of a non-Hamiltonian ray dynamics in optical microcavities will be identified soon (S. Bittner, B. Dietz-Pilatus and U. Kuhl, H.-J. Stöckmann, private communications) and that quantum chaos in open systems will remain a fascinating research topic in the future.

5 Acknowledgements

The author would like to thank all her co-authors and colleagues for inspiring discussions over the past years. Sincere thanks go to the Alexander von Humboldt Stiftung (Feodor-Lynen Fellowship 2002–2004) and the Deutsche Forschungsgemeinschaft (Emmy-Noether Group 2006-2011 and Research Group FG 760, 2007-2010) for ongoing support.

References

1. *Mesoscopic Electron Transport*, edited by L. L. Sohn, L. P. Kouwenhoven, and G. Schön (Kluwer Academic Publishers, Dordrecht, 1997)
2. K. S. Novoselov et al.: Nature (London) **438**, 197 (2005)
3. Y. Zhang et al.: Nature (London) **438**, 201 (2005)
4. D. Goldhaber-Gordon et al.: Nature (London) **391**, 156 (1998)
5. S. M. Cronenwett et al.: Science **281**, 540 (1998)
6. A. M. Chang, H. U. Baranger, L. N. Pfeiffer, and K. W. West: Phys. Rev. Lett. **73**, 2111 (1994)
7. M. Hentschel and M. Vojta: Opt. Lett. **26**, 1764 (2001)
8. S.-Y. Lee, J.-W. Ryu, T.-Y. Kwon, S. Rim, and C.-M. Kim: Phys. Rev. A **72**, 061801(R) (2005)
9. S.-B. Lee et al.: Phys. Rev. A **75**, 011802 (2007)
10. J. Wiersig: J. Opt. A: Pure Appl. Opt. **5**, 53 (2003)
11. M. Hentschel, Dissertation, TU Dresden, Germany (2001)
12. M. Hentschel and K. Richter, Phys. Rev. E **66**, 056207 (2002)
13. M. Hentschel, H. Schomerus, and R. Schubert: Europhys. Lett. **62**, 636 (2003)
14. H. Schomerus, J. Wiersig, and M. Hentschel: Phys. Rev. A **70**, 012703 (2004)
15. T. Harayama, S. Sunada, and K. Ikeda: Phys. Rev. A. **72**, 013803 (2005)
16. H. G. L. Schwefel et al.: J. Opt. Soc. Am. B **21**, 923 (2004)
17. S. Shinohara and T. Harayama: Phys. Rev. E **75**, 036216 (2007)
18. T. Tanaka et al.: Phys. Rev. Lett. **98**, 033902 (2007)
19. M. Hentschel, T.-Y. Kwon, M. Belkin, and F. Capasso: in preparation
20. J. Wiersig and M. Hentschel: Phys. Rev. Lett. **100**, 033901 (2008)
21. M. Hentschel and H. Schomerus: Phys. Rev. E **65**, 045603(R) (2002)
22. F. Goos and H. Hänchen: Ann. Phys. (Leipzig) **1**, 333 (1947)
23. K. Artmann: Ann. Phys. (Leipzig) **8**, 270 (1951)
24. J. Wiersig and M. Hentschel: Phys. Rev. A **73**, 031802(R) (2006)
25. S.-Y. Lee et al.: Phys. Rev. Lett. **93**, 164102 (2004)
26. Ch.-M. Kim et al.: Appl. Phys. Lett. **92**, 131110 (2008)
27. R. Audet et al.: Appl. Phys. Lett. **91**, 131106 (2007)
28. T. Ben-Messaoud and J. Zyss: Appl. Phys. Lett. **86**, 241110 (2005)
29. G. D. Chern et al.: Appl. Phys. Lett. **83**, 1710 (2003)
30. H. Schomerus and M. Hentschel: Phys. Rev. Lett. **96**, 243903 (2006)
31. H. E. Tureci and A. D. Stone: Opt. Lett. **27**, 7 (2002)
32. E. G. Altmann, G. Del Magno, and M. Hentschel, Europhys. Lett. 84, 10008 (2008).

Nonlinear Transport Properties of Electron Y-Branch Switches

Lukas Worschech, David Hartmann, Stefan Lang, D. Spanheimer, Christian R. Müller, and Alfred Forchel

Technische Physik, Universität Würzburg, Am Hubland, 97074 Würzburg, Germany
lukas.worschech@physik.uni-wuerzburg.de

Abstract. Nonlinear transport properties of Y-branched semiconductor nanostructures were studied in the non-linear transport regime. The Y-branch switches (YBSs) were defined by electron beam lithography and wet etching techniques in modulation doped GaAs/AlGaAs heterostructures. Due to capacitive couplings between different branches, YBSs can be used as amplifiers, inverters, bistable switches and compact logic gates with functional integration. Room temperature operation of a logic gate with gain solely based on coupled YBSs without external load resistors is reviewed.

1 Introduction

Semiconductor nanostructures are often referred to as mesoscopic systems as they are small compared to a macroscopic scale but large with respect to single atoms [1, 2]. Mesoscopic conductors show several transport properties very different from those of diffusive conductors due to their reduced dimensionality. In this framework, two important length scales are the Fermi wavelength λ_F and the mean free path l_{mfp}, which is limited by inelastic scattering [3]. A well known effect of mesoscopic conductors is the conductance quantization in units of the conductance quantum $2e^2/\mathrm{h}$ [4, 5], which can be observed in quantum-point contacts with a geometrical width and length smaller than λ_F and l_{mfp}, respectively.

In the last decades, mesoscopic structures were intensely studied as regards ballistic and phase-coherent effects in the linear transport regime with electron transport close to the Fermi energy. Recently, non-linear properties of mesoscopic structures robust up to room temperature were reported. In the non-linear regime, with differences of the electro-chemical potentials between different contacts exceeding by far the Fermi energy, capacitive couplings between different sections of complex shaped nanostructures influence the transport properties significantly [6, 7]. Selfswitching and rectification can be observed in the non-linear transport of nanoelectronic devices. Here we review studies of different logic elements based on electron Y-branch switches (YBSs).

A YBS consists of a narrow source, which splits along a branching section into two drains. By a lateral electric field electrons can be guided into one of the drains without creating a barrier in the other branch. Such a directional switching allows to observe switching voltages smaller than the thermal voltage kT/e [8, 9, 31, 32], which typically represents a fundamental limit for barrier controlled switches like the field-effect transistor. The YBS shows several advantages for functional integration in nanoelectronic circuits. In a YBS the switching is complementary. YBSs driven in the non-linear ballistic regime show a pronounced rectification [10, 11]. Due to selfswitching, YBSs are especially interesting for low power applications.

2 Fabrication techniques

The authors have realized YBSs monolithically integrated in modulation doped GaAs/AlGaAs heterostructures. Narrow channels were defined by electron beam lithography and wet etching techniques. Figure 1 shows a scheme of a typical layer sequence in which a quantum wire was defined. A two-dimensional electron gas (2DEG) was formed at the GaAs/AlGaAs heterointerface. The 2DEG is locally separated from the ionized Si donors by a 20 nm thick, undoped AlGaAs spacer [12]. Trenches, typically 100–200 nm wide and 100 nm deep function as barriers between outer electron reservoirs and define also narrow channels. The outer 2DEGs serve as gates as demonstrated for quantum-wire transistors [13–15].

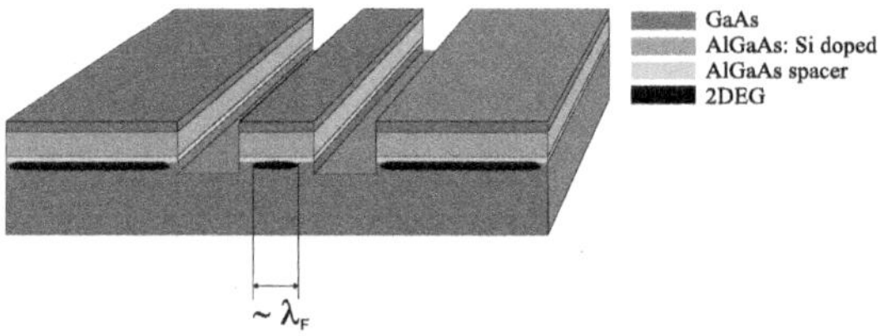

Fig. 1. Sketch of a GaAs/AlGaAs heterostructure in which two etched trenches were defined, which confine electrons to a quantum wire with a width comparable to λ_F. Parts of the AlGaAs are doped with Si. Ionized Si donors are locally separated from the 2DEG, which forms at the GaAs/AlGaAs interface. Unconstrained 2DEG at the two sides of the quantum wire are used as side-gates.

3 Quantum capacitance and self-gating

A gate can efficiently control a nearby channel via the field effect. A change of the electro-chemical potential of the gate, $\mu_{\rm qg}=-eV_{\rm qg}$, with e the electron charge and $V_{\rm qg}$ the voltage applied at the gate, leads to $\Phi_{\rm qd}$, a change of the electrostatic potential in the channel. The leverage factor η_{qg} often called gate efficiency can be estimated by taking into account the circuit diagram depicted in Fig. 2. In low dimensional conductors, it is useful to describe the reduced screening of the electric field by the density of states (DOS). For that purpose Luryi has introduced the concept of the quantum capacitance $D=e^2 \times DOS$ [16, 30]. Therefore, $\eta_{\rm qg}=(\Phi_{\rm qd}-\mu_{\rm qd})\,/\,(\mu_{\rm qg}-\mu_{\rm qd})=\left(1+\frac{D_{\rm qd}}{C_{\rm qd}}+\frac{D_{\rm qd}}{D_{\rm qg}}\right)^{-1}$. As a result, a small quantum capacitance of the channel and a large geometrical capacitance of the gate, $C_{\rm gd}$, enhance the gate efficiency. The geometrical capacitance per unit length is proportional to the dielectric constant, $C \approx \epsilon \approx 10^{-10}\,{\rm F/m}$. The quantum capacitance is $D_{\rm qd} = 4e^2/hv_F$, with $v_F \approx 10^5\,{\rm m/s}$ the Fermi velocity. For $D_{\rm qg}$ of the gate much larger than $D_{\rm qd}$, the gate efficiency is $\eta_{\rm qg} \approx 1/11$, a value typically observed for quantum wires and quantum-point contacts [7, 28, 29]. Interestingly, a simple estimation shows that a quantum wire can also serve as efficient gate for another quantum wire as $D_{\rm qg} \approx D_{\rm qd}$ leads to a gate efficiency $\eta_{\rm qg} \approx 1/12$, a value similar to that of metallic gates with a large DOS.

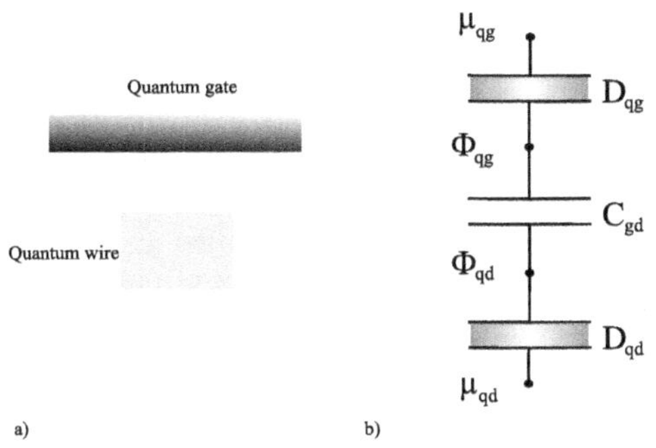

Fig. 2. (**a**) Cross section scheme of a quantum wire controlled by a gate. (**b**) Corresponding capacitive circuit used for the description of the gate efficiency. The quantum capacitance D describes the role of electric-field screening.

The question arises now of whether or not gates, which are leaky coupled to the channel, are still efficient. Typically, gate leakage currents reduce the gate efficiency. We will first discussion such a scenario for a quantum wire controlled by leaky side gates. As shown in Fig. 3, the contact between the gate and the trench as well as the contact between the trench and the quantum wire form two barriers. A corresponding circuit diagram is presented in Fig. 3(b), which can be simplified to a network reflecting two mesoscopic dipols each defined by a resistor, which is shunted by two quantum capacitances and a geometrical capacitance (Fig. 3(c)). In the circuit diagram the upper path describes the electro-static potentials along the cross section shown adjacently in the left side, whereas the lower path reflects the electro-chemical potentials. In this case the gate efficiency reads as $\eta_{qg} = \left(1+\frac{R_1}{R_2}\right)^{-1}\left(1+\frac{D_{\rm qd}}{D_{C_{B2}}}+\frac{D_{\rm qd}}{D_B}\right)^{-1}$

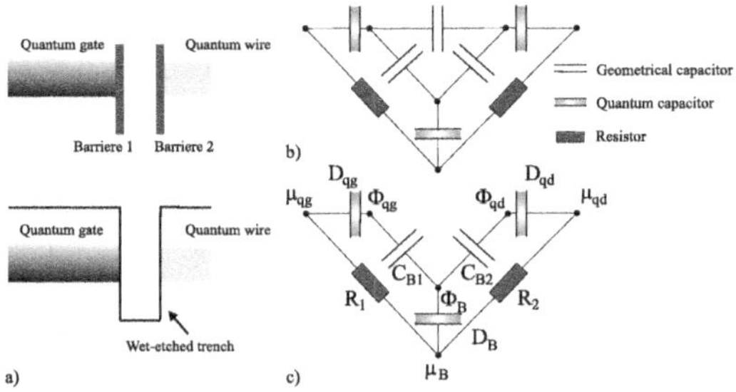

Fig. 3. (**a**) Cross section scheme of a quantum wire with a nearby quantum gate separated by an etched trench (two-barrier-system). (**b**) Equivalent capacitive circuit for the system. (**c**) Simplified network.

Similarly, the gate efficiency of a YBS can be derived. As shown in Fig. 4 a lateral field due to a difference of the electro-chemical potentials in the gates, $\mu_{\rm gl} - \mu_{\rm gr}$, leads to a difference of the electro-static potentials in the drains. Based on the circuit diagram shown in Fig. 4, $(\Phi_{\rm bl} - \Phi_{\rm br}) / (\mu_{\rm gl} - \mu_{\rm gr}) = C_g / (C_g + D_b + 2C_{lr})$. Most interestingly, also a difference of the electro-chemical potentials at the drains leads to a switching field with a self-gating efficiency $\eta_b = D_b / (C_g + D_b + 2C_{\rm lr})$ [8].

4 Self-gating in a Y-branch switch at room temperature

As described above, in a miniaturized YBS the switching operation is not only dependent on geometrical capacitances but also sensitive to quantum capacitances. Particularly in three terminal systems like the YBS the quantum capacity effect is very interesting. Due to an intrinsic parallel connection of quantum capacities self-gating can dominate the switching. Reitzenstein et. al. [6] have observed self-gating in a YBS at cryogenic temperatures. With increasing temperature, the thermal energy kT of the system rises and the self gating effect disappears, because the thermal energy exceeds the switching energy $E_s = \hbar v_F / e L_i < 3$ meV (typical switching length of $L_i = 200\,\text{nm}$).

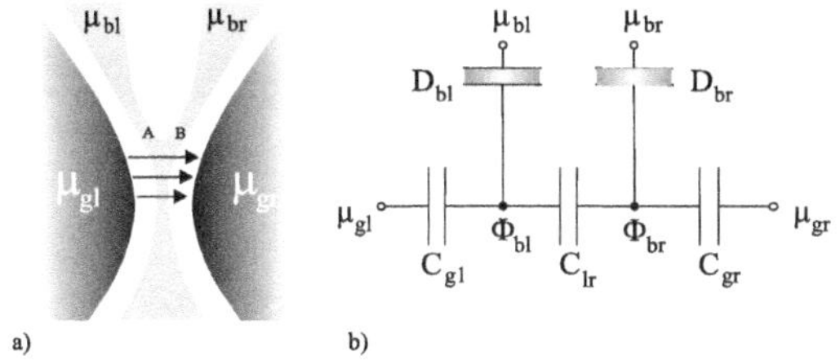

Fig. 4. (**a**) Schematic sketch of a Y-branch structure. The lateral electric field of the sidegates (indicated by the *arrows*) leads to a difference in the electrochemical potentials of the two points A and B. (**b**) Capacitive network of the structure.

In the following, it is demonstrated that self-gating can also be observed at room temperature provided the YBS is very small. The investigated structure was based on a GaAs/AlGaAs heterostructure with a 2DEG only 30 nm below the surface. The upper part of Fig. 5 shows a YBS, which is controlled by four side gates, together with the experimental setup of the measurement.

The bias voltage V_{bias} was applied to the drains in series with resistors ($R_b = 10\,\mathrm{M}\Omega$) with the stem used as ground. For the measurement V_{gl} and V_{gr} serve as input voltages ($V_{\mathrm{gl}} = V_{\mathrm{gl,u}}$ and $V_{\mathrm{gr}} = V_{\mathrm{gr,u}}$). The voltage drop across the branches (V_{bl} or V_{br}) was detected as output signal. By sweeping the side-gate voltages in the push-pull fashion ($\delta V_{\mathrm{gl}} = -\delta V_{\mathrm{gr}}$) the bias voltage was increased from $\Delta V_{\mathrm{bias}} = 0$ up to 0.7 V and the voltages at the drain were detected. The lower part of Fig. 5 shows the output signal (V_{bl} or V_{br}) as a function of the difference of the sidegate voltages $\Delta V_g = V_{\mathrm{gl}} - V_{\mathrm{gr}}$.

In Fig. 6, the maximum differential voltage gain $g_{\mathrm{max}} = (d\Delta V_b)/d\Delta V_g))_{\mathrm{max}}$ is plotted as a function of the forward bias. Interestingly, $g_{\mathrm{max}}(V_{\mathrm{bias}})$ rises superlinearly. This behavior was predicted theoretically [8] as well as reported experimentally [6] indicating a self-gating between the branches. For this, the two switching parameters γ_l and γ_r were introduced which describe the conductance of the left and the right branch, respectively, and depend on the lateral electrical field. These switching parameters do not only take into account voltage differences between the sidegates, but also between the branches.

$$\gamma_l = \tanh\left(\frac{\eta_g(V_g - V_{\mathrm{wp}}) + \eta_b \Delta V_b}{V_S}\right) \tag{1}$$

$$\gamma_r = \tanh\left(\frac{\eta_g(V_g + V_{\mathrm{wp}}) + \eta_b \Delta V_b}{V_S}\right) \tag{2}$$

with η_g being the sidegate efficiency, η_b the gate efficiency between the branches and V_S the switching voltage. The working point voltage V_{wp} defines the condition at which $V_{\mathrm{bl}} = V_{\mathrm{br}}$. For a symmetrical Y-branch structure, $V_{\mathrm{wp}} = 0$ i.e. $\gamma = \gamma_l = \gamma_r$. Subsequently, the currents through the branches can be expressed by

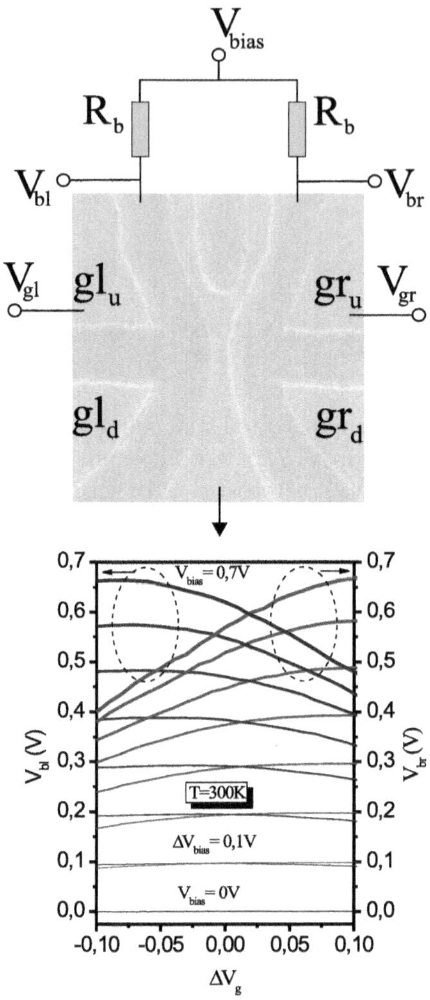

Fig. 5. *Upper part*: Electron microscope image of a YBS. The minimal branch width is 47 nm. *Lower part*: Voltage drop at the different branches as a function of the difference of the side gate voltages in push-pull mode. Self-gating increases with increasing bias voltage.

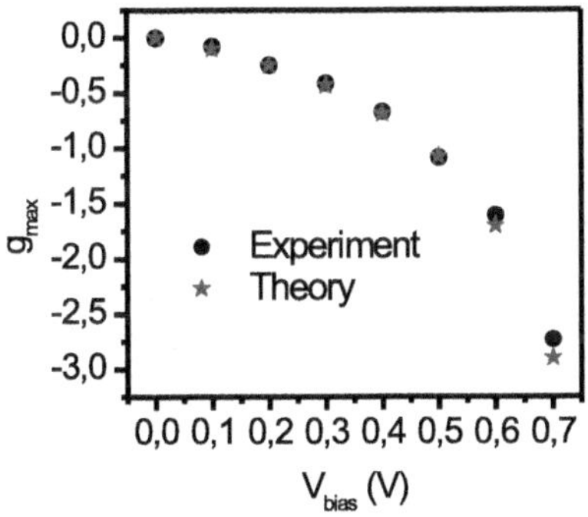

Fig. 6. Maximum differential voltage gain versus V_{bias}.

$$I_l = \frac{G}{2}(1+\gamma)V_{\text{bl}} = \frac{V_{\text{bias}} - V_{\text{bl}}}{R_b} \tag{3}$$

$$I_r = \frac{G}{2}(1-\gamma)V_{\text{bl}} = \frac{V_{\text{bias}} - V_{\text{br}}}{R_b} \tag{4}$$

Equations 3 and 4 finally lead to

$$\Delta V_b = V_{\text{bias}} \left(\frac{1}{1 + GR_b(1+\gamma)} \frac{1}{GR_b(1-\gamma)} \right) . \tag{5}$$

Equation 5 can be solved numerically for calculating the maximum voltage gain $g_{\max} = (d\Delta V_b)/d\Delta V_g))_{\max}$ by considering Eq. (2). A good agreement between the theoretical data and the experimental values was found with $V_{\text{wp}} = 0$ for $\eta_g/V_S = 3.6\,\text{V}^{-1}$ and $\eta_b/V_S = -4.1V^{-1}$. The maximum conductivity of the structure was $G = 1.9 \times 10^{-8}\,\Omega^{-1}$. The differential voltage gain rises quadratically with the forward bias and reaches values with $|\text{d}(\Delta V_b)/\text{d}(\Delta V_g)| > 1$ at room temperature. The ratio between η_g and η_b indicates, that voltage differences between the sidegates dominate the switching process for low bias voltages i.e. small voltage differences δV_b between the branches. However, with increasing bias voltages, self-gating becomes more pronounced and the switching parameter is dominated by the value of the factor $\eta_b V_b$ due to $|\eta_b/\eta_g| > 1$. The sidegate efficiency at room temperature is smaller compared to the gating parameters in Y-branches at helium temperature [6] with $\eta_g/V_S = 10.0\,\text{V}^{-1}$ but the value for the self-gating efficiency is more than a decade larger than $\eta_b/V_S = -0.36\,\text{V}^{-1}$.

5 The Y-branch as logic device

An ideal digital switch is a bistable device where the output voltage jumps instantaneously from a logical high-level (H) to a logical low-level (L) and vice versa as a function of the input voltage [17]. The switching is induced by constant voltages at static inputs or by pulses at dynamical inputs. However, both states are stable. The functionality of a Schmitt-Trigger (ST) was recently demonstrated based on a Y-branch structure with the external setup shown in Fig. 7(a) [18, 19]. Hereby, an external coupling between one branch and the opposite gate was used and the bias voltage V_{bias} was applied to the branches over two external resistors $R_{\mathrm{bl}} = R_{\mathrm{br}} = 10\,\mathrm{M}\Omega$. The ST-characteristics of the device is shown in Fig. 7(c), with the output signal $V_{\mathrm{out}} = V_{\mathrm{bl}}$ remaining in the low state as long as the input voltage V_{gl} is smaller than the threshold voltage $V_{L\rightarrow H}$. When this critical value is reached, the output jumps immediately to the H-state, and remains there, until the input voltage becomes smaller than a second threshold $V_{H\rightarrow L}$, with $V_{H\rightarrow L} < V_{L\rightarrow H}$. The observed switching process is related to a capacitive interplay between the sidegates and the self-induced switching of the branches. An equivalent circuit is shown in Fig. 7, with the branches and the stem each one represented by two parallel n-channel field-effect transistors with different threshold voltages and conductivities. Voltages applied to the gates control the electrical width of the channel and, in first order only, of the branch nearby. The gate efficiency depends on the capacitive coupling between the gate and the channel. The external coupling between the left branch and the right sidegate hereby enhances the controllability of the conductivity of the right branch significantly and bistability can be observed. Such a bistability is shown in Fig. 7(c). The output voltages V_{bl} and V_{br} are switching abruptly with the input voltage V_{gl} and shows hystereses between $V_{\mathrm{gl}} = 0.365\,\mathrm{V}$ and $V_{\mathrm{gl}} = 0.550\,\mathrm{V}$.

In the non-linear transport regime, the experimental results revealed that the Y-branch as a compact switching device shows the functionality of a ST. The interplay between the intrinsic coupling of the branches and the external coupling of the left branch to the right sidegate leads to a feedback loop and enables bistable switching.

Based on three electron YBS without side gates, a fully integrated logic NOR gate was realized. The structure was based on a modulation-doped GaAs/AlGaAs heterostructure with a 2DEG situated approximately 30 nm below the surface. For the lateral definition of the YBSs, the upper layers were locally removed by applying mask technology and wet etching techniques. For good insulation between the conducting regions of the YBSs, the etch depth must be larger than 40 nm. For YBS ballistic rectification has been reported even at room temperature [10, 20–23]. Based on the transistor-like behavior of the structure, the current and voltage amplification can be reached and nanoelectronic devices and logic gates were realized [24–27, 33].

By applying a voltage to the one branch, the conductivity of the other branch and the stem is controlled due to a lateral electric field build up in the

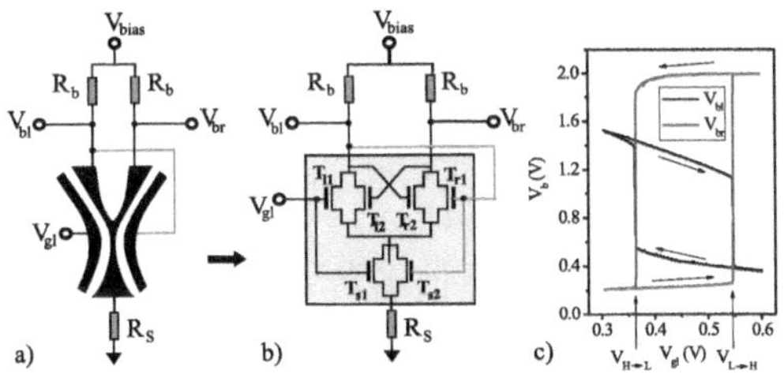

Fig. 7. (**a**) Schematic representations of a YBS as a bistable device (Schmitt-Trigger) with external feedback. V_{gl} served as input while the voltage drops at the branches V_{bl} and V_{br} were measured. V_{bias} was kept at a constant value of 2 V. (**b**)Equivalent circuit of a YBS switch with each branch and the stem represented by two field effect transistors. (**c**) Bistable switching with external feedback. The hysteretic loop is plotted for V_{bl} and V_{br} versus V_{gl}.

branching region. The upper part of Fig. 8(b) shows a scanning electron microscope (SEM) image of a YBS with a 250 nm long and 90 nm wide branching region. The constriction in the gate branch was about 80 nm.

A sketch of the electric setup used for the logic NOR gate is shown in Fig. 8a. The concept of the logic NOR gate is based on a typical n-type MOSFETs architecture with a load transistor. The voltages V_x and V_y were applied to the gate branches of the two YBSs which were used as inputs of the logic NOR gate. The stems of both YBSs were grounded. The gate branch of the third YBS, the so called load transistor, was coupled to the bias voltage V_{bias}. All measurements were performed in the dark at room temperature.

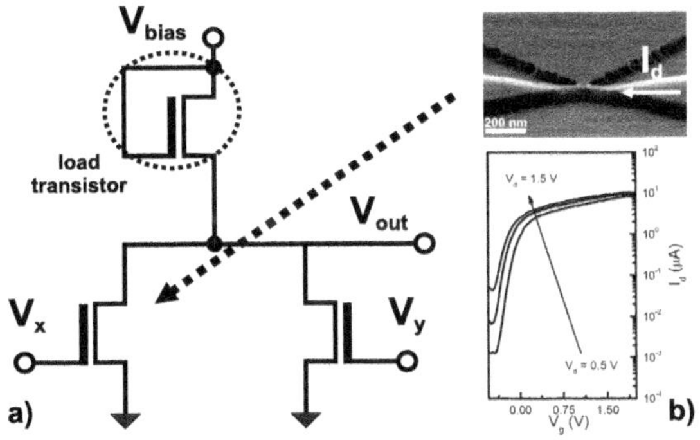

Fig. 8. (**a**) Sketch of the electric setup used for the logic NOR-gate and (**b**) SEM image of a studied YBS together with the transfer characteristics for the drain voltages $V_d = 0.5$, 1.0, and 1.5 V.

The lower part of Fig. 8(b) displays the transfer characteristics of a YBS and points out the transistor-like behavior of the device. Therefore, the drain current I_d was measured versus the gate branch voltage V_g. The drain voltage V_d was applied to the other branch and I_d is shown for $V_d = 0.5$, 1.0, and 1.5 V. Independent of V_d, the branching region of the YBS is depleted for $V_g < -0.6$ V and no drain current is detected. With increasing V_g, I_d increases monotonically and high drain currents up to 10 μA are reached for $V_d \geq 0.5$ V. A maximum transconductance exceeding 10 μA/V is observed for $V_g = 0.0$ and $V_d = 1.0$ V.

Based on the transfer characteristics of the YBSs, the lower input level of the input voltages V_x and V_y was defined as -0.7 V. The upper input level of V_x and V_y was set to -0.2 V, where the saturation regime is reached even for moderate drain voltages. For the NOR-gate function, the bias voltage was set to 10 V.

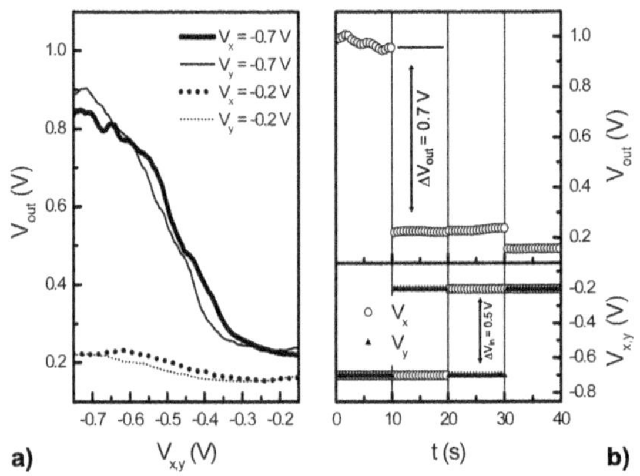

Fig. 9. (**a**) Transients of the output voltage in push-fix mode and (**b**) output voltage for different configurations of V_x and V_y for $V_{\text{bias}} = 10$ V.

The transients of V_{out} are shown in Fig. 9(a) for different configurations of V_x and V_y. As one can clearly see, V_{out} has a maximum value of 0.9 V only if both inputs are equal $V_x = V_y = -0.7$ V. For $V_x = V_y = -0.2$ V or at least one input equal -0.2 V, minimum output voltages in the range of 0.17–0.24 V are reached. Therefore, the output corresponds to the truth of a logic NOR gate. Figure 9(b) displays a similar measurement, whereas the applied input voltages are shown in the lower part and the resulting output voltage is shown in the upper part. Clearly, V_{out} reaches its maximum value only for $V_x = V_y = -0.7$ V.

A minimum voltage difference ΔV_{out} of 0.7 V between the high level and the low levels of V_{out} is observed in Fig. 9(b). Therefore, as $\Delta V_{\text{in}} = 0.5$ V, ΔV_{out} is 1.4 times larger than the voltage difference between the input levels.

In addition, maximum differential voltage gain of −3.9 and −4.3 was extracted for dV_{out}/dV_x with $V_y = -0.7\,\text{V}$ and dV_{out}/dV_y with $V_x = -0.7\,\text{V}$, respectively.

6 Acknowledgements

In summary, we have realized nanoelectronic Y-branch switches and studied several non-linear transport properties. Selfgating, amplification and bistable switching were observed and exploited for logic circuits. We are grateful for financial support through SUBTLE (European Commission), FORNEL (Bayerische Forschungsstiftung), nanoQUIT (BMBF) and the state of Bavaria. Expert sample preparation by Monika Emmerling is gratefully acknowledged.

References

1. Y. Imry: *Directions in Condensed Matter Physics*, World Scientific Publ., Singapore (1986)
2. S. Datta: *Electronic Transport in Electronic System*, Cambridge University Press, Cambridge (1995)
3. D. K. Ferry, S. M. Goodnick: *Transport in Nanostructures*, Cambridge University Press, Cambridge (1997)
4. B. J. van Wees, H. van Houten, C. W. J. Beenakker, J. G. Williamson, L. P. Kouwenhoven, D. van der Marel, C. T. Foxon: Phys. Rev. Lett. **66**, 848 (1988)
5. D. A. Wharam, T. J. Thornton, R. Newbury, M. Pepper, H. Ahmed, J. E. F. Frost, D. G. Hasko, D. C. Peacock, D. A. Richtie, G. A. C. Jones: J. Phys. C: Solid State Phys. **21**, L209 (1988)
6. S. Reitzenstein, L. Worschech, P. Hartmann, M. Kamp, A. Forchel: Phys. Rev. Lett. **89**, 226804 (2002)
7. L. Worschech, D. Hartmann, S. Reitzenstein, A. Forchel: J. Phys.: Condens. Matter **17**, R775 (2005)
8. J.-O. J. Wesström: Phys. Rev. Lett **82**, 2564 (1999)
9. T. Palm: Appl. Phys. Lett. **60**, 237 (1992)
10. L. Worschech, H. Q. Xu, A. Forchel, L. Samuelson: Appl. Phys. Lett. **79**, 3287 (2001)
11. H. Q. Xu: Nat. Mater. **4**, 649 (2005)
12. H. L. Störmer, R. Dingle, A. C. Gossard, W. Wiegmann: Proceedings of the 14th International Conference on the Physics of Semiconductors, Edinburgh: edited by B. L. H. Wilson, 557 (1978)
13. A. D. Wieck, K. Ploog: Appl. Phys. Lett. **56** 928 (1990)
14. J. Nieder, A. D. Wieck, P. Grambow, H. Lage, D. Heitmann, K. von Klitzing: Appl. Phys. Lett. **57** 2695 (1990)
15. H. Okada, K. Jinushi, N. Wu, T. Hashizume, H. Hasegawa: Jpn. J. Appl. Phys. **34**, 1315 (1995)
16. S. Luryi: Appl. Phys. Lett. **52**, 501 (1988)
17. P. Horowitz, W. Hill: *The Art of Electronics*, Cambridge University Press, Cambridge (1989)

18. S. Reitzenstein, L. Worschech, P. Hartmann, A. Forchel: Phys. Rev. Lett. **82**, 1980 (2003)
19. D. Hartmann, S. Reitzenstein, L. Worschech, A. Forchel: Electron. Lett. **41**, 303 (2005)
20. H. Q. Xu: Appl. Phys. Lett. **78**, 2064 (2001)
21. J. Mateos, B. G. Vasallo, D. Pardo, T. González, E. Pichonat, J.-S. Galloo, S. Bollaert, Y. Roelens, A. Cappy: IEEE Electron Device Lett. **25**, 235 (2004)
22. L. Rashmi, B. Bednarz, Hackens, G. Farhi, V. Bayot, I. Huynen: Solid State Comm. **134**, 217 (2005)
23. D. Wallin, I. Shorubalko, H. Q. Xu, A. Cappy: Appl. Phys. Lett. **89**, 092124 (2006)
24. H. Q. Xu, I. Shorubalko, D. Wallin, I. Maximov, P. Omling, L. Samuelson, W. Seifert: IEEE Electron Device Lett. **25**, 164 (2004)
25. S. Reitzenstein, L. Worschech, A. Forchel: IEEE Electron Device Lett. **25**, 462 (2004)
26. C. R. Müller, L. Worschech, D. Spanheimer, A. Forchel: IEEE Electron Device Lett. **27**, 208 (2006)
27. C. R. Müller, L. Worschech, P. Höpfner, S. Höfling, A. Forchel: IEEE Electron Device Lett. **28**, 859 (2007)
28. D. K. De Vries, P. Stelmaszyk, A. D. Wieck: J. Appl. Phys. **79**, 8087 (1996)
29. Ya. M. Blanter, F. W. J. Hekking, M. Büttiker: Phys. Rev. Lett. **81**, 1925 (1998)
30. M. Russ, C. Meier, A. Lorke, D. Reuter, A. D. Wieck: Phys. Rev. B **73**, 115334 (2006)
31. D. Grimm, P. Venezuela, A. Latge: Phys. Rev. B **71**, 155425 (2005)
32. K.-K. Voo, S.-C. Chen, C.-S. Tang, C.-S. Chu: Phys. Rev. B **73**, 035307 (2006)
33. V. I. Talyanskii, M. R. Graham, H. E. Beere: Appl. Phys. Lett. **88**, 083501 (2006)

Rectification Through Entropic Barriers

Gerhard Schmid, P. Sekhar Burada, Peter Talkner, and Peter Hänggi

Department of Physics, University of Augsburg, Universitätsstr. 1, 86135, Augsburg, Germany
Gerhard.Schmid@physik.uni-augsburg.de

Abstract. The dynamics of Brownian motion has widespread applications extending from transport in designed micro-channels up to its prominent role for inducing transport in molecular motors and Brownian motors. Here, Brownian transport is studied in micro-sized, two dimensional periodic channels, exhibiting periodically varying cross sections. The particles in addition are subjected to a constant external force acting alongside the direction of the longitudinal channel axis. For a fixed channel geometry, the dynamics of the two dimensional problem is characterized by a single dimensionless parameter which is proportional to the ratio of the applied force and the temperature of the environment. In such structures entropic effects may play a dominant role. Under certain conditions the two dimensional dynamics can be approximated by an effective one dimensional motion of the particle in the longitudinal direction. The Langevin equation describing this reduced, one dimensional process is of the type of the Fick-Jacobs equation. It contains an entropic potential determined by the varying extension of the eliminated transversal channel direction, and a correction to the diffusion constant that introduces a space dependent diffusion. We analyze the influence of broken channel symmetry and the validity of the Fick-Jacobs equation. For the nonlinear mobility we find a temperature dependence which is opposite to that known for particle transport in periodic energetic potentials. The influence of entropic effects is discussed for both, the nonlinear mobility, and the effective diffusion constant. In case of broken reflection symmetry rectification occurs and there is a favored direction for particle transport. The rectification effect could be maximized due to the optimal chosen absolute value of the applied bias.

1 Introduction

The phenomenon of entropic transport is ubiquitous in biological cells, ion channels, nano-porous materials, zeolites and microfluidic devices etched with grooves and chambers. Instead of diffusing freely in the host liquid phase the Brownian particles frequently undergo a constrained motion. The geometric restrictions to the system's dynamics results in entropic barriers and regulate the transport of particles yielding important effects exhibiting peculiar properties. The results have prominent implications in processes such as catalysis, osmosis and particle separation [1–12] and, as well, for the noise-induced transport in periodic potential landscapes that lack reflection symmetry (Brownian ratchet systems) [13–15] or Brownian motor transport occurring in arrays of periodically arranged asymmetric obstacles, termed "entropic" ratchet devices [16–20]. Motion in these systems can be induced by imposing different concentrations at the ends of the channel, or by the presence of external driving forces supplying the particles with the energy necessary to proceed. The study of the kinetics of the entropic transport, the properties of transport coefficients in far from equilibrium situations and the possibility for transport control mechanisms are pertinent objectives in the dynamical characterization of those systems.

Because the role of inertia for the motion of the particles through these structures can typically be neglected the Brownian dynamics can safely be analyzed by solving the Smoluchowski equation in the domain defined by the available free space upon imposing the appropriate boundary conditions. Whereas this method has been very successful when the boundaries of the system possess a rectangular shape, the challenge to solve the boundary value problem in the case of nontrivial, corrugated domains represents a difficult task. A way to circumvent this difficulty consists in coarsening the description by reducing the dimensionality of the system, keeping only the main direction of transport, but taking into account the physically available space by means of an entropic potential. The resulting kinetic equation for the probability distribution, the so called Fick-Jacobs (FJ) equation, is similar in form to the Smoluchowski equation, but now contains an entropic term. The entropic nature of this term leads to a genuine dynamics which is distinctly different from that observed when the potential is of energetic origin [21]. It has been shown that the FJ equation can provide a very accurate description of entropic transport in channels of varying cross-section [21–24]. However, the derivation of the FJ equation entails a tacit approximation: The particle distribution in the transversal direction is assumed to equilibrate much faster than in the main (unconstrained) direction of transport. This equilibration justifies the coarsening of the description leading in turn to a simplification of the dynamics, but raises the question about its validity when an *external force is applied.* To establish the validity criterion of a FJ description for such biased diffusion in confined media is, due to the ubiquity of this situation, a subject of primary importance.

Our objective with this work is to investigate in greater detail the FJ-approximation for biased diffusion and to study rectification due to the asymmetry of a geometrical confinement. We will analyze the biased movement of Brownian particles in 2D periodic, but asymmetric channels of varying cross-section. On the basis of our numerical and analytical results we recapitulate the striking and sometimes counterintuitive features [21], which arises from entropic transport and which are different from those observed in the more familiar case with energetic, metastable landscapes [25].

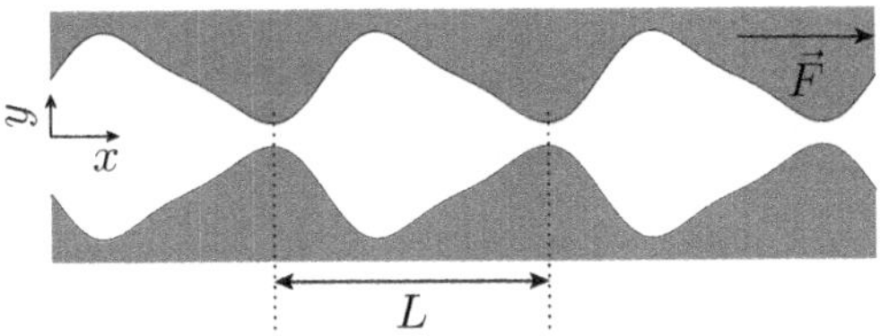

Fig. 1. Schematic diagram of a channel confining the motion of forced Brownian particles. The half-width ω is a periodic function of x with periodicity L.

2 Diffusion in confined systems

Transport through pores or channels (like the one depicted in Fig. 1) may be caused by different particle concentrations maintained at the ends of the channel, or by the application of external forces acting on the particles. Here we will exclusively consider the case of force driven transport. The external driving force is denoted by $\boldsymbol{F} = F\boldsymbol{e}_x$. It points into the direction of the channel axis. In general, the dynamics of a suspended Brownian particles is overdamped [26] and well described by the Langevin equation in dimensionless variables [23],

$$\frac{\mathrm{d}\boldsymbol{r}}{\mathrm{d}t} = \boldsymbol{f} + \boldsymbol{\xi}(t)\,, \tag{1}$$

where t is dimensionless time, $\boldsymbol{r}$ corresponds to the position vector of the particle (given in units of the period length L), $\boldsymbol{\xi}$ to Gaussian white noise with $\langle \boldsymbol{\xi}(t) \rangle = 0$ and $\langle \boldsymbol{\xi}(t)\boldsymbol{\xi}(t') \rangle = 2\delta_{i,j}\delta(t-t')$ for $i,j = x,y,z$ and with the dimensionless force

$$\boldsymbol{f} = f\boldsymbol{e}_x \text{ and } f = \frac{LF}{k_\mathrm{B}T}\,. \tag{2}$$

The dimensionless parameter f characterizes the force as the ratio of the work which it performs on the particle along a distance of the length of the period L and the thermal energy $k_\mathrm{B}T$.

The boundary of the 2D periodic channel which is mirror symmetric about its axis is given by the periodic functions $y = \pm\omega(x)$, i.e. $\omega(x+1) = \omega(x)$ for all x, where x and y are the Cartesian components of $\boldsymbol{r}$. Except for a straight channel with $\omega = const$ there are no periodic channel shapes for which an exact analytical solution of Eq. (1) and the corresponding Fokker-Planck equation with boundary conditions is known [27, 28]. Approximate solutions though can be obtained on the basis of an one dimensional diffusion problem in an effective potential. Narrow channel openings, which act as geometric hindrances in the full model, show up as entropic barriers in this one dimensional approximation [21–23, 29–32]. This approach is valid under conditions that will be discussed below in some detail.

3 Transport in periodic channels with broken symmetry

3.1 The Fick-Jacobs approximation

In the absence of an external force, i.e. for $\boldsymbol{f} = 0$, it was shown [29–32] that the dynamics of particles in confined structures (such as that of Fig. 1) can be described approximatively for $|\omega'(x)| \ll 1$ by the FJ equation, with a spatial dependent diffusion coefficient:

$$\frac{\partial P(x,t)}{\partial t} = \frac{\partial}{\partial x} D(x) \left(\frac{\partial P(x,t)}{\partial x} - \frac{\omega'(x)}{\omega(x)} P(x,t) \right) , \tag{3}$$

obtained from the full 2D Smoluchowski equation upon the elimination of the transversal y coordinate assuming fast equilibration in that direction. Here $P(x,t) = \int_{-\omega(x)}^{\omega(x)} \mathrm{d}y \, P(x,y,t)$ denotes the marginal probability density along the axis of the channel. We note that for three dimensional channels an analogue approximate Fokker-Planck equation holds in which the function $\omega(x)$ is to be replaced by $\pi\omega^2(x)$ (area of cross-section). The prime refers to the derivative of the function with respect to its argument, i.e. $\omega'(x) = \mathrm{d}\omega/\mathrm{d}x$. In the original work by Jacobs [29] the 1D diffusion coefficient $D(x)$ is constant and equals the bare diffusion constant which is unity in the present dimensionless variables. Later, Zwanzig [30] and Reguera and Rubí [31] proposed different spatially dependent forms of the 1D diffusion coefficient which allows for an extended regime of validity of the FJ-description.

Reguera and Rubí [31] put forward this form of the 1D diffusion coefficient:

$$D(x) = \frac{1}{(1 + \omega'(x)^2)^{\gamma}} , \tag{4}$$

where $\gamma = 1/3$ for 2D structures and $\gamma = 1/2$ for 3D systems. The right hand side of Eq. (4) can be considered as a resummation of Zwanzig's original perturbational result [30].

In the presence of a constant force F along the direction of the channel the FJ equation (3) can be recast into the form [21–23, 31]:

$$\frac{\partial P}{\partial t} = \frac{\partial}{\partial x} D(x) \left(\frac{\partial P}{\partial x} + \frac{\mathrm{d}A(x)}{\mathrm{d}x} P \right) \tag{5}$$

with the dimensionless free energy $A(x) := E - S = -f\,x - \ln\omega(x)$. In physical dimensions the energy is $\tilde{E} \equiv k_{\mathrm{B}}TE = -F\tilde{x}$ ($\tilde{x} = xL$) and the dimensional entropic contribution is $\tilde{S} \equiv k_{\mathrm{B}}TS = k_{\mathrm{B}}T \ln\omega$. For a periodic channel with broken reflection symmetry this free energy assumes the form of a tilted periodic, ratchet-like potential. In the absence of a force the free energy is purely entropic and Eq. (5) reduces to the FJ equation (3). On the other hand, for a straight channel the entropic contribution vanishes and the particle is solely driven by the external force.

3.2 Transport characteristics

Key quantities of particle transport through periodic channels are the average *particle current*, or equivalently the nonlinear *mobility*, and the *effective diffusion coefficient.* For a particle moving in a one dimensional tilted *energetic* periodic potential the heights ΔE of the barriers separating the potential wells provide an additional energy scale apart from the work of the force FL and the thermal energy $k_B T$. Hence, at least two dimensionless parameters, say $\Delta E/(k_B T)$ and $FL/(k_B T)$ govern the transport properties of these systems. In contrast, as already noted in the context of the full 2D model the transport through channels is governed by the single dimensionless parameter $f = FL/(k_B T)$ [21–23]. This, of course, remains to hold true in the one dimensional approximation which models the transversal spatial variation in terms of an entropic potential.

For any non negative force the average particle current in periodic structures can be obtained from mean-first-passage time analysis [21–23, 34, 35], i.e. the average particle current $\langle \dot{x} \rangle$ is given as ratio of period length L and the mean-first-passage time $\langle T \rangle$ for a particle to overcome one period length, i.e. in dimensionless units: $\langle \dot{x} \rangle = 1/\langle T \rangle$.

The nonlinear mobility $\mu(f)$ is defined by

$$\mu(f) = \frac{\langle \dot{x} \rangle}{f} \,. \tag{6}$$

Consequently, one can obtain the following Stratonovich formula for the nonlinear mobility [21–23]

$$\mu(f) = \frac{1 - \exp(-f)}{f \displaystyle\int_0^1 \mathrm{d}z \, I(z,f)} \,, \tag{7}$$

where

$$I(z,f) := \frac{h^{-1}(z)}{D(z)} \exp(-f\, z) \int_{z-1}^{z} \mathrm{d}\tilde{z} \, h(\tilde{z}) \exp(f\, \tilde{z}) \,, \tag{8}$$

depends on the dimensionless position z, the force f and the shape of the tube given in terms of the half width $\omega(x)$ and its first derivative.

The effective diffusion coefficient of the movement alongside the channel axis is defined as the asymptotic behavior of the variance of the position

$$D_{\text{eff}} = \lim_{t \to \infty} \frac{\langle x^2(t) \rangle - \langle x(t) \rangle^2}{2t} \,. \tag{9}$$

It is related to the first two moments of the first passage time $\langle T \rangle$ and $\langle T^2 \rangle$ by the expression [34–36]:

$$D_{\text{eff}} = \frac{\langle T^2 \rangle - \langle T \rangle^2}{2\, \langle T \rangle^3} \,. \tag{10}$$

After some algebra it can be transformed to read

$$D_{\text{eff}} = \frac{\displaystyle\int_0^1 \mathrm{d}z \int_{z-1}^{z} \mathrm{d}\tilde{z}\, \mathcal{N}(z,\tilde{z},f)}{\left[\displaystyle\int_0^1 \mathrm{d}z\, I(z,f)\right]^3}, \tag{11}$$

where

$$\mathcal{N}(z,\tilde{z},f) := \frac{D(\tilde{z})}{h(z)}\,\frac{h(\tilde{z})}{D(z)}\,[I(\tilde{z},f)]^2 \exp(-f\,z + f\,\tilde{z}). \tag{12}$$

The predicted dependence of the average particle current and the effective diffusion coefficient was compared with 2D Brownian dynamic simulations performed by a numerical integration of the Langevin equation (1), within the stochastic Euler-algorithm. The shape of the exemplarily taken 2D channel is described (in dimensionless units) by

$$\omega(x) := \sin(2\pi x) + 0.25\,\sin(4\pi x) + 1.12. \tag{13}$$

For the considered channel configuration, the widest opening of the channel is by a factor of 116 larger than the width at the narrowest openings, i.e. at the bottlenecks. One may therefore expect strong entropic effects for these channels. The particle current and effective diffusion coefficient were derived from an ensemble-average of about $3 \cdot 10^4$ trajectories:

$$\langle \dot{x} \rangle = \lim_{t\to\infty} \frac{\langle x(t) \rangle}{t}, \tag{14}$$

and Eq. (9), respectively.

Transport in one dimensional periodic *energetic* potentials behaves very differently from one dimensional periodic systems with *entropic* barriers [21]. The fundamental difference lies in the temperature dependence of these models. Decreasing temperature in an energetic periodic potential decreases the transition rates from one period to the neighboring by decreasing the Arrhenius factor $\exp\{-\Delta V/(k_{\mathrm{B}}T)\}$ where ΔV denotes the activation energy necessary to proceed by a period [25]. Hence decreasing temperature leads to a decrease in mobility. For a one dimensional periodic system with an entropic potential, a decrease of temperature leads to an increase of the dimensionless force parameter f and consequently to an increase of the mobility, cf. Fig. 2.

The reduction of dimensionality leading to the FJ equation relies on the assumption of equilibration in the transversal direction which results in an almost uniform distribution of the transversal positions y at fixed values of the longitudinal coordinate x. One can formulate criteria determining whether the FJ equation approximatively describes the stationary state of the considered problem [23]. For channels with varying width the narrow positions confine the positions of the particles. From there they are dragged by the force and

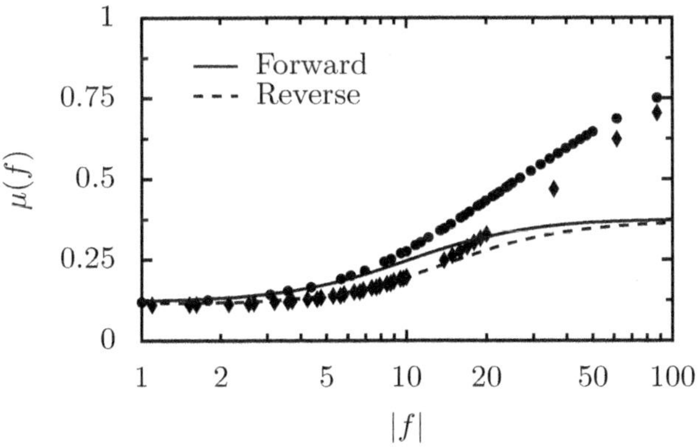

Fig. 2. The numerically simulated (*symbols*) and analytically calculated (cf. Eq. (7) – lines) dependence of the absolute value of the nonlinear mobility $\mu(f)$ *vs.* the dimensionless force $f = FL/k_B T$ is depicted for a 2D channel with the scaled half-width given by $\omega(x) = \sin(2\pi x) + 0.25\sin(4\pi x) + 1.12$; for transport in positive x-direction, i.e. positive f: *circles and solid line*, for negative f-values: diamonds and *dashed line*. For the linear response regime, i.e. small $|f|$, the nonlinear mobility for forward and backward transport converge to each other.

– at the same time – they perform a diffusive motion until the channel narrows again. The required uniform distribution in the transversal direction can only be achieved if the diffusional motion is fast enough in comparison to the deterministic drift under the influence of f. Therefore the time scale of equilibration in transversal direction must be short compared to the time it takes to drag a particle from the narrow position to the position with largest channel width. The latter requirement leads to an estimate of the minimal forcing above which the FJ description is expected to fail in providing an accurate description of the transport properties in the long time limit [22, 23]. Detailed analysis demonstrates that, for the considered asymmetric channel, cf. Fig. 1, the FJ description holds for larger force value when forcing towards the negative x-direction than for forcing in the positive x-direction, cf. Fig. 2. The dependence of the nonlinear mobility on the direction of the forcing which arises due to the asymmetry of the shape of the channel walls is addressed in Sect. 3.3.

Another interesting effect can be observed for the effective diffusion if looked as a function of the force f. Already the expression for the effective diffusion (11) which follows rigorously from the FJ equation displays a maximum as a function of f which may even exceed the value 1 of the bare diffusion, cf. Fig. 3. For $f \to \infty$ the effective diffusion approaches the bare value 1. If one decreases the force to finite but still large values then the stationary distribution acquires a finite width in the transversal direction with a "crowded" region in front of the narrowest place of the channel Refs. [22, 23]. The transport becomes more noisy and consequently the effective diffusion exceeds the

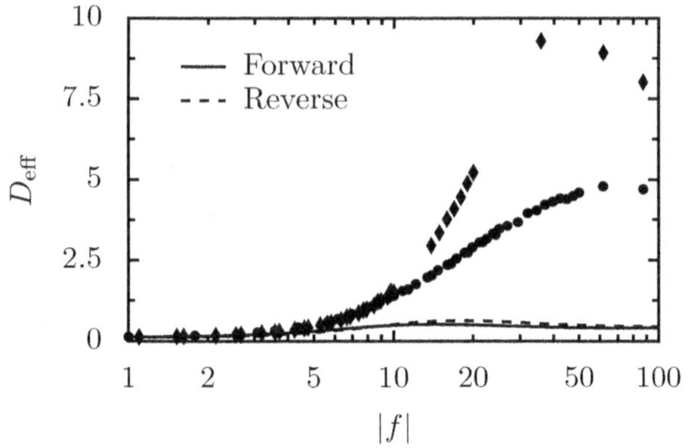

Fig. 3. The numerically simulated (*symbols*) and analytically calculated (cf. Eq. (11) – lines) dependence of the effective diffusion coefficient D_{eff} is depicted *vs.* the dimensionless force $f = FL/k_{\mathrm{B}}T$ for two channels in 2D. For both channels the scaled half-width is given by $\omega(x) = \sin(2\pi x) + 0.25\sin(4\pi x) + 1.12$; transport in positive x direction ($f > 0$): *circles and solid line*; transport in negative x direction ($f < 0$): *diamonds and dashed line.*

bare value 1. On the other hand if one starts at $f = 0$ the entropic barriers diminish the diffusion such that the effective diffusion is less than bare diffusion. Consequently, somewhere in between there must be a value of f with maximal effective diffusion [21–23]. For the considered 2D channel defined by Eq. (13) the value of the force at the maximal effective diffusion is outside the regime of validity of the FJ equation for both forcing directions. The numerical simulations give a much more pronounced peak of the effective diffusion. These observations lead us to the conclusion that entropic effects increase the randomness of transport through a channel and in this way decrease the mobility and increase the effective diffusion. A similar enhancement of effective diffusion was found in titled periodic *energetic* potentials [33–36].

3.3 Rectification in asymmetric channels

Due to different focussing towards the narrowest width of the channel, i.e. the bottlenecks, which is a consequence of the broken reflection symmetry of the channel, the nonlinear mobility depends on the direction of the constant bias and not only on its absolute value, cf. Fig. 2. Moreover, within the FJ description an asymmetric shape of the channel walls leads to a ratchet-like free energy landscape facilitating the transport in one direction. The rectification thereby not only depends on the channel geometry but also on f. To quantify the rectification effect, we define the quantity

$$\alpha = \frac{|\mu(f) - \mu(-f)|}{\mu(f) + \mu(-f)}\,, \tag{15}$$

as a rectification-measure.

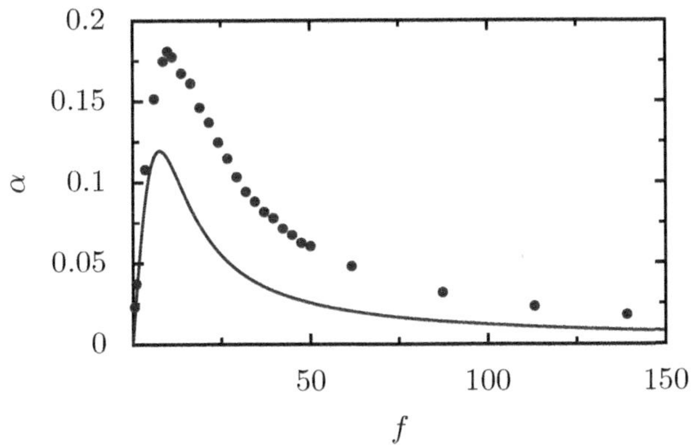

Fig. 4. The numerical simulated dependence of the rectification measure α, cf. Eq. (15), giving the relative discrepancy of the nonlinear mobilities in positive and negative x-direction on the force value f is depicted by the *symbols*. There is an optimal forcing value for which the rectification for the given 2D channel, cf. Eq. (13), is maximal. Within the FJ description (*solid line*) the same qualitative behavior could be observed.

In Fig. 4 the dependence of the rectification measure α on the force f is depicted. Interestingly, there is an optimal value for f where the rectification is maximum. Due to the favoring of transport in one direction for finite f-values there is rectification, whereas for $f \to 0$ (linear response) and $f \to \infty$ (corresponding to a flat channel geometry [22, 23]) the nonlinear mobilities for forward and backward propagation equal each other, i.e. $\mu(f) = \mu(-f)$. Consistently, within the FJ approximation the qualitative behavior could be observed.

4 Conclusions

In summary, we demonstrated that transport phenomena in periodic channels with varying width exhibit some features that are radically different from conventional transport occurring in energetic periodic potential landscapes. The most striking difference between these two physical situations lies in the fact that for a fixed channel geometry the dynamics is characterized by a single parameter $f = FL/(k_B T)$ which combines the external force F causing a drift, the period length L of the channel, and the thermal energy $k_B T$, which is a measure of the strength of the acting fluctuating forces. Transport in periodic energetic potentials depends, at least, on one further parameter which is the activation energy of the highest barrier separating neighboring periods. This leads to an opposite temperature dependence of the mobility. While the mobility of a particle in an energetic potential increases with increasing temperature the mobility of a particle in a channel of periodically varying width decreases. The incorporation of the spatial variation of the channel width as an entropic potential in the FJ equation allows a qualitative understanding of the dependence of the transport properties on the channel geometry. In channels without a mirror symmetry about a vertical axis rectification favoring transport in one channel direction occurs. An optimal forcing regime could be found for which the rectification effect is maximal.

The effective diffusion exhibits a non monotonic dependence versus the dimensionless force f. It starts out at small f with a value that is less than the bare diffusion constant, reaches a maximum with increasing f and finally approaches the value of the bare diffusion from above.

Under certain conditions, the two dimensional Fokker-Planck equation governing the time dependence of the probability density of a particle in the channel can be approximated by one dimensional Fokker-Planck equation: the approximated equation is termed the Fick-Jacobs equation; it contains an entropic potential and a position dependent diffusion coefficient. In principle the FJ equation describes both the transient behavior of a particle and also the stationary behavior of the particle dynamics which is approached in the limit of large times. In this paper we demonstrated the suitability of the FJ approximation on describing the biased Brownian motion in periodic channels with broken symmetry where rectification takes place. Although we restricted our discussion to two dimensional channels, a generalization of the presented methods to three dimensional pores with varying cross section is straight forward.

5 Acknowledgements

The authors acknowledge D. Reguera and J.M. Rubí (Universidad de Barcelona) for fruitful discussions. This work has been supported by the DFG via research center, SFB-486, project A10, the Volkswagen Foundation (project I/80424), and by the German Excellence Initiative via the *Nanosystems Initiative Munich* (NIM).

References

1. L. Liu, P. Li, S. A. Asher: Entropic trapping of macromolecules by mesoscopic periodic voids in a polymer hydrogel, Nature **397**, 141 (1999)
2. Z. Siwy, I. D. Kosinska, A. Fulinski, C. R. Martin: Asymmetric diffusion through synthetic nanopores, Phys. Rev. Lett. **94**, 048102 (2005)
3. A. M. Berezhkovskii, S. M. Bezrukov: Optimizing transport of metabolites through large channels: Molecular sieves with and without binding, Biophys. J. **88**, L17 (2005)
4. B. Hille: *Ion Channels of Excitable Membranes* (Sinauer, Sunderland, 2001)

5. R. M. Barrer: *Zeolites and Clay Minerals as Sorbents and Molecular Sieves* (Academic Press, London, 1978)
6. T. Chou, D. Lohse: Entropy-driven pumping in zeolites and biological channels, Phys. Rev. Lett. **82**, 3552 (1999)
7. C. Kettner, P. Reimann, P. Hänggi, F. Müller: Drift Ratchet, Phys. Rev. E **61**, 312 (2000)
8. S. Matthias, F. Müller: Asymmetric pores in a silicon membrane acting as massively parallel Brownian ratchets, Nature **424**, 53 (2003)
9. B. Ai, L. Liu: Current in a three-dimensional periodic tube with unbiased forces, Phys. Rev. E **74**, 05114 (2006)
10. W. D. Volkmuth, R. H. Austin: DNA electrophoresis in microlithographic arrays, Nature **358**, 600 (1992)
11. G. I. Nixon, G. W. Slater: Saturation and entropic trapping of monodisperse polymers in porous media, J. Chem. Phys. **117**, 4042 (2002)
12. R. Chang, A. Yethiraj: Dynamics of chain molecules in disordered materials, Phys. Rev. Lett. **96**, 107802 (2006)
13. P. Hänggi, F. Marchesoni, F. Nori: Brownian motors, Ann. Physik (Berlin) **14**, 51 (2005)
14. R. Astumian, P. Hänggi: Brownian motors, Phys. Today **55**, 33 (2002)
15. P. Reimann, P. Hänggi: Introduction to the physics of Brownian motors, Appl. Phys. A **75**, 169 (2002)
16. I. Derenyi, R. Astumian: Ac separation of particles by biased Brownian motion in a two-dimensional sieve, Phys. Rev. E **58**, 7781 (1998)
17. T. A. J. Duke, R. H. Austin: Microfabricated sieve for the continuous sorting of macromolecules, Phys. Rev. Lett. **80**, 1552 (1998)
18. A. V. Oudenaarden, S. Boxer: Brownian ratchets: Molecular separations in lipid bilayers supported on patterned arrays, Science **285**, 1046 (1999)
19. M. Kostur, L. Schimansky-Geier: Numerical study of diffusion induced transport in 2D systems, Phys. Lett. **265**, 337 (2000)
20. C. Keller, F. Marquardt, C. Bruder: Separation quality of a geometric ratchet, Phys. Rev. E **65**, 041927 (2002)
21. D. Reguera, G. Schmid, P. S. Burada, J. M. Rubí, P. Reimann, P. Hänggi: Entropic transport: Kinetics, scaling and control mechanisms, Phys. Rev. Lett. **96**, 130603 (2006)
22. P. S. Burada, G. Schmid, D. Reguera, J. M. Rubí, P. Hänggi: Biased diffusion in confined media: Test of the Fick-Jacobs approximation and validity criteria, Phys. Rev. E **75**, 051111 (2007)
23. P. S. Burada, G. Schmid, P. Talkner, P. Hänggi, D. Reguera, J. M. Rubí: Entropic particle transport in periodic channels, Biosystems **93**, 16 (2008)
24. I. D. Kosinska, I. Goychuk, M. Kostur, G. Schmid, P. Hänggi: Rectification in synthetic conical nanopores: A one-dimensional Poisson-Nernst-Planck modeling, Phys. Rev. E **77**, 031131 (2008)
25. P. Hänggi, P. Talkner, M. Borkovec: Reaction rate theory: Fifty years after Kramers, Rev. Mod. Phys. **62**, 251 (1990)
26. E. Purcell: Life at low Reynolds-number, Am. J. Phys. **45**, 3 (1977)
27. H. Risken: *The Fokker-Planck Equation* (Springer, Berlin, 1989)
28. P. Hänggi, H. Thomas: Stochastic processes: Time-evolution, symmetries and linear response, Phys. Rep. **88**, 207 (1982)
29. M. Jacobs: *Diffusion Processes* (Springer, New York, 1967)

30. R. Zwanzig: Diffusion past an entropic barrier, J. Phys. Chem. **96**, 3926 (1992)
31. D. Reguera, J. M. Rubí: Kinetic equations for diffusion in the presence of entropic barriers, Phys. Rev. E **64**, 061106 (2001)
32. P. Kalinay, J. K. Percus: Corrections to the Fick-Jacobs equation, Phys. Rev. E **74**, 041203 (2006)
33. G. Costantini, F. Marchesoni: Threshold diffusion in a tilted washboard potential, Europhys. Lett. **48**, 491 (1999)
34. P. Reimann, C. Broeck, H. Linke, P. Hänggi, J. M. Rubí, A. Pérez-Madrid: Giant acceleration of free diffusion by use of tilted periodic potentials, Phys. Rev. Lett. **87**, 010602 (2001)
35. P. Reimann, C. Broeck, H. Linke, P. Hänggi, J. M. Rubí, A. Pérez-Madrid: Diffusion in tilted periodic potentials: Enhancement, universality, and scaling, Phys. Rev. E **65**, 031104 (2002)
36. B. Lindner, M. Kostur, L. Schimansky-Geier: Optimal diffusive transport in a tilted periodic potential, Fluct. Noise Lett. **1**, R25 (2001)

Part VI

Characterization of Materials and Devices

Microstructure Tomography – An Essential Tool to Understand 3D Microstructures and Degradation Effects

Alexandra Velichko[1] and Frank Mücklich[2]

[1] Functional Materials, Saarland University, Postbox 151150, 66041Saarbrücken, Germany
a.velitchko@matsci.uni-sb.de
[2] Functional Materials, Saarland University, Postbox 151150, 66041 Saarbrücken, Germany
muecke@matsci.uni-sb.de

Abstract. In materials science the 3D morphology of the structure is the key to understand the relationships between the manufacturing parameters and the properties of the material. The question is at which conditions two-dimensional characterization and the application of the stereological relations can provide sufficient information and when 3D analysis is indispensible or offers additional insights. Correct description of the 3D grain (particle, pore, object, etc.) size distribution is one of the most important requirements for the microstructure characterization. Taking this problem as an example it will be shown for which grain shapes 2D image analysis with stereological estimates delivers reliable results for the 3D situation and when the microstructural tomography is absolutely necessary. The problem will be discussed assuming grain shapes of different complexity (sphere – equiaxial polyhedra – non-equiaxial polyhedra – non-convex grain shapes).

It will be shown that the 3D characterization of non-convex, irregular and interconnected grains is absolutely indispensable in order to quantify spatial parameters, like connectivity or particles density. Growth mechanisms will be characterized using euclidic distance transformation. A detailed 3D image analysis and simulations enable the comprehensive quantitative evaluation of local microstructure degradation effects.

1 Introduction

The aim of the quantitative microstructural analysis is to provide adequate microstructural information to clarify the correlations between manufacturing, microstructure, and properties. The pure empirical correlations between manufacture and properties are not sufficient anymore for the tailor-made materials. Thus, the complete description of the spatial microstructure is required.

One of the most important parameters of the quantitative microstructural analysis still remains the grain size and the grain volume density (N_V), as it determines fundamental material properties.

In order to use a stereological method for the estimation of (N_V), the model assumption has to be adapted to the structure configuration. The structure can be assumed as a spatial system of non-overlapping, random, convex polyhedra [1, 2]. The complete space filling can be simulated by assuming a random mosaic [3]. A third alternative, which is to a large extent independent of restrictive model assumptions, is the tomographical imaging of the spatial structure [4]. FIB nanotomography provides extended possibilities for the analysis even of very complex three dimensional morphologies on the scale from micron down to several nanometers.

The growth and development of such structures and thus the influence of the processing parameters can be quantitatively characterized.

2 Basic characteristics of the microstructure

The basics of the microstructural characterization are summarized in the theorem of Hadwiger [5]. It says that all features of one particle, which possess the special properties regarding its size, shape, topology, etc., can be represented as a linear combination of following four parameters: V – the volume, S – the surface, M – the integral of the mean curvature and K – the integral of the total curvature of the particle [6, 7].

The densities of the particle features are used to characterize the components of the microstructure. Volume fraction (V_V) directly reflects the materials composition, i.e. the phase structure. The density of the surface area (S_V) depends primarily on the kinetic aspects of the material fabrication, and influences significantly the mechanical (e.g. Hall Petch effect) and physical properties [8]. Density of the integral of mean curvature (M_V) characterizes the geometrical arrangement (i.e. shape) of the second phase. Density of the integral of total curvature (K_V) is mainly influenced by the nucleation velocity [2].

The volume fraction (V_V), the specific surface area (S_V) and the specific integral of the mean curvature (M_V) can be calculated from the 2D microstructural images according to the stereological equations (see Table 1) [2].

While the parameters V_V, S_V and M_V can be obtained from 2D microstructure images under condition that they are isotropic microstructures or that for the anisotropic microstructures the analysis of several sections offers a solution, the total curvature K_V, as well as other topological parameters, such as Euler number (χ_V), particle number (N_V) and connectivity (C_V), cannot be calculated from the elementary stereological method. Thus, 3D images provide remarkable information gain in comparison with 2D. The four basic characteristics are independent from each other; it means that they describe on an abstract level various aspects of the complex microstructural arrangements.

Table 1. Stereological equations. Parameters from area, lineal and point count methods can be used to estimate spatial data such as V_V, S_V and M_V, where as K_V is only accessible via 3D measurement. For details see [2].

Spatial structure	Area method	Lineal analysis	Point count method
V_V	$= A_A$	$= L_L$	$= P_P$
S_V	$= \frac{4}{\pi} L_A$	$= 2P_L$	
M_V	$= 2\pi\chi_A$		
K_V			

Hence, the microstructure can be considered to be completely characterized, when all four basic parameters are known, as every further specific microstructural characteristic can be described by combining these four parameters. However, this is not trivial for the definition of easily implementable structural parameters like, for example, the grain size, which therefore is to be discussed in some detail.

3 Determination of the 3D grain size distribution from 2D micrographs

If the structure is assumed as a spatial system of non-overlapping, random, convex polyhedra [1] stereological methods can be applied to calculate the size distribution. Analysis of the simulated microstructural models can be used if the real microstructure can be approximated by complete space filling random mosaics [3].

3.1 Spherical particles

Already in 1925 Wicksell [9] has suggested the analytical solution for the calculation of the 3D size distribution of the spherical particles from the random 2D sections. This results in the estimation of the average particle number per unit volume (N_V) and the distribution function of the sphere diameter ($F_V(u)$, with sphere diameter u) from the measurement of the average number of the 2D sections through the spheres per unit area (N_A) and the distribution function of the section diameters ($F_A(s)$, with circle section diameter s) on the planar section. The integral equation

$$N_A \left(1 - F_A(s)\right) = N_V \int_s^{\infty} p(u,s)\, \mathrm{d}F_V(u)\ , s \geq 0 \tag{1}$$

connects the values $(N_V, F_V(u))$ of the spatial microstructure with the values $(N_A, F_A(s))$ from the 2D analysis. Further details and the solution are described in [6]. This 3D estimate provides very good results, as long as the assumed sphere shape is statistically correct, as, for example, when measuring spherical precipitations in a matrix, like spheroidal graphite in cast iron or isolated pores.

3.2 Isometric polyhedra and non-isometric prismatic grains

Another possibility to determine spatial microstructure parameters for particles of different elongation but principally the same shape (Fig. 1a) is the generalization of Wicksell's corpuscle problem [10].

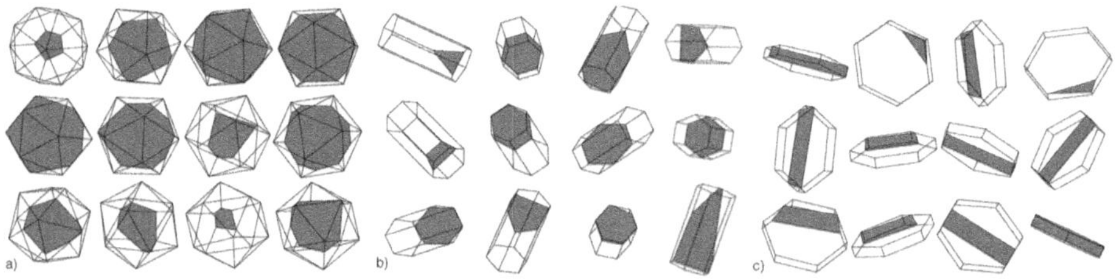

Fig. 1. Simulation of the stereology for (**a**) isometric polyhedra and non-isometric (**b**) elongated and (**c**) flat prisms. Size and degree of elongation are determined as bivariate distributions.

A further step to generalize the problem has been taken in the direction of non-isometric particle systems. In those particle systems for which the orientation distribution of the particles is unknown, the trivariate distribution of the size, the shape, and the direction is determined. The model solution of such a problem has been realized for the hexagonal needle-like grain shape in silicon nitride ceramics (Fig. 1b) [11].

It can be shown for the example of elongated prisms of the silicon nitride grains, that only extremely few of the 2D grain sections can display the actual shape of the 3D-prisms. The higher the elongation degree, the lower is this probability, so that subjective estimation leads to significant misinterpretations (see also Fig. 1c) [10].

The integral equation takes into consideration next to the particle size its shape parameter, e.g. elongation in 2D (t) and 3D (ν):

$$N_A F_A(s,t) = N_V \int_0^\infty \int_0^\infty K(u,\nu,s,t)\,\mathrm{d}F_V(u,\nu) \qquad (2)$$

The function $K(u,\nu,s,t)$ includes the shape assumption for the particles. It is acquired from the numerical simulation and analysis of all possible sections through this particle of the certain shape.

3.3 Single-phase polyhedral structures

Even when grain shape (Fig. 2a) strongly deviate from the ideal above mentioned shapes and the analytical solution is not possible it might be described by spatial well known mosaics, e.g. Poisson-Voronoi mosaic (Fig. 2b–c). Thus, by means of 2D measurements for single-phase polyhedral structures further 3D parameters can be determined when the verification of the correlation with this mosaic is done [1].

The reason for that is the correlation (4) which has been reliably examined by means of computer simulation for the Poisson-Voronoi mosaic between the values point density per line length (P_L) and particle density per unit area (N_A) (which are actually independent) which corresponds to the one which has empirically been found for real single-phase structures and which is specified in DIN 50 601 (now DIN EN ISO 643) (3).

$$lg\left(\frac{N_A}{\mathrm{mm}^{-2}}\right) = 3.9 - 2 \cdot lg\left(\frac{100\ mm^{-1}}{P_L}\right) \tag{3}$$

$$N_A = 0.6887\ P_L^2 \tag{4}$$

In this context, it is especially advantageous, that one can easily verify whether the structure to be examined can be described by means of a Poisson-Voronoi mosaic or not (i.e. whether it is a sufficiently homogeneous structure). The values for the point density N_V in the mosaic (corresponds to the number of grains per volume) calculated from experimentally determined values for P_L and N_A, should only be sufficiently close to one another.

$$P_L = 1.455\ N_V^{1/3} \tag{5}$$

$$N_A = 1.458\ N_V^{2/3} \tag{6}$$

When this criterion for the existence of a Poisson-Voronoi mosaic is fulfilled, values for the average number of grains in the volume N_V and the average spatial grain size can easily be given (see [1]).

We can conclude, that stereological methods can be successfully applied as long as the 3D microstructure consists of convex, non-overlapping grains with a defined shape.

4 Analysis of the 3D tomographical images

The 3D analysis is especially indispensable, when not periodic, not convex and not symmetric structures have to be characterized. Such structures, depending on their measurement and shape, can appear significantly different in 2D as in 3D. And thus flawed conclusions about the material properties can be derived.

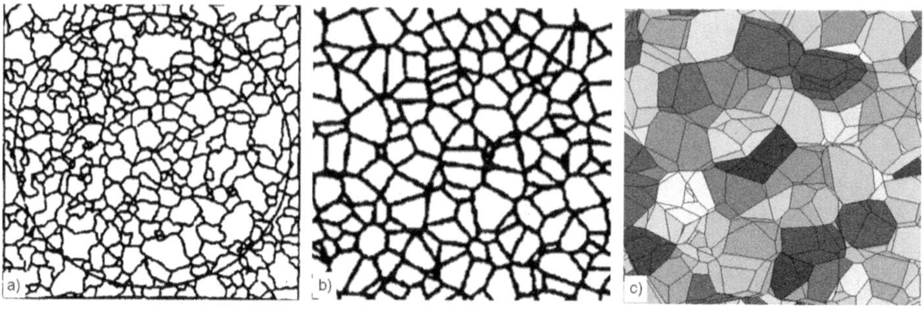

Fig. 2. Microstructure and simulation of the model. (**a**) Grain structure, standardized in accordance with DIN 50 601 (now DIN EN ISO 643); (**b**) result of the computer simulation of a planar section of the mosaic and (**c**) of the complete spatial Poisson-Voronoi mosaic [1].

4.1 Characterization of the interconnected microstructures

When particles are non-isolated in the structure, not only the particle density (N_V), but also the particle connectivity (C_V) has to be taken into consideration. This is important when, for example, the ductility of the bulk material decreases with increasing connectivity of a brittle phase, as is the case in Al/Si alloys [12], or the higher connectivity of a phase with high thermal conductivity leads to a rise of the effective thermal conductivity of the material (e.g. cast iron with flake graphite Fig. 3b).

As the particles in Fig. 3a and b are connected to each other, the particle density (N_V) is very low. In an extreme case, such as, for example, with open-pored foam (Fig. 3c), only one single connected particle per unit volume is measurable. Therefore it is often much more important for connected structures, to quantitatively describe the connectivity of the phase rather than the spatial particle density. Thus, the 3D Euler number (χ_V) is an appropriate quantity.

4.2 Microstructure development studied on non-convex particles

Single 3D non-convex particle can be cut by a plane so that it produce several 2D sections, so that no precise result can be achieved by means of the stereology.

Combining tomographic analysis with the analysis of the chemical composition and structure, 3D images can be used for the characterization of the microstructure development processes. The analysis of the growth mechanism of the complex particles is shown for the example of the irregular graphite precipitates in cast iron. Depending on the chemical composition of the melt and the processing, the graphite shape can change between extreme cases: spherical, which is promoted by additions of Mg and/or the absence of oxygen, sulfur and phosphorus; and lamellar, where according to [13] O and S

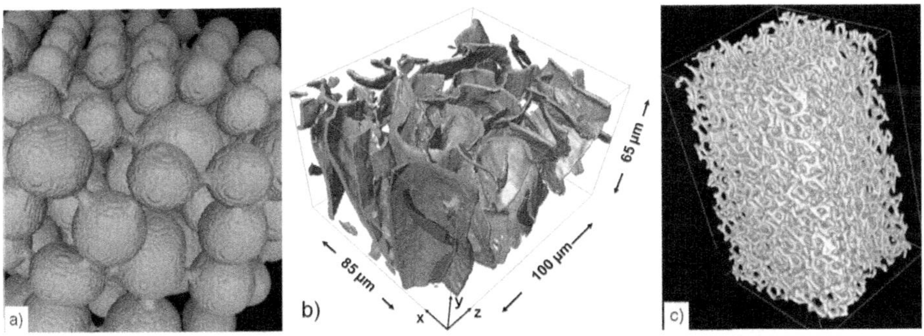

Fig. 3. Partially connected (**a**) *spherical-shaped* particles of sintered copper [2] and (**b**) flake graphite particles of the eutecticum in a cast iron [4], as well as (**c**) open-pored Ni foam [7]. The connectivity of the structure increases from the *left* to the *right*.

induce flake morphology by adsorption onto graphite. By means of the FIB-nanotomography, it became possible to reconstruct the complete graphite precipitations from serial sections [4]. Figure 4 illustrates the particles of three typical graphite types. They have very different 3D parameters which are, however, specific for the respective type: *V, S, M, K*, as well as the shape parameters; i.e. they can clearly be distinguished from each other [14].

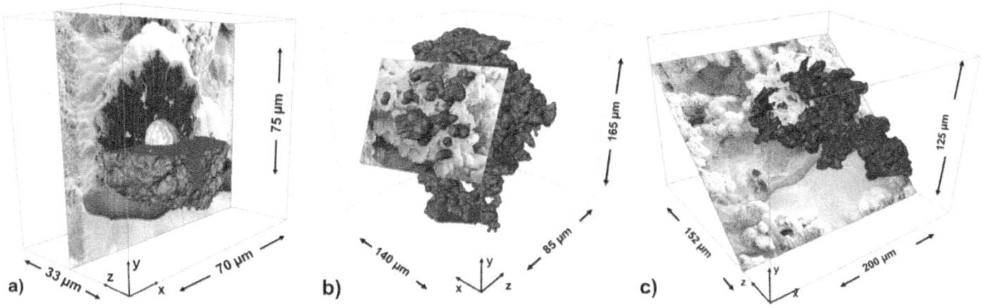

Fig. 4. Non-convex, irregular graphite particles in cast iron: (**a**) spheroidal graphite, (**b**) temper graphite and (**c**) vermicular graphite. *White* impurities in nodular graphite hamper the regular graphite growth and induce surface roughness.

Nuclei, pores and inclusions have an important influence on the growth mechanisms of the graphite. Nuclei were found almost in every particle. The radially distributed inclusions in the nodular graphite interrupt regular graphite growth and cause surface roughness (Fig. 4a). The distribution of the amount of the graphite voxels in 3D with regard to the distance from their nucleus was characterized with the help of the Euclidean distance transformation (EDT). The transformation assigns each voxel the gray value, which is proportional to its distance from the nucleus (or several nuclei) as it is shown on the Fig. 5.

Assuming that the nucleus is the origin of the particle, the growing process can be characterized with the help of the EDT. Figure 6 depicts the distribution curves for three nodular graphite (SG) and two vermicular graphite (CG) particles.

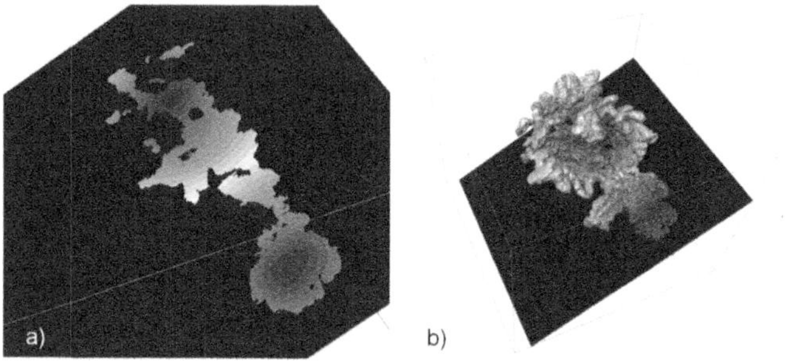

Fig. 5. Demonstration of the Euclidean distance transformation (EDT) on the reconstructed oblique 2D section and for the complete vermicular graphite particle.

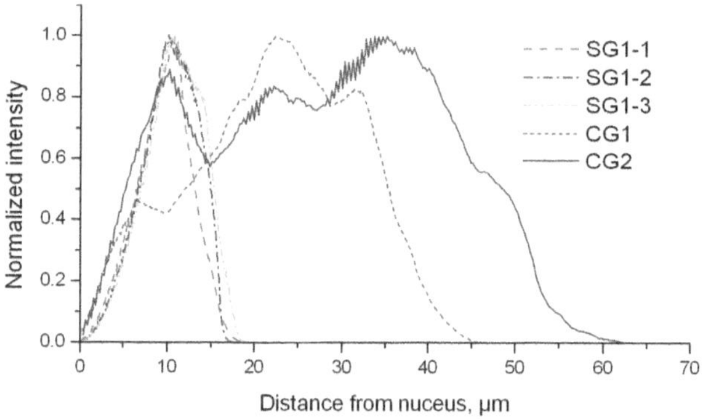

Fig. 6. Normalized distribution of the intensity of the voxel grey values for different graphite particles. Each value corresponds to the distance from the particle nucleus.

The distribution curves for two vermicular graphite particles in the Fig. 6 follow at the beginning the distribution for the nodular shape, which indicates its initial formation as graphite nodules (Fig. 4a). The integration of this distribution curves provides the quantitative characterization of the 3D particle arrangement in regard to their growth mechanisms.

3D quantitative characterization of the growth mechanisms supports the observation of crystal structure of different graphite morphologies.

4.3 Study of the microstructure degradation

The development of the new art of contact materials with long-term stability can succeed only with better basic understanding of the degradation processes called electro erosion. The employment of the FIB-nanotomography and thus available three dimensional chemical (FIB+EDX) [15] and structural material analysis (FIB+EBSD) [16] is of essential importance. A major advantage of FIB tomography is the possibility to perform site-specific investigations like in micro scale arc erosion craters [17] or in crack propagation zones [18].

Here the microstructure of the initial state of the Ag/SnO_2 material (Fig. 7a) and of the switched (after plasma discharge) state (Fig. 7b) was three dimensionally analyzed with the help of the FIB-nanotomography.

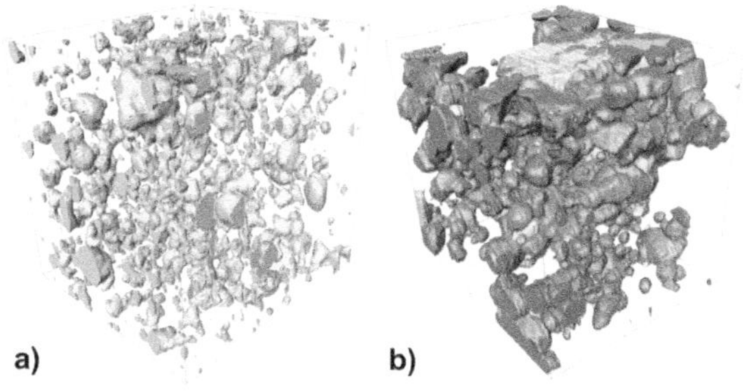

Fig. 7. Microstructure (**a**) of the initial state and (**b**) of the switched state approx. 70 µm below the surface. Interconnected oxide agglomerates and pores were built. 3D reconstruction.

4.3.1 Determination of the distance between the oxides

Besides size and shape of the particles, their dispersity in the volume plays an important role for the properties of a polyphase material. The oxide particles in Ag/SnO_2 initial state are mainly separated from each other (disperse state). The oxides in the switched state are agglomerated around the pores.

3D image analysis method such as morphological operations can be used for the characterization of the spatial distances and so the particle-free volume regions. By means of dilation every particle is gradually dilated in all space directions by voxel addition. After each step (Fig. 8a), the number of particles is counted.

Figure 8b shows exemplarily the number of particles which were unified after each dilation step. Such a curve progression makes possible the quantitative description of the qualitatively recognizable agglomeration.

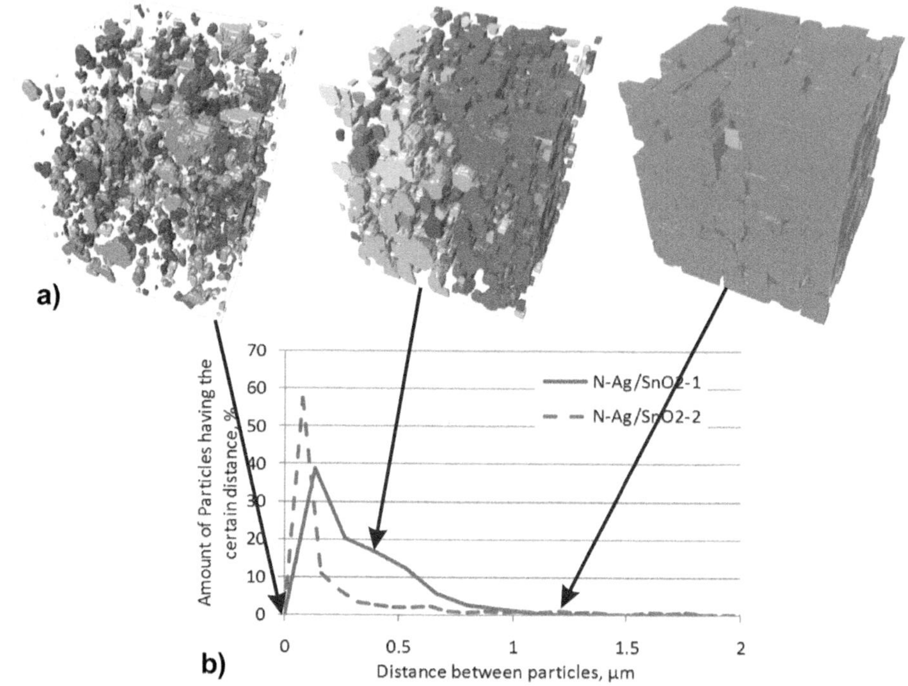

Fig. 8. Amount of particle which have certain distance to each other for Ag/SnO_2 in the initial state and in the "switched" state after exposure to a plasma discharge. In the initial state a quite homogeneous dispersion of the oxide particles prevails, while the large amount of particles growing together already after one dilation step in the switched state describes the relatively high agglomeration.

The method that is based on morphological transformations is also particularly qualified for particles with varying sizes, shapes (also non-convex), and orientations.

4.3.2 3D Simulation of material properties

Estimation of the thermal and electrical conductivity was performed on the tomographic data using simulation software GeoDict (Fraunhofer ITWM), which allows the use of even anisotropic voxels. Results were acquired in three spatial directions: x, y, and z. Mean values and standard deviations are summarized in the Table 2.

The effective conductivity of the initial state with random distributed SnO_2 inclusions can be approximated by various models [19]. For the switched state the influence of the local agglomeration (Fig. 9) can be studied.

Table 2. Simulated electrical and thermal conductivities using tomographic data.

	Ag/SnO_2 – initial state	Ag/SnO_2 – switched
Electrical conductivity, σ, 10^6 S/m	56.4 ± 0.7	40.1 ± 1.5
Thermal conductivity, λ, W/mK	370 ± 4	272 ± 9

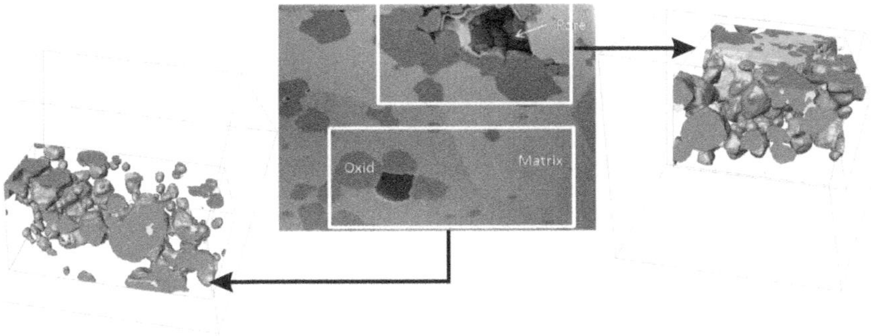

Fig. 9. Two regions of the switched state used for the simulation of the local properties of contact material. Electrical conductivity of the *upper* region with pore and oxide agglomerate is equal to $(24.1 \pm 1.4)\ 10^6$ S/m and of the *lower* region with homogenic distribution of the oxides is equal to $(46 \pm 9)\ 10^6$ S/m.

5 Conclusions

It has been shown, that different possibilities are available for the adequate quantification of the spatial microstructure which is relevant for the material properties. Focusing on the practical problem, one can have recourse to different methods. Often the field features analysis of the three basic microstructural parameters which are accessible from 2D and which can be extrapolated to 3D, is sufficient.

For uniform, convex grain shapes, well-elaborated stereological solutions are available, which lead to an unambiguous size and, if necessary, size-shape-distribution in the material volume. For the homogeneous polyhedral structure, one can particularly easily verify and, when applicable, make use of the assumption of a spatial Poisson-Voronoi mosaic.

More complex and especially non-convex grain populations have to be imaged by means of a tomography of the microstructure, which allows an extensive spatial analysis without limiting shape assumptions, as is possible with the help of the nanotomography with combined FIB/SEM.

The examples of the microstructure development, agglomerate formation and material degradation, have been discussed. The consequences for the material properties caused by these local microstructural effects were estimated.

The combination of the FIB microstructural tomography with the modern 3D analysis and simulation methods provide new possibilities of the site-specific characterization of materials with complex microstructures.

References

1. F. Mücklich, J. Ohser, G. Schneider: Z. Metallkd. **88**, 27–32 (1997)
2. J. Ohser, F. Mücklich: *Statistical Analysis of Microstructures in Materials Science* (John Willey and Sons, England, 2000)
3. C. Lautensack, T. Sych: Image Anal Stereol. **25**, 87–93 (2006)
4. A. Velichko, C. Holzapfel, F. Mücklich: Adv. Eng. Mat. **9**, 39–45 (2007)
5. H. Hadwiger: *Vorlesungen über Inhalt, Oberfläche, und Isoperimetrie*, (Springer Verlag, Berlin, 1957)
6. J.C. Russ, R.T. Dehoff: *Practical Stereology*, (Kluwer Academic/Plenum Publishers, New York, 2000)
7. H. Schumann, H. Oettel: *Metallographie, 14th Ed.* (Wiley-VCH, Weinheim, 2005)
8. G. Herzer: IEEE Trans. Magn. **26**, 1397–1402 (1990)
9. S.D. Wicksell: Biometrica. **17/18**, 84–89, 152–172 (1925/26)
10. J. Ohser, F. Mücklich: Adv. Appl. Prob. **27**, 384–396 (1995)
11. F. Mücklich, J. Ohser, S. Blank, D. Katrakova, G. Petzow: Z. Metallkud. **90**, 8, 557–561 (1999)
12. F. Lasagni, A. Lasagni, E. Marks, C. Holzapfel, F. Mücklich, H.P. Degischer: Acta Mat. **55**, 3875–3882 (2007)
13. D.D. Double, A. Hellawell: Growth structure of various forms of graphite in *The Metallurgy of Cast Iron*, B. Lux, I. Minkoff, F. Mollard (Eds.), Georgi Publ. St Saphorin (1975), pp. 509–525
14. A. Velichko, C. Holzapfel, A. Siefers, K. Schladitz, F. Mücklich: Acta Mater. **56**, 1981–1990 (2008)
15. F. Lasagni, A. Lasagni, M. Engstler, H.P. Degischer, F. Mücklich: Adv. Eng. Mat. **10**, 62–66 (2008)
16. J. Konrad, S. Zaefferer, D. Raabe: Acta Mater. **54**, 5, 1369–1380 (2006)
17. N. Jeanvoine, C. Holzapfel, F. Soldera, F. Mücklich: Pract. Metallography **43**, 107–119 (2006)
18. C. Holzapfel, W. Schäf, M. Marx, H. Vehoff, F. Mücklich: Scripta Mat. **56**, 697–700 (2007)
19. L. Weber, J. Dorn, A. Mortensen: Acta Mater. **51**, 3199–3211 (2003)

Profiling of Fiber Texture Gradients by Anomalous X-ray Diffraction

M. Birkholz[1], N. Darowski[2] and I. Zizak[2]

[1] IHP, Innovations for High Performance, Im Technologiepark 25, 15236 Frankfurt (Oder), Germany
[2] Hahn-Meitner-Institut, BESSY II, Albert-Einstein-Str. 15, 12489 Berlin, Germany

Abstract. Preferred crystallographic orientation or texture is a typically observed phenomenon in polycrystalline thin films. In addition, texture was revealed in numerous x-ray diffraction studies to increase with layer thickness. The phenomenon is rather significant for the optimized preparation of thin films, but was difficult to measure so far. A method is presented that allows for texture profiling by exploiting the anomalous variation of the x-ray attenuation coefficient in the vicinity of an elemental absorption edge. The study reports the application of the technique to thin ZnO:Al films by measuring with wavelengths below and above the Zn K edge. Large texture gradients between 0.03 and 0.3 mrd/nm were revealed to arise in these samples. Anomalous diffraction is concluded to enable the determination of texture gradients as required in many thin film projects.

1 Why are texture profiles of interest?

Modern technological devices to a large extent rely on polycrystalline thin films with thickness in the nanometer and micrometer range. This holds, for instance, for all microelectronic and most optoelectronic systems or sensoric and superconducting layers, to mention only a few important examples. Many of those layer systems develop characteristic microstructures that can be classified by the ratio of deposition temperature over melting temperature as described in structure zone models [1]. Next to its technological significance the growth of polycrystalline layers is of basic physical interest, since it may be regarded as a continuous disorder-to-order transition.

An order parameter of decisive relevance for the proper functioning of polycrystalline materials is the degree of preferred orientation. The phenomenon is also denoted as texture and accounts for the spatial distribution of crystallographic orientations [2–4]. In an ideal powder sample the orientation distribution is random, i.e. the end points of normal vectors of a particular lattice plane (hkl) would be equally distributed over the surface of the unit sphere. In a textured sample, however, the distribution of normal vectors $\langle hkl \rangle$ would exhibit pronounced patterns indicative for growth and processing conditions, to which the polycrystal has been subjected previously. The experimentally determined texture strength is given in units of mrd, which stand for multiples of a random distribution and specifies the fraction of sample volume oriented within a certain range of solid angle.

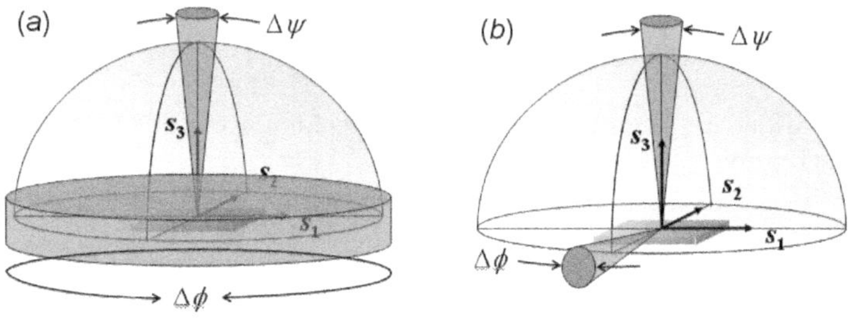

Fig. 1. Schematic representation of the two most common texture types in thin films given in the sample reference frame: (**a**) in case of a fiber texture most normal vectors of a particular lattice plane $\{hkl\}$ are confined to a small region of tilt angles $\Delta\psi$ around the substrate normal, while the in-plane orientation is random, (**b**) whereas for a biaxial texture a second set of normal vectors $\{uvw\}$ – being perpendicular to $\{hkl\}$ – is moreover restricted to a certain azimuthal range $\Delta\phi$.

Well pronounced textures were found to occur in thin films, for a review see [5], with most of them may be categorized into two different types: (1) for so-called fiber textures the orientation of a certain low-indexed lattice plane $\{hkl\}$ is preferentially parallel to the substrate plane and most lattice plane vectors $\langle hkl \rangle$ are confined to small tilt angles around the substrate normal being denoted by $\mathbf{s}_3$, see Fig. 1(a). In this case the crystallographic orientation of the grain ensemble assumes a preferential orientation parallel to one fundamental direction of the layer, being $\mathbf{s}_3$. (2) In contrast, in biaxial textures also the in-plane orientation of crystallites denoted by azimuth ϕ becomes locked with respect to the sample reference frame unit vectors $\mathbf{s}_1$ and $\mathbf{s}_2$. Such effects are accordingly observed in nearly epitaxial growth processes, where the crystallographic axes of the layer tend to align along those of the substrate, Fig. 1(b).

Tailoring the texture on demand is not only an interesting fundamental question in the study of structure-property relations, but became an important task in recent thin film technology, too. In the case of anisotropic oxides compounds, for instance, intended for dielectric devices [6–9] like surface acoustic wave (SAW) propagators, the polar axis must be aligned perfectly along the substrate normal, see Fig. 2(a). The figures of merit of thin film SAW devices using either AlN or ZnO as the active layer were found to directly scale with the degree of texture. Another example is given by high-T_c

superconducting cables that are being developed as oxide multilayer systems deposited on metallic ribbons [10–12]. The adjustment of the biaxial texture in $YBa_2Cu_3O_7$ layers turned out as the decisive prerequisite for achieving sufficiently large critical currents. In fact, the technical tailoring of thin film texture currently is the *conditio sine qua non* for a large-scale introduction of HTSC cable technology.

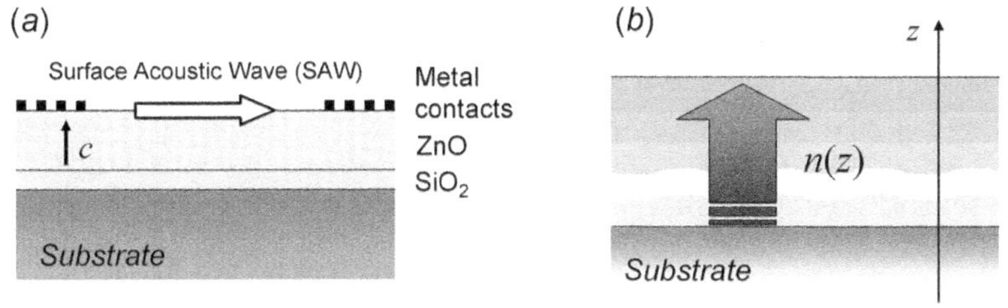

Fig. 2. (a) Sketch of a thin-film surface acoustic wave (SAW) device. (b) Model architecture of a thin film with a non-homogeneous fibre texture. The complete film is composed of individual layers at different height z that exhibit a linearly increasing order parameter $n(z)$.

Interestingly, it has been revealed that the degree of texture is often subjected to an evolution during thin film growth [13] and that most pronounced textures are only obtained after the layer has achieved a certain thickness. The process of texture optimization would thus deserve not only information about the average texture in the layer, but also on the "velocity", by which texture evolves with thickness, see Fig. 2(b). Therefore, thin film growers require information about the texture profile or the texture gradient in order to optimize the deposition process.

Textures of polycrystalline films may be determined by either electron microscopy techniques or x-ray diffraction (XRD) procedures. The latter have the advantage of averaging over volume sizes relevant to the macroscopic behavior of the sample and are thus often the method of choice. The determination of texture gradients by XRD, however, is not straightforward, because different depths of a specimen contribute to a diffraction peak. Techniques that allow for the adequate deconvolution of diffraction intensity have recently been developed intending to analyze non-homogeneous textures in the framework of kinematical theory [14, 15]. In Birkholz's approach different x-ray wavelengths have to be applied in order to probe different depths of the sample. For this purpose the strong variations of x-ray absorption in the vicinity of an elemental absorption edge are advantageously exploited, which is generally denoted as anomalous diffraction. It will be presented in the following the application of the technique to some strongly textured thin films. The focus will be on fiber textures, but the technique may easily be extended to biaxial textures. Samples of c axis fiber-textured ZnO:Al layers will serve as example systems, for which the probing x-ray energies were chosen to lie below and above the Zn K edge of 9.67 keV.

2 Conceptual approach

Fiber textures from thin film samples may be determined by x-ray diffraction in a reflective experimental set-up as shown in Fig. 3. For this purpose the integral intensity I_{HKL} or I_H of the associated Bragg peak is measured from symmetrical $\theta/2\theta$ scans for a set of tilt angles ψ that account for the inclination of the scattering vector Q with respect to s_3. In case of an ungraded or homogeneous fiber texture the intensity distribution $I_H(\psi)$ is a direct measure of the texture distribution $T_H(\psi)$ as is realized from the expression of Bragg reflection intensity in kinematical theory [5]

$$I_H(\psi) = SCF \cdot \lambda^3 Lp \left|F_H\right|^2 m_H T_H(\psi) \frac{1}{2\mu} A(t) \tag{1}$$

In Eq. (1) SCF, Lp, F_H and m_H stand for the instrumental scaling factor, Lorentz-polarization and structure factor and the multiplicity of the reflection, while μ, t and A account for the linear attenuation coefficient, layer thickness and absorption factor, having $A = (1 - exp(-2\mu tk))$, respectively. The configuration factor k indicates the geometry of the measurement and becomes $k = 2/(\sin\theta\cos\psi)$ for the case considered here. The $T_H(\psi)$ distribution accounts for the volume density of lattice planes (HKL) at inclination angle and has to be used in normalized form to cover all crystallites in the sample. Describing the texture distribution by the shape of the $\cos^n\psi$ function, a normalization factor of $(n+1)/(2\pi)$ has to be taken into account. The $\cos^n\psi$ distribution obeys the convenient property that the concentration of lattice planes – in mrd units – along the direction of the fiber pole directly scales with the order parameter n. For instance, a $T_H = (3+1)/(2\pi)\cos^3\psi$ distribution yields a texture strength of 4 mrd in the fiber pole at $\psi = 0$, while $n = 0$ gives the random distribution or the powder case having $T_H = 1$ for all ψ.

If graded fiber textures are considered, a depth-dependent distribution function $T_H(\psi, z)$ has to be introduced. It has been shown that the kinematical intensity of the texture peak can be described by the finite Laplace transform of the texture distribution $\pounds^t(T_H)$ [15], where the transformation is from z space to the penetration depth coordinate $(\mu k)^{-1}$ and the general expression for the Bragg peak intensity becomes

$$I_H(\psi) = SCF \cdot \lambda^3 Lp \left|F_H\right|^2 m_H \frac{k}{2} \pounds^t \left\{T_H(\psi)\right\} \tag{2}$$

For $T_H = (n+1)/(2\pi)\cos^n\psi$ functions the depth dependence might be reliably modeled by regarding the order parameter n as a z-dependent polynomial, see Fig. 2(b). Analytical solutions become possible by restricting the

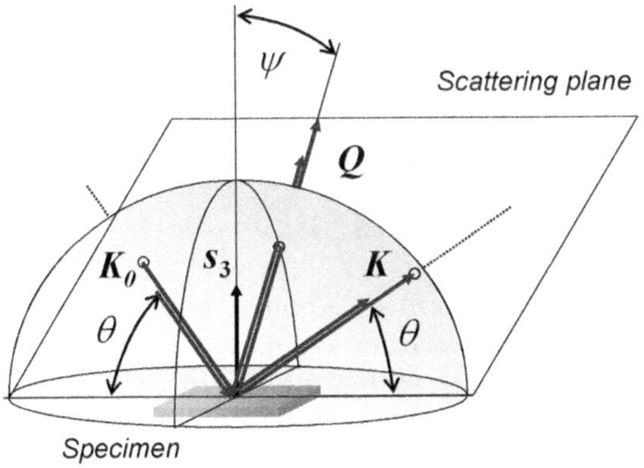

Fig. 3. Experimental set-up for $\theta/2\theta$ measurements under various tilt angles ψ and for different wavelengths, with the latter being symbolized by incoming and exiting wave vectors $\mathbf{K}_0$ and $\mathbf{K}$ of different color and magnitude. For non zero tilts, $\psi \neq 0$, the scattering vector $\mathbf{Q} = \mathbf{K} - \mathbf{K}_0$ deviates from the direction of the substrate normal $\mathbf{s}_3$.

polynomial degree to the most simple case, i.e. by setting $n = n_0 + n_1 z$. In the linear case the finite Laplace transform is solved to give

$$\pounds^t \{T_H(\psi)\} = \widetilde{N} \cos^{n_0} \psi \frac{1}{M} \widetilde{A}(t) \tag{3}$$

where $M = \mu k - n_1 \ln(\cos\psi)$ is valid and the abbreviations

$$\widetilde{N} = \frac{1}{2\pi}\left[n_0 + n_1 t\left(\frac{1}{Mt} + \frac{1}{1-\exp(Mt)}\right) + 1\right] \tag{4}$$

$$\widetilde{A}(t) = 1 - \left(\cos^{n_1 t}\psi\right)\exp(-\mu t k) \tag{5}$$

are used that stand for the generalized absorption factor and generalized normalization factor in the graded texture case. It should be noted the fiber texture gradient n_1 is of dimensionality mrd per unit depth and appears as the product with film thickness $n_1 t$, which is this quantity to be derived from the regression of experimental data.

This formalism represents the basis from which the problem of experimentally determining fiber texture gradients may be tackled. Our ansatz relies on the distinct penetration of x-rays into a sample for different wavelengths. Intending the determination of both n_0 and n_1, at least two $I_H(\psi)$ distributions have to be measured at two different wavelengths λ_1 and λ_2. Favorably, both wavelengths are chosen to lie slightly below and above of an elemental absorption edge present in the sample, since this assures sufficiently different penetration depths of the probing x-rays. The two intensity distributions obtained may then be subjected to a numerical regression by two equations of type (2) – the first for λ_1 and the second for λ_2 – in order to derive the parameters n_0 and n_1. This procedure principally allows for the determination of a fiber texture gradient from a single sample without requiring a set of thickness-controlled samples.

3 Experiments and results

The presented formalism has been applied to study the evolution of fiber texture in thin ZnO:Al layers, from a sample series of which a significant increase of texture strength with increasing thickness has recently been revealed [7, 9]. Thin films of ZnO and Al-doped ZnO typically crystallize in the hexagonal wurzite structure and exhibit a pronounced $00L$ fiber texture with the c axis of most crystallites orienting along the substrate normal s_3. The effect is suitably investigated by determining the integral intensity of the 00.2 Bragg peak being the dominating reflection in a symmetrical $\theta/2\theta$ scan. The samples investigated here were prepared by reactive magnetron sputtering from a Zn-2wt%Al alloy target.

Highly accurate μt products (x-ray attenuation coefficient $\times$ thickness) of these samples were derived from Rutherford backscattering [9] that allowed for the determination the projected thickness $t_{pro} = \mu t/(\rho \mu_m)$. The projected thickness is the thickness the layer would have in case its density would compare to the ZnO bulk value of $\rho = 5675$ kg/m^3 [16]. This seemingly complicated procedure was necessary, because in many polycrystalline layers, the effective thickness exceeds the projected one due to inclusion of voids and other excess volume [5]. Values of t_{pro} of the investigated samples amounted to 118, 246 and 413 nm.

Diffraction experiments were performed at Bessy II beam line KMC-2 with an energy resolution $\Delta E/E$ of about 10^{-4} [17] and beam energies set to 8.048 and 10 keV. The higher beam energy is sufficiently above the Zn K edge to ensure that EXAFS oscillation in the mass attenuation coefficient μ_m are damped out [18] and to use the free atom value. Intensity $\theta/2\theta$ scans of 00.2 Bragg peaks were collected from each sample for both energies and inclinations set to 0, 10, 20, 30, 45 and 60°. Figure 4 schematically displays the performed scans in the $(2\theta, \psi)$ plane. Measured intensity data were corrected for by the Lorentz-polarization factor in appropriate form for in-plane polarized synchrotron radiation, $Lp = (\sin\theta \sin 2\theta)^{-1}$.

Lp-corrected Bragg peaks were fitted by pseudo Voigt function profiles yielding the desired integral intensity distributions $I_H(\psi)/Lp$ with estimated standard deviations on the order of 1..2% for maximum intensities at $\psi = 0$. Those values were subsequently divided by $\left(\lambda^3 \left|F_H\right|^2 m_H\right)$ and subjected to a non-linear regression in terms of Eqs. (2) and (3). Effects due to the polar space group of ZnO were estimated to be of minor significance and were thus neglected by usage of an averaged structure factor F_H. The normal scattering factors f for Zn and O were set to 24.56 and 5.768, respectively, as adequate

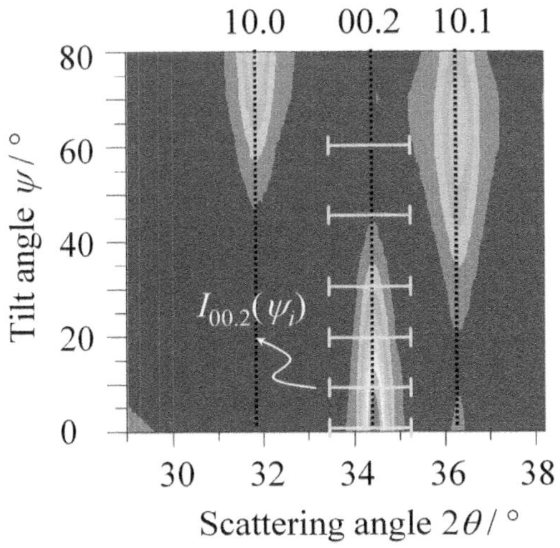

Fig. 4. Representation of the $\theta/2\theta$ scans performed in the $(2\theta, \psi)$ plane for 8 keV measurements. Integral intensities of 00.2 Bragg peaks of ZnO:Al layers were determined by scanning θ and 2θ along the white lines of constant tilt angle ψ.

for the ZnO 00.2 reflection at $\sin\theta/\lambda = 1.92$ nm^{-1}. Anomalous scattering factors f', f'' as well as mass attenuation coefficients μ_m for both x-ray wavelengths were taken from the Berkeley x-ray server [19]. The regression of corrected intensities $I_H(\psi)/(\lambda^3 Lp\,|F_H|^2\, m_H)$ was performed for each sample to both energies simultaneously by assuming n_0 and $n_1 t$ to be independent of energy. Only the scaling factors decomposed into distinct values $SCF(E_1)$ and $SCF(E_2)$. The resulting fits showed an excellent agreement of measured data and the modeling of the texture gradient through Eqs. (2) to (5).

The values obtained by the regression for n_0 and $n_1 t$ are given in Fig. 5 as a function of sample thickness. It is realized from this figure that $n_1 t$ shows a falling tendency, while the opposite trend is observed for n_0. This reflects an increase of preferred orientation with increasing thickness of ZnO:Al layers, albeit the grading of texture is diminishing. Both results are fully in accordance with previous studies [7, 9] justifying the reliability of the approach. The inset of the figure depicts the course of n_1, the values of which were derived from the $n_1 t/t_{pro}$ ratio.

For the thinnest sample, having $t_{pro} = 118$ nm, the texture gradient n_1 approaches a value of 0.3 mrd/nm averaged over the full depth of the layer, which means that every three nanometers the texture order n increases by 1 mrd. This is an unexpected and remarkable large value emphasizing the strong increase of fiber texture with increasing thickness in these samples. Our investigation shows that the strongest texture gradient is observed close to the layer-substrate interface. This is an interesting result, since it contrasts

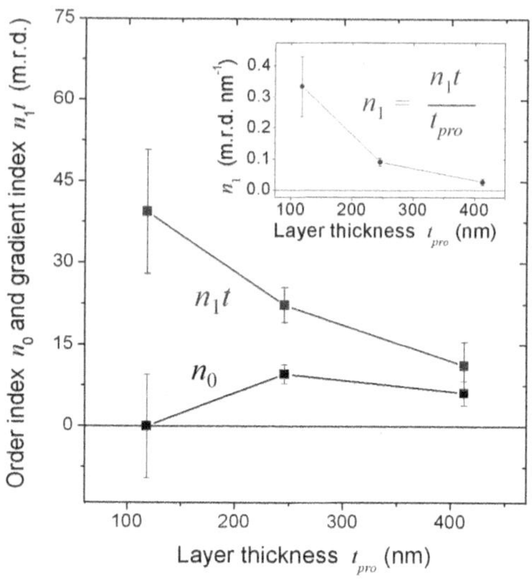

Fig. 5. Quantitative results and estimated standard deviations for the initial texture index n_0 and the product of thickness and texture gradient n_1t for the investigated samples. The inset shows the pure texture gradient n_1 in units of mrd/nm.

to those typically obtained for residual stress gradients, where the strongest gradients were mostly observed in the vicinity of the layer surface.

An on-going discussion is concerned with the relation between fiber texture and residual stress. The question is of practical and theoretical interest, since the occurrence of texture renders the usual technique of residual stress analysis obsolete and alternatives had to be developed [20–22]. In addition, a possible interaction between both may explain the often observed evolution of microstructural properties during the growth of thin film [23–26]. It can be expected that the discussion will benefit from the availability of quantitative results of fiber texture gradients as they may be determined by the concept presented in this work.

4 Conclusions

It can be concluded that anomalous diffraction has successfully been applied for the quantitative determination of fiber texture gradients. The technique may operate on a single thin film specimen and lifts the burden of preparing complete sample series of different thickness or to rely on in situ XRD techniques. Large texture gradients on the order of 0.03 to 0.3 mrd/nm have been uncovered for thin ZnO:Al films, with the strongest texture increase close to the interface with the substrate. The presented concept is completely general and may equally be applied to other material systems and other distributions than of the $\cos^n \psi$ type. It can be expected that the quantitative determination of texture gradients will deliver valuable experimental input to extend the spare data base of microstructure gradients as required for a perspective full *in silico* modeling of thin film growth.

5 Acknowledgements

We thank Daniel Chateigner, Christoph Genzel, Maarten Heyn, Peter Zaumseil and Harald Beyer for helpful discussion and Frank Fenske for leaving these otherwise already fruitful samples for this investigation.

References

1. E. S. Machlin: The Relationship Between Thin Film Processing and Structure, (Giro Press, Croton-on-Hudson, 1995).
2. H.-J. Bunge: *Texture Analysis in Materials Science* (Butterworth, London, 1982).
3. H.-R. Wenk and P. van Houtte: Rep. Prog. Phys. **67**, 1367 (2004).
4. D. Chateigner: Combined Analysis: Structure-texture-microstructure-phase-stresses-reflectivity determination by x-ray and neutron scattering, www.ecole.ensicaen.fr/~chateign/texture/combined.pdf (2005).
5. M. Birkholz: Thin Film Analysis by X-ray Scattering (Wiley-VCH, Weinheim, 2006).
6. V. Bornand, I. Huet, J. F. Bardeau, D. Chateigner and P. Papet: Integr. Ferroelectr. **43**, 51 (2002).
7. M. Birkholz, B. Selle, F. Fenske and W. Fuhs: Phys. Rev. B **68**, 205414 (2003).
8. J. Ricote, R. Poyato, M. Alguero, L. Pardo, M. L. Calzada and D. Chateigner: J. Am. Ceram. Soc. **86**, 1571 (2003).
9. F. Fenske, B. Selle and M. Birkholz: Jpn. J. Appl. Phys. **44**, L662 (2005).
10. P. Berdahl, R. P. Reade, J. Liu, R. E. Russo, L. Fritzemeier, D. Buczek and U. Schopp: Appl. Phys. Lett. **82**, 343 (2003).
11. J. D. Budai, W. Yang, N. Tamura, J.-S. Chung, J. Z. Tischler, B. C. Larson, G. E. Ice, C. Park and D. P. Norton: Nat. Mat. **2**, 487 (2003).
12. Y. Iijima, K. Kakimoto, Y. Yamada, T. Izumi, T. Saitoh and Y. Shiohara: MRS Bulletin **29**, 564 (2004).
13. A. van der Drift: Philips Res. Rep. **22**, 267 (1967).
14. J. T. Bonarski: Prog. Mat. Sc. **51**, 61 (2006).
15. M. Birkholz: J. Appl. Cryst. **40**, 735 (2007).
16. W. Martienssen: Functional Materials – Semiconductors, in W. Martienssen and H. Warlimont (eds.) Springer Handbook of Condensed Matter and Materials Data, p. 575 (Springer, Berlin, 2005).
17. A. Erko, I. Packe, W. Gudat, N. Abrosimov and A. Firsov: A Graded Crystal Monochromator at BESSY II, in A. K. Freund et al. (eds.) SPIE conf. proc. **4145**, p. 122 (2000).

18. F. Decremps, F. Datchi, A. M. Saitta, A. Polian, S. Pascarelli, A. Di Cicco, J. P. Itié and F. Baudelet: Phys. Rev. B **68**, 104101 (2003).
19. B. L. Henke, E. M. Gullikson and J. C. Davis: Atomic Data and Nuclear Data Tables **54**, 181 (1993).
20. C. Genzel and W. Reimers: Phys. Stat. Sol. (a) **166**, 751 (1998).
21. A. Saerens, P. Van Houtte, B. Meert and C. Quaeyhaegens: J. Appl. Cryst. **33**, 312 (2000).
22. P. Scardi and Y. H. Dong: J. Mater. Res. **16**, 233 (2001).
23. B. Rauschenbach and J. W. Gerlach: Cryst. Res. Technol. **35**, 675 (2000).
24. P. Reinig, F. Fenske, W. Fuhs, V. Alex and M. Birkholz: J. Vac. Sc. Technol. A **20**, 2004 (2002).
25. J. Almer, U. Lienert, R. L. Peng, C. Schlauer and M. Odén: J. Appl. Phys. **94**, 697 (2003).
26. M. Birkholz, C. Genzel and T. Jung: J. Appl. Phys. **96**, 7202 (2004).

Film Production Methods in Precision Optics

Hans K. Pulker

Thin Film Technology Group, Institute of Ion Physics and Applied Physics, University of Innsbruck, Technikerstraße 25, A-6020, Innsbruck, Austria
Hans.Pulker@uibk.ac.at

Abstract. Basic PVD processes are either fast running but low in particle energy (evaporation), or exactly the other way round (sputtering) or have disturbing side effects (arc ablation). After presentation of the basic processes, it is shown how they are improved to become effective in reliable and fast deposition of high quality optical films. This is performed by adding ions and plasma species or combining magnetic fields and operate the various hybrid processes in continuous or pulsed modes. Characteristic data of the processes are presented as well as properties of metal oxide films produced with energetic reactive processes.

1 Introduction

Films of metals, chemical compounds and composites can be formed on surface of solid substrates by wet and dry, chemical and physical technologies. Coating is carried out on air, in chemical controlled atmosphere, or under vacuum. Among the various technologies, physical vapour deposition PVD processes, operated under vacuum are the preferred ones for film production in optics. Although these processes are based principally on purely physical effects, chemical reactions, by adding reactive gas into the coating chamber, are also performed to deposit stoichiometric compound films. In modern ion and plasma processes, input of energy into the growing films by collision and momentum transfer of activated and kinetically enhanced ions, atoms and molecules stimulate chemical reactions and cause densification improving besides optical quality also structural and mechanical film properties and their environmental stability. All the processes are used to deposit coatings in the thickness range between few nanometers up to some microns. Single films and multilayers are deposited homogeneously or with graded composition.

2 PVD-methods

2.1 General considerations

In this paper exclusively PVD-technologies performed under vacuum will be considered, since these technologies are predominant in optical and interference optical film production. In PVD there are three fundamental mechanisms to transfer the coating material into the vapour phase: Evaporation, Sputtering, Arc ablation.

Comparing these fundamental mechanisms, speed of ablation, kinetic particle energy and number of electrically charged vapour species are found to be very different. Evaporation is a thermal process. That means the number of predominantly to exclusively neutral atoms or molecules evaporating from a heated surface depends on the temperature. The average energy of the evaporating species $E_{kin} = 3/2\,\mathrm{kT}$ is about 0.1 eV for temperatures between 1000 and 2000 Kelvin. Evaporation is a fast process and the velocity of the vapour atoms is around 10^5 cm/s.

Sputtering is compared to evaporation, a completely different process with material erosion from a solid target by momentum transfer mechanisms of the impinging energetic noble gas ions generated in gas discharge plasma or in an ion gun. The ablated material may contain different amounts of ions. Fundamental sputtering is a rather slow process with a rate of only about $5 * 10^{-4}$ g/cm^{2}s compared to about $1 * 10^{-4}$ g/cm^{2}s for evaporation. However, the particle energy is high with 1 up to 40 eV and the molecular velocity is also high with values between 5 and $10 * 10^5$ cm/s.

The exploding plasma plume in electrical arc and pulsed laser ablation is the fastest material transfer process. The vapour phase contains a large amount of single and multiple ionized material. The particle energy is high with values of $E_{kin} = 10$ to 100 eV. The velocity of the ablated species may be higher than $50 * 10^5$ cm/s. Disadvantage in both processes are the droplets.

A growing number of hybrid coating process variants based on these three fundamental transfer mechanisms, are in practical application. The hybrids may influence chemical reactions and modify kinetic and other energetic conditions in a coating process. In all cases the vaporized/ablated coating material is transported through a reduced, differently composed atmosphere and condensed on generally rotated, non-heated or heated substrates where finally the coating is formed [1–3]. On their way to the substrates the vaporized coating material species experience collisions with the residual gas molecules changing their kinetic energy and the direction of flight. Number of collisions is dependent on gas pressure and distance source to substrate.

Technical vacua in the range between 10^{-2} and 10^{-4} mbar are sufficient to deposit a coating. However, for e.g. high quality evaporated metal films, residual pressures lower than 10^{-6} mbar and precisely controlled gas compositions are required to avoid unwanted chemical reactions and other film contamination. Although many reactive gas processes run in the 10^{-4} to 10^{-3} mbar range, the starting pressure, before adding the reactive gas e.g. O_2, should be lower than 10^{-6} mbar to obtain clean and reproducible conditions in the actual working range. In all the reactive gas processes exactly and rapidly working gas regulation systems, operating in agreement with applied deposition rate, are very important.

PVD processes are carried out in sealed, cylindrical, cubic or some other way shaped stainless steel chambers. The coating chambers are equipped with all required installations, as e.g. vapour sources, substrate holders, heating devices, gas/vapour control and rate and film thickness measurement devices for optimum deposition conditions. In situ broad brand on line monitoring systems are applied for controlling optical thicknesses. Different pumping systems are used to evacuate a coating chamber. Most common are mechanical displacement pumps combined with baffled synthetic oil operated diffusion pumps. Cleaner vacua are achieved with turbo molecular pumps or, even better, refrigerator cryo pumps. They pump all gases and show the highest pumping speed of all high vacuum pumps. Modern optics coating systems, are automated to a high degree and generally installed in humidity controlled and dust free clean rooms. Evaporation based PVD technologies are summarized in Table 1. Sputter and other ablation based PVD technologies are listed in Table 2.

Table 1. Evaporation based PVD Technologies.

Conventional evaporation	Pohl, Pringsheim	1912
reactive mode	Auwärter	1952
Activated reactive evaporation	Heitmann	1972
Ion beam assisted deposition	Ebert	1982
non reactice/reactive mode	Martin, Macleod et al.	1983
Ion plating non reactive	Berghaus	1939
reactive mode	Mattox	1963
Reactive low voltage ion plating	Moll, Pulker	1985
Advanced plasma source	Matl	1991
	Beisswenger	1994
	Zöller	1997

2.2 Conventional evaporation

The simplest and gentlest process is conventional evaporation. Metals and stable dielectrics (chemical compounds) are evaporated under vacuum from resistively heated boats or today by electron beam heating from an e-gun and deposited on unheated or heated substrates. A schematic of an evaporation plant is shown in Fig. 1.

Table 2. Sputter and other Ablation based PVD Technologies.

Gas discharged sputtering	Grove	1852
	Plücker	1858
dc non reactive	Wehner	1955
dc, rf, reactive mode	Laegreid, Wehner	1961
	Schwartz	1963
planar magnetron	Clarke	1973
	Chapine	1974
rf, mf, pulsed	Schiller	1993
	Sproul	1994
mf,dc,dc-pulsed twin cathodes, reactive mode	Schiller, Bräuer	1995
Ion beam sputtering	Tilsch, Tschudi, Harper	1977
reactive, non reactive, dual mode	Wei (Litton)	1979
	Barnes	1987
Electric arc deposition	Boxman	1996
reactive, non reactive, pulsed mode	Martin, Netterfield, Bendavid	2001
Laser deposition	Brinsmaid	1957
pulsed, reactive, non reactive	Reisse	1994

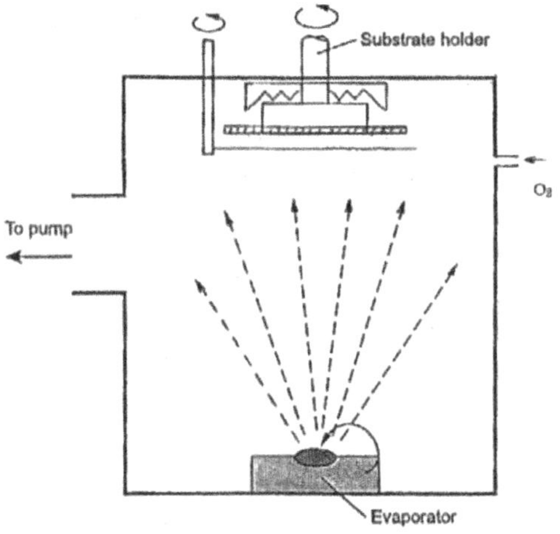

Fig. 1. Main features of a set-up for evaporative deposition.

As consequence of the low particle energy in evaporation adherence, hardness and abrasion resistance of such coatings is often poor. Their microstructure is generally columnar or spongy and less dense with rough surface. Metal oxide layers deposited with this technology show generally remarkable residual optical absorption caused by chemical degradation during vaporization. Most of the oxygen liberated in the degradation disappears in the vacuum pumps. The absorption problem could only be overcome by controlled addition of oxygen to the residual gas in course of the deposition to involve chemical reactions repairing stoichiometry during condensation and film growth. Deposition on heated substrates influences positively film quality. This important technology is termed reactive evaporation (RE) [4]. Increasing quality requirements particularly for lowest-loss He-Ne-laser mirrors stimulated further development of reactive evaporation to activated reactive evaporation (ARE).

2.3 Activated reactive evaporation (ARE)

Gas discharge activation of oxygen resulted apparently in higher chemical reactivity and clearly lower optical losses in the oxide films [5]. However the mechanical properties and the environmental stability of the less dense films was still unsatisfactory since deposition occurred with thermal energies of coating material species and gas atoms/molecules which come to only about 0.1 eV. The discharge tube for oxygen activation used by Heitmann is shown in Fig. 2.

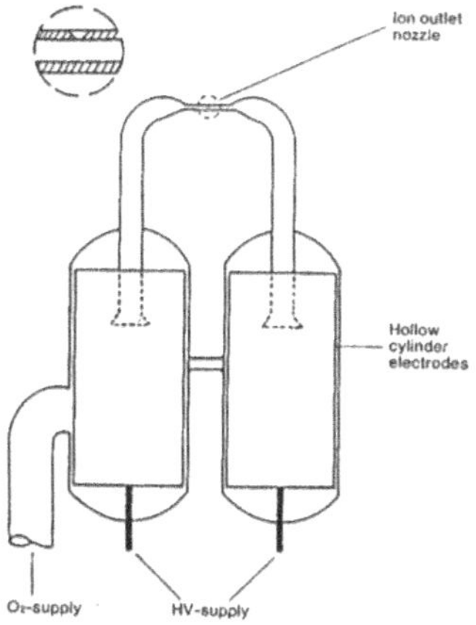

Fig. 2. Heitmann – Source (1972) Discharge tube for activation of the reaction gas.

2.4 Ion beam assisted deposition (IBAD)

Bombardment of a growing film with energetic gas species, improves adherence, density and hardness. Controlled ion beam bombardment with reactive species (IBAD) during film growth, improves additionally stoichiometry and surface smoothness and therefore the optical properties of oxide films deposited onto unheated substrates [6]. A schematic of the plant is shown in Fig. 3. Such techniques can also be applied to clean substrates before coating or more rarely afterwards to densify or improve other properties. Particle energies of around 10 eV until up to more than 100 eV are generally applied.

The mechanism responsible for densification is momentum transfer, shown by molecular-dynamic calculations [7]. Hot cathode ion guns (Kaufman type) are

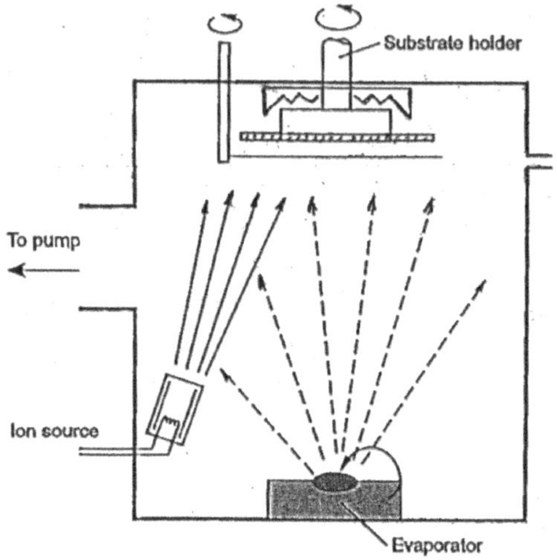

Fig. 3. Main features of a set-up for ion beam-assisted deposition.

often used but are replaced today by cold cathode installations, besides other advantages more safe in possible film contamination. Some round and rectangular broad beam ion sources are commercially available. Larger substrate surfaces can, however, hardly be bombarded uniformly, so that intermittent treatment of rotating substrates is generally applied.

2.5 Ion plasma plating (IPP)

This term encompasses different evaporation based ion and plasma technologies for supplying energy uniformly and simultaneously over larger areas to films growing generally on unheated substrates. Depositions can be performed in non-reactive and reactive modes. Very dense cold plasmas are typically in such processes. All molecular and atomic partners in the coating process are ionized and activated to varying degrees. The energy input during film formation raises the substrate temperature. The total thermal load, however, can be kept low enough to coat most plastic materials without damage. The energetic deposition improves adherence, changes microstructure, increases film density and with it all related properties. The plant to perform deposition is shown schematically in Fig. 4. Ion plating was invented 1937 and patented by Berghaus [8] but not applied at this time. In 1963 the diode ion plating was reinvented by Mattox [9, 10]. It was not used for optical coatings.

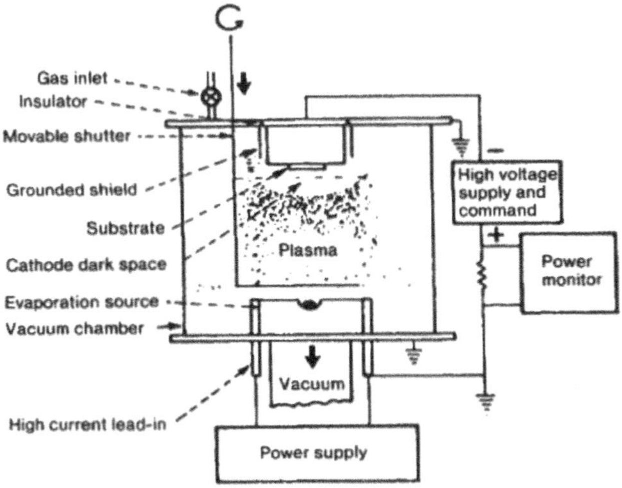

Fig. 4. Schematic representation of a simple diode type ion plating system.

2.6 Reactive low voltage ion plating (RLVIP)

The reactive low ion plating process, invented by Balzers AG [11] in the early 80s is performed in the BAP 800 batch box coater shown schematically in Fig. 5a. The starting materials, metals or sub oxides form electrically conducting melts. Effective ionization and activation of the vaporized coating material atoms and the admitted reactive gas molecules O_2 is performed mainly by a low voltage about 40 V high current about 50 A electron beam of a non-self-sustaining thermionic arc discharge plasma directed to one of the anodic crucibles of the e-beam evaporators. The negative self-biasing potential of the insulated substrates and the repulsive forces of the positive ions from the anodic crucible mainly determine the impact energy of the film forming species. This system is used in precision optics for production of the environmental stable long-and short-pass-filters, various high quality line filters and special fluorescence filters.

2.7 Advanced plasma source (APS)

The advanced plasma source APS ion plating technology was developed at Leybold Optics [12, 13] and is applied in a batch box coater, shown schematically in Fig. 5b. A high density plasma source, consisting of a large area indirectly heated LaB_6-cathode, a cylindrical anode tube and a solenoid magnet, is placed in the centre of the coating chamber bottom next to the electron beam evaporators. The solenoid magnet surrounding the source forces the electrons of an argon plasma in axial directions. Due to the magnetic field the

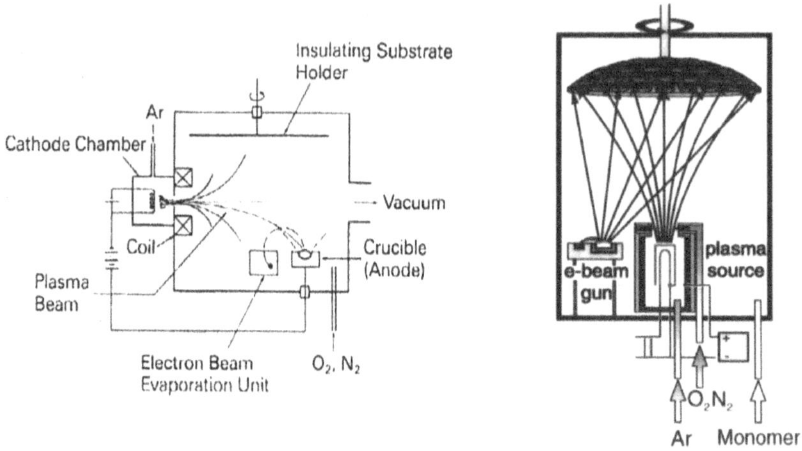

a) Balzers/Evatec BAP 800 b) Leybold APS 900

Fig. 5. (**a**) and (**b**) Industrial ion plating systems.

electrons move in the direction to the substrate holder. They spiral along the field lines thus bringing the plasma into the coating chamber. A ring shower for the reactive gas is placed on the top of the anode tube. Directly blown into the plasma the reactive gas is ionized and activated. Since dense plasma fills the total volume of the chamber, the evaporate becomes also partly ionized. Chemical reactions are improved and also ion plating takes place because of the negative self-biasing potential of about -100V of the substrates. In this process the generation of the plasma is completely separated from evaporation. The process is used in ophthalmics for a.r. coatings but also in precisions optics to produce e.g. highest quality line filters and other high quality interference optical film components.

2.8 Pulsed laser deposition (PLD)

A light beam can easily be brought into an evacuated coating chamber and focused on a coating material target. Regular and non-linear absorption cause strong heating. A plasma plume is formed at the target spot. The ablated material, highly ionized with high kinetic energy up to some 10 eV, is deposited on the substrate opposite the target. Reactive deposition is possible. CO_2-, NdYAG- and excimer-laser are applied. They produce power densities on the target up to $10^{10}\,\mathrm{W/cm^2}$. Unfortunately the ablated material contains also droplets, degrading film quality. This problem can partially be overcome by using pulsed irradiation with pulse duration of 20–200 ns and repetition rates between 40 and 150 Hz [14, 15]. Congruent ablation of alloys and composite materials can be achieved with laser ablation. A schematic of this technology is shown in Fig. 6.

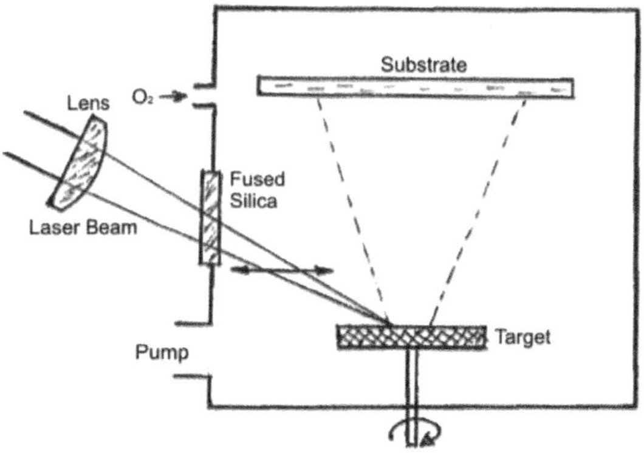

Fig. 6. Schematic of a pulse laser ablation System.

2.9 Electrical arc ablation

A metallic cathode in an evacuated and electrically grounded coating chamber is operated in a low-voltage high-current mode. 1 to 10 µm sized, fast travelling cathode spots are formed in the discharge with duration of about 10^{-6} s. The current density in the spots is between 10^6 and 10^8 A/cm^2. The trace of the travelling arc spots can be influenced by magnetic and other means. The plasma ablated material contains high concentration of positive ion i.e., single and multiple charged atoms, with energy of 5–50 eV. Reactive deposition in presence of reactive gas, e.g. oxygen, is possible [16, 17]. The problem are droplets or even macro particles, which appear independent on mode of operation, dc or pulsed, in the ablated material with sizes of 0.1–100 µm. Defect free optical film of remarkable quality can only be obtained with expensive filtered arc technology shown schematically in Fig. 7.

2.10 Sputter ablation

In sputtering atoms from a solid target are ejected with kinetic energies of 1 to some 10 eV, initiated by momentum transfer processes of bombarding energetic gas ions or atoms from a gas discharge plasma (gas discharge sputtering) or of energetic ions from an ion beam source (ion beam sputtering). Reactive processes are possible in both cases and very good optical and mechanical film quality can be achieved even when films are deposited on unheated substrates. Particularly after the invention of planar magnetron cathodes with improved sputter rate, gas discharge sputtering has become an important technology in coating plane and large substrate areas [18]. The schematic of a planar magnetron is shown in Fig 8. For precision optical film deposition newly designed ion beam sputter systems are known, [19–21] which enable production of highest quality laser mirrors, broad – and narrow-band interference filters, beam splitters and coatings for EUV optics.

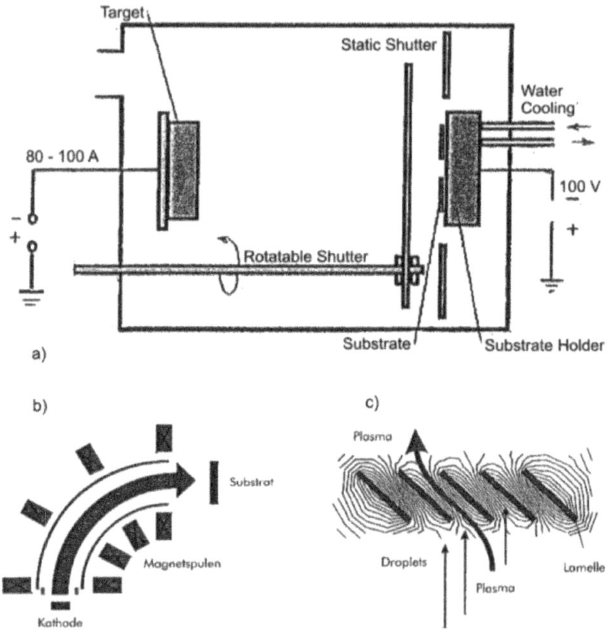

Fig. 7. (**a**) Schematic of an electrical arc source deposition plant and magnetic devices (**b**) and (**c**) for particle (droplets) separation.

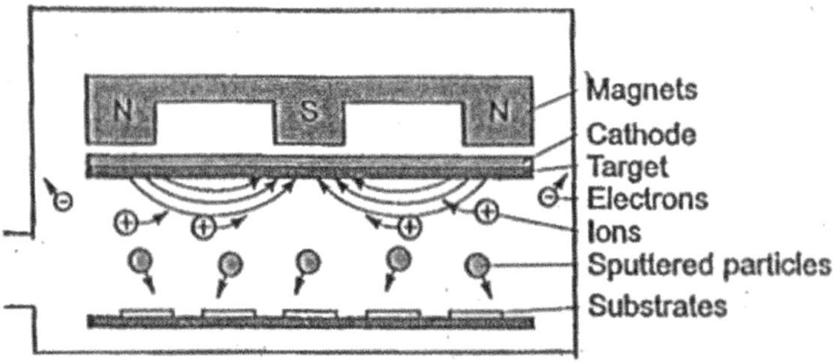

Fig. 8. Schematic of a planar magnetron cathode.

2.11 Gas discharge magnetron sputtering

Chemical compound films are generally produced in reactive mode from metal targets. In reactive dc magnetron sputtering of metal oxide films the chamber wall act as anode. Its conducting surface area, however, decreases with sputter time because of an insulating overcoat. This causes irreproducible and inhomogeneous plasma conditions negatively influencing film quality. The problem can be eliminated by reactive mid-frequency magnetron sputtering. As shown in Fig. 9 two magnetron cathodes are operated by one mid-frequency power supply with harmonic or rectangular pulses. Both targets are switched alternatively as cathode and then as anode in the gas discharge. By this way neutralization of surface charges in the oxidized areas of the target surface occurs before critical break through field strength is attained. Activation frequencies range between 10 and 40 kHz. Generally 40 kHz are applied. This relatively new technique is called twin or dual magnetron sputtering [22, 23]. Besides the more technical advantages of this technology, the modified plasma conditions of this discharge produces intense high energetic ions consequently resulting in dense non-columnar, homogeneous oxide films of excellent quality [24]. Optical and mechanical properties are similar to that of films obtained in IBAD and ion plating.

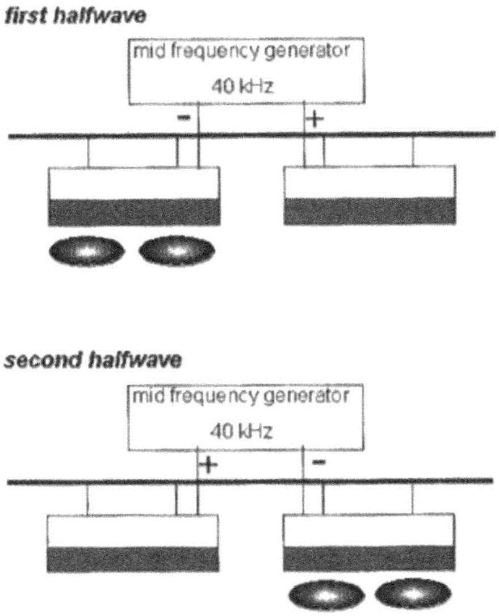

Fig. 9. Operation principle of a mid-frequency operated double magnetron system.

2.12 Ion beam sputtering (IBS)

Ion beams from various ion sources operating in clean vacua between in dc or rf mode with 45° impact angle on the target perform sputtering. In the dual beam technique a second, lower energetic ion beam from an additional ion source is applied. This beam is generally directed at normal incidence on the growing film for better oxidation and further densification, similar to ion beam assisted deposition. Ion beam sputtering is known as a slow process but resulting in highest quality optical coatings [25]. Recently productivity and economics of IBS was successfully optimized in replacing traditional ion sources by modern broad beam devices extending area of homogeneous distribution up to 330 mm diameter and increasing deposition rate from about 0.03 nm/s to 0.1 nm/s. Film thickness monitoring could also be improved by broad band in situ transmittance measurement [19]. The sputtered material arrive at the unheated substrate with energies between 20 an 100 eV . The energy input into the growing film in clean, reactive, low-pressure environment, results in dense, pure, highly adherent, amorphous and stoichiometric layers with excellent optical properties and high environmental stability. An ion beam sputtering system is shown in Fig. 10. An overview concerning characteristic data of the various PVD technologies discussed in this paper is assembled in Table 3.

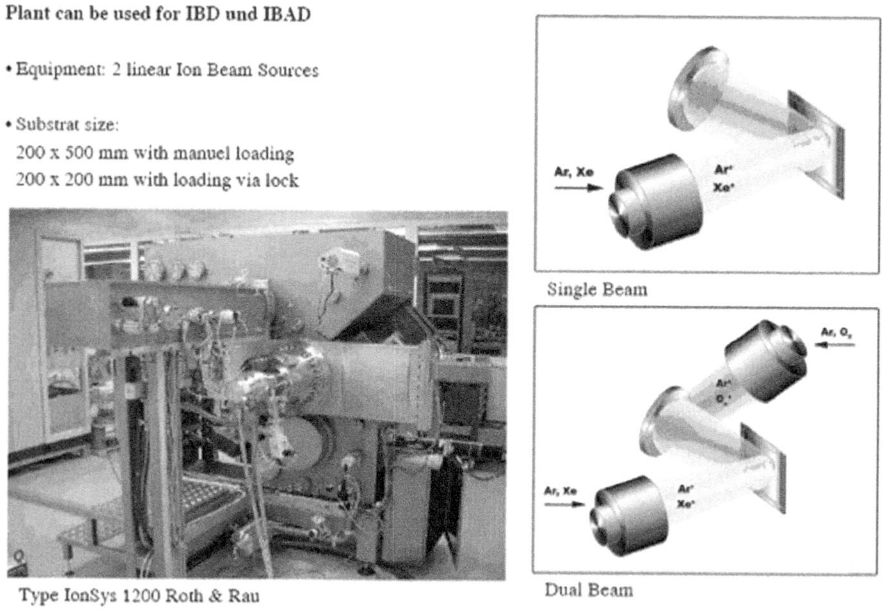

Fig. 10. Principle and plant of ion beam sputtering.

3 Thin film properties

The film growth at the surface of a solid substrate is strongly dependent on chemistry and topography of the surface, on surface temperature, gas pressure, deposition rate, angle of incidence, and on the energy of the impinging film forming and bombarding atoms and molecules. The dissipated energy per incident and adhering atom/molecule is composed of the loss of kinetic energy

Table 3. Characteristic data of various PVD deposition technologies.

	Power density on target	E_{kin} of particles	Beam velocity	Deposition rate	Current density on substrate
e-gun evaporation	$10^4\,Wcm_{met}^{-2}$ $10^3\,Wcm_{diel}^{-2}$	<0.3 eV	$10^3\,ms^{-1}$	High	
Reactive low voltage ion platting	$>10^4\,Wcm^{-2}$	15–16 eV	$10^4\,ms^{-1}$	High	$0.2–1\,mAcm^{-2}$
Magnetron sputtering	$10\,Wcm^{-2}$	1–15 eV	$10^4\,ms^{-1}$	Medium	$0.1\,mAcm^{-2}$
Ion beam sputtering	$>10\,Wcm^{-2}$	5–100 eV	$>10^4\,ms^{-1}$	Low	$0.5–1\,mAcm^{-2}$
Pulsed laser ablation	$10^7–10^{10}\,Wcm^{-2}$	10–100 eV	$>10^4\,ms^{-1}$	High	$>1\,mAcm^{-2}$
Arc source ablation	$10^7–10^{10}\,Wcm^{-2}$	20–100 eV	$>10^4\,ms^{1}$	High	$>1\,mAcm^{-2}$

and of the liberated heat of condensation and, in case of reactive deposition, also of the heat of reaction. Modern energetic, continuous or pulsed ion and plasma supported, hybrid coating technologies provide this additional energy to the substrate/film surface. This causes activation and improved chemical reaction, surface and volume diffusion and peening effects which densify film microstructure and smoothen film surface topography. Measures for improved chemical reaction are extremely important for the proper production of high quality metal oxide films. Even traces of non-oxidized metal atoms or metal suboxides in the film causes optical absorption which is not acceptable in precision optical film components such as laser mirrors with required low losses and high damage threshold. However, with all energetic deposition technologies unwanted possibly occurring local energy spikes may cause dissociative degradation of film composition resulting in optical absorption. Therefore the dose of energy input must be carefully evaluated. With most energetic coating processes, substrate heating prior to deposition is not required since preceding ion bombardment create a clean and activated surface with no problems concerning film adherence. The energy input during deposition results, with many metal oxides, in a homogeneous, constrained high density amorphous film structure and atomically flat film surface. The high density amorphous film structure can be seen in TEM micrographs [26] shown in Fig. 11.

Fourier transform of diffraction pattern showed nearest neighbors at a distance of about 1 nm. After heat treatment 4 h at 350°C the film relaxed slightly in density [27, 28], however, remained amorphous with a slightly increased distance of order from 1 nm to about 1.5 nm [26] Fig. 12.

Figure 13 shows a comparison of the classical structural zone model of Movchan and Demchishin with a new zone model in which substrate

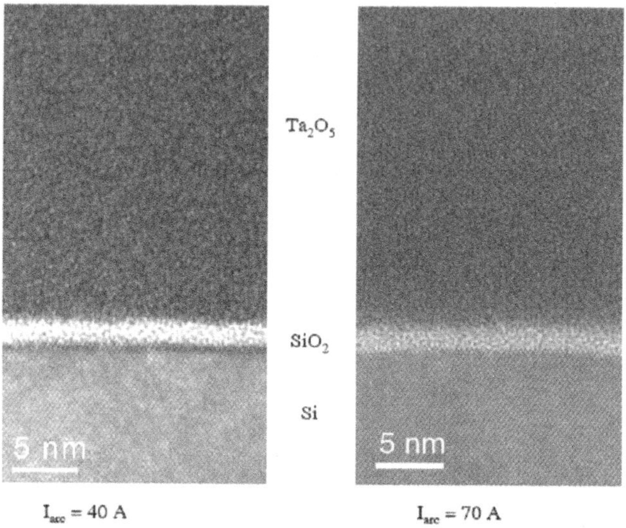

Fig. 11. TEM micrographs of RLVIP-Ta_2O_5-films on Si-wafer (U. Kaiser, 2004).

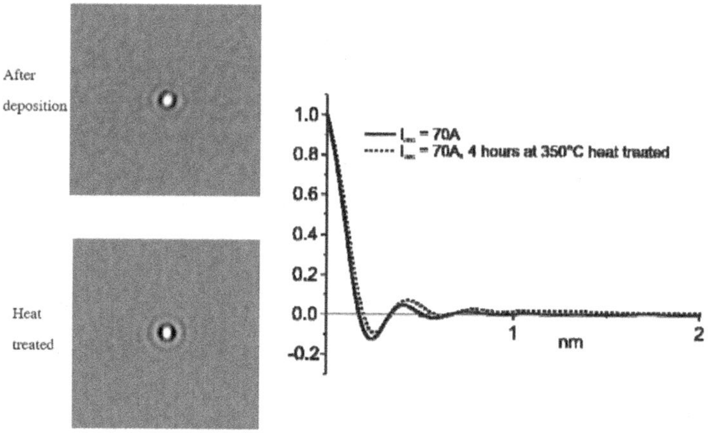

Fig. 12. Electro diffraction pattern of RLVIP-Ta_2O_5-films (U. Kaiser, 2004).

temperature is replaced mainly by kinetic particle energy of the film forming and film bombarding species.

Besides a refractive index equal or close to the value of the bulk material, the dense films show compressive stress. There is no water vapour absorption to observe and the films behave like a diffusion barrier. Their characteristics are environmentally stable and remain practically unchanged with time. The film properties depending on the density can be influenced and optimized by careful dose regulation of the energy input during film formation [29]. A reduction of density and with it of compressive stress is important not only

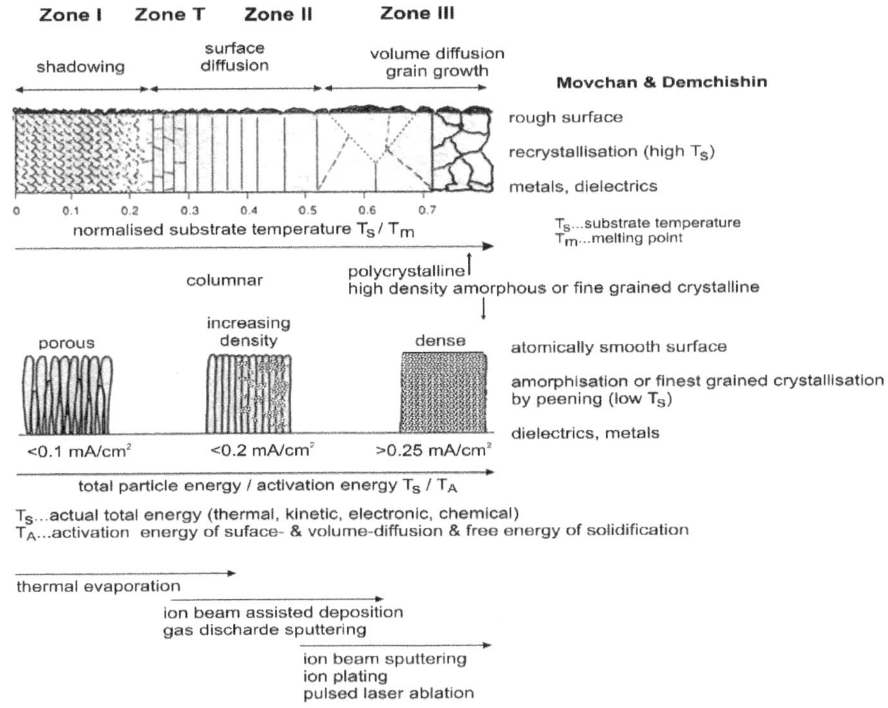

Fig. 13. New extended structure zone model.

for mechanical film stability but also for the stability of lowest residual optical absorption. In chemical compound films high stress seems to lower the resistance against colour centre formation by local non-stoichiometry when exposed to short wave radiation. Figure 14 shows the principle dependences of some film properties on energy input during film production. According Fig. 14 the values of all properties shown in graphic first increase nearly linear with rising energy input. Density, refractive index, and stress reach a plateau and then drop slightly by onset of film density relaxation. Absorption, however, experience a further steady increase because of bombardment induced chemical degradation. Higher gas pressure reduces particle energy and with it film density and residual optical absorption. The influence of energy input on refractive index of some often applied metal oxide films produced by reactive low voltage ion plating is shown in Tables 4 and 5.

With RLVIP metal oxide films, deposition at arc current $I_{\mathrm{arc}} = 40\,\mathrm{A}$ and relatively high oxygen partial pressure between $10–20 * 10^{-4}\,\mathrm{mbar}$ results in a high optical film quality with $k < 10^{-4}$ and in a lowered film density. Consequently the compressive film stress is drastically reduced from -600 to $-500\,\mathrm{MPa}$ to about -100 to $-50\,\mathrm{MPa}$. Whereas the refractive indices, listed in Table 5, are only slightly decreased by about 1–1.5%. The resulting lowered

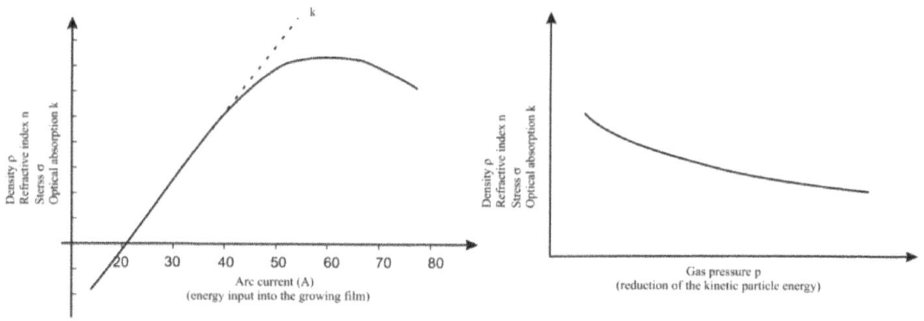

Fig. 14. RLVIP-film properties as function of energy input (*left*) and as function of gas pressure (*right*).

Table 4. Arc current variation 20–60 A in RLVIP-metal oxide film deposition. Deposition rate: 0.1–0.7 nm/s, $p_{\mathrm{Ar}} = 4 * 10^{-4}$ mbar, $p_{O_2} = 5\text{–}11 * 10^{-4}$ mbar, unheated glass substrates.

Arc current I_{arc}	RLVIP - Film refractive indices n550						
	SiO_2	Al_2O_3	HfO_2	ZrO_2	Ta_2O_5	Nb_2O_5	TiO_2
20 A	1.46	1.61	2.05	2.15	2.16	2.27	2.38
40 A	1.48	1.66	2.13	2.19	2.24	2.40	2.55
60 A	1.49	1.68	2.14	2.20	2.24	2.40	

Table 5. Oxygen pressure variation at constant Ar-pressure and constant arc current in RLVIP-film deposition. Deposition rate: 0.1–0.7 nm/s, $p_{\mathrm{Ar}} = 4 * 10^{-4}$ mbar, $p_{O_2} = 5\text{–}11 * 10^{-4}$ mbar, unheated glass substrates.

Arc current I_{arc}	O_2-pressure	RLVIP - Film refractive indices n550				
		Al_2O_3	HfO_2	ZrO_2	Ta_2O_5	Nb_2O_5
40 A	$10 * 10^{-4}$ mbar	1.66	2.13	2.194	2.25	2.38
40 A	$20 * 10^{-4}$ mbar	1.66	2.10	2.187	2.24	2.36

density, however, remained high enough to prevent the negative influence of water vapour sorption.

4 Conclusions

Conventional deposition processes are improved by various modifications resulting in new sometimes hybrid technologies. Optimised technologies are highly required to solve the increasing quality demands in optical thin films. Precision optical coatings are requested for various small and large area thin film products. New rapidly growing fields of application in information technology, medicine, energy conversion and production technology require excellent surfaces with environmentally stable, highly complex and multifunctional optical coatings. Development and experience in manipulation of new, possibly also graded or composite materials, for reliable high quality film production applicable in the visible but also in UV and IR, are of enormous importance.

References

1. H.K. Pulker: Coatings on Glass, 2nd rev. ed. (Elsevier, Amsterdam, 1999)
2. N. Kaiser, H.K. Pulker (Eds.): Optical Interference Coatings (Springer, Berlin, Heidelberg, 2003)
3. B. Schultrich: VIP Lexikon Beschichtungsverfahren, Vakuum in Forschung und Praxis, Start Vol. **18** No. 1 2006 to be continued
4. M. Auwärter: Austrian Pat. Nr. 192650 (1952), US Pat. No. 2.920.002 (1960)
5. W. Heitmann: Appl. Optics **10**, 2414 (1971)
6. J. Ebert: Proc. SPIE Technol. Meeting, Los Angeles, paper 325-04 Jan. (1982)
7. K.H. Müller: J. Vac. Sci. Technol. **A4**, 461 (1986)
8. B. Berghaus: German Pat. Nr. 683414 Klasse 12 Gruppe 37 (1939)
9. D. Mattox: Sandia Corp. Rep. SC-DR-281-63 (1963)
10. D. Mattox: J. Vac. Sci. Technol. **10**, 47 (1973)
11. H.K. Pulker, E. Moll, W. Haag: US Pat. No. 4.619.748 (priority 1985)
12. K. Matl, W. Klug, A. Zöller: Mat. Sci. Eng. A **140**, 523 (1991)
13. A. Zöller: Vacuum's Best 08, 21–25 (2008)
14. D. Brinsmaid, G. Koch, W. Keenan, W. Parson: KODAK US Pat. No. 2.784.115, Application Made May 4 (1953)
15. G. Reisse, S. Weissmantel, B. Keiper, B. Steiger: Thin Films Proc. DGM Informationsgesellschaft Verlag, TATF, 320 (1994)
16. R. Boxman, P. Martin, S. Sanders (Eds.): Handbook of Vacuum Arc Science, (Noyes, New York, 1996)
17. P.S. Martin, A. Bendavid: Review of the Filtered Vac. Arc Process and Materials Deposition, Thin Sold Films **394**, 1–15 (2001)
18. P.J. Clarke: US Pat. No. 3.711.398 (1973)
19. H. Ehlers, M. Lappschies, N. Beermann, D. Ristau: Optik & Photonik **3**, 41–46 (2007)
20. M. Scherer: Photonik **3**, 8 (2007)
21. P. Gawlitza, St. Braun, A. Leson, S. Lipfert, M. Nestler: Vak in Forschung u. Praxis **19**, 2, 37–43 (2007)
22. S. Schiller, K. Goedicke, V. Kirchhoff, T. Kopte: Proc. SVC 38th Ann. Technol. Conf., Chicago, 288 (1995)
23. J. Szczyrbowski, G. Teschner, G. Bräuer: Proc. SVC 38th Ann. Technol. Conf., Chicago, 288 (1995)
24. G. Bräuer: Vak. in Forschung u. Praxis, **20**, 2, 30–32 (2008)
25. D.T. Wei: Appl. Opt. **28**, 2813–2817 (1989)
26. U. Kaiser: University of Ulm, Germany, unpublished results (2004)
27. A. Hallbauer, D. Huber, A. Kunz, H. K. Pulker: Proc. ICCG-6, Dresden, Starke & Sachse, Grossenhain, Saxonia, 321–326 (2006)
28. S. Yulin: Fraunhofer IOF, Jena, unpublished results (2006)
29. H. K. Pulker, S. Schlichtherle: Advances in Optical Thin Films, St. Etienne (F), Proc. SPIE, Vol. 5250, C. Amra, N. Kaiser, H. A. Macleod (Eds.), 1–11 (2003)

Advanced Metrology for Next Generation Transistors

Alain C. Diebold

College of Nanoscale Science and Engineering, University at Albany, NY 12203, Albany, USA,
adiebold@uamail.albany.edu

Abstract. This paper overviews advances in optical and electrical measurements for advanced transistor materials and processes. Optical measurements such as ellipsometry require advances in both optical methods and optical measurement technology [1, 2]. Although laboratory based optical measurements are capable of measuring into the VUV, the availability of in-line systems is a recent development. The Cody Lorentz model is used to describe the optical response of high k materials and advance traditional measurements such as ellipsometry [1]. Electrical measurements provide a well recognized means of quantifying defect states in transistor film stacks. Advanced charge pumping methods extend capacitance – voltage measurements to high k materials [3]. Non-traditional optical approaches such as second harmonic generation which confirm the observation of trapped charge induced changes in the silicon dielectric function by ellipsometry are also discussed [4, 5]. The paper ends with a discussion of impact of quantum confinement on the optical measurement of thin SOI films [6, 7].

1 Introduction

As transistor dimensions shrink, development of new measurements must keep pace with equivalent scaling methods such as high k, metal gate, fully depleted SOI, and stress enhanced mobility. Recent advances in optical methods have enabled measurements of several of these key materials and process areas. New optical models for the dielectric function enable thickness metrology for high k materials. New phenomena such as charge build-up in the high k – interface oxide stack continue to require advances in both electrical and physical measurements.

The first section of this paper covers optical models for high k films. The second section covers optical models for thin metal films. The third section covers the use of charge pumping methods to enable capacitance – voltage measurements of high k films. The fourth section discusses observation of trapped charge in high k films using optical methods. The measurement of ultra-thin SOI is discussed in the last section. In that section we review the impact of quantum confinement on the dielectric function of ultra-thin semiconductor films.

2 Optical models for high K

When compared to silicon dioxide, high K dielectric layers are more absorbing at UV wavelengths and thus require optical models that account for the absorption and reflect the band-gap [1]. There are many approaches to optically modeling the substrate – interface layer – high k film stack. Each of these options impacts the overall precision of the measurement. Due to the thickness of each film in the stack, it is very difficult to separate the contribution of the interface layer and the high-k film. Thus the simplest approach is model the two layers as one medium returning a single optical response. The optimum optical model should represent the Si/interfacial layer/high-k film stack and not just provide a measure of total film thickness. In order to measure the high-k and interface thickness separately, an accurate optical model for each film is required. It is possible to model the high k film using non-physical optical models such as a summation of several damped oscillators. The multi oscillator approach has too many variables and will average the noise contribution over several of these variables. A better optical model such as the Cody Lorentz model uses as few variables as possible [1].

Initial attempts to model the optical response of high k films used the Tauc-Lorentz and the Generalized Tauc-Lorentz models [1, 8, 9]. The issue with these models is that they explicitly restrict 2 to be zero for energies less than the band gap. Contributions to the joint DOS due to defects, as well as intra-band absorption, will cause the imaginary part of the dielectric function to have non-zero values below the band gap. High-k films typically contain defects due oxygen vacancies and process induced problems. These defects, in turn, will contribute localized states below the band edge and thus optical absorption.

The Tauc-Lorentz model was first applied to Ta_2O_5 by Richter and Nguyen [8] and then to Ta_2O_5 and ZrO_2 [9]. The onset of optical absorption of both Ta_2O_5 ($\sim$ 4.2 eV) and ZrO_2 ($\sim$4.8 eV; band gap $\sim$ 5 to 5.5 eV) are below HfO_2 [9]. Thus these materials absorb light in visible wavelength range and do not require VUV ellipsometry. The dielectric constant of these materials is significantly changed after the amorphous films are annealed and become polycrystalline [9]. The Tauc Lorentz model was also applied to HfO_2 [10]. Again the dielectric function shows significant change due to crystallization which was modeled using two Tauc-Lorentz functions [10].

Contributions to absorption (from the localized wave functions) below the band gap are known to follow Urbach's exponential form [1]:

$$\alpha(E) = \alpha_0 * \exp\{\gamma'(E - E_0)/kT\} \tag{1}$$

where α is the absorption coefficient, γ is a constant, and T is the absolute temperature [1].

As a matter of comparison, we note that Tauc derived his expression for the imaginary part of the dielectric function based on the assumption of parabolic bands and a constant momentum matrix element [1, 2]:

$$\epsilon_2(E) = A\frac{(E - E_g)^2}{E^2} \tag{2}$$

The Cody expression for ϵ_2 is based on the assumption of parabolic bands, but with a constant dipole matrix element [11, 12]. This results in a different form for the imaginary part of the dielectric function:

$$\epsilon_2(E) \propto (E - E_g)^2 \tag{3}$$

In the high-k films of interest, the band gap is $\sim 6.0\,\text{eV}$, the difference in absorption is much more important because ϵ_2 scales as E^{-2} close to Eg for the Tauc-Lorentz model. Because absorption above the band gap is greater in magnitude for the Cody-Lorentz model, it more accurately models high optical absorption and enables inclusion of the interfacial layer.

The Cody-Lorentz model for the dielectric function is:

$$\begin{aligned} \epsilon_2(E) &= \frac{E_1}{E}\exp\left[\frac{(E - E_t)}{E_\mu}\right], 0 < E \le E_t \\ \epsilon_2(E) &= G(E)L(E) = G(E)\frac{AE_0\Gamma E}{[(E^2 - E_0^2)^2 + \Gamma^2 E^2]}, E > E_t. \end{aligned} \tag{4}$$

where E_t is a transition energy between the Urbach Tail and band-to-band transitions, and E_μ represents the extent of broadening [12].

Here, G(E) is the Joint Density-of-States and is a result of the constant dipole approach:

$$G(E) = \frac{(E - E_g)^2}{(E - E_g)^2 + E_p^2}. \tag{5}$$

The denominator, with the extra term, E_p (a transition energy that separates the absorption onset behavior from the Lorentzian behavior), was added to allow the DOS to converge with increasing energy. At high photon energies, G(E) will approach 1, and ϵ_2 will assume a pure Lorentzian form [1, 12]. As shown in Fig. 1, the Cody-Lorentz model, which accounts for absorption below the band edge, has proven to be a useful model for high-k films. Cody-Lorentz accounts for absorption due to defect states a few tenths of an eV below the band gap. The impact of an interfacial layer between the high k film and silicon substrate is clearly observed in Fig. 1 [1]. The change in optical response when increasing amounts of silicon are added to Hf oxide are shown in Fig. 2 [1].

Optical models provide a means of measuring film thickness, but are not well suited to observation of certain defect states. The data inversion method

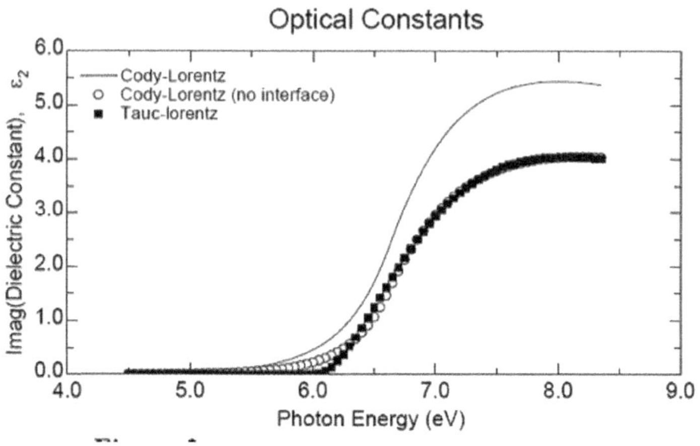

Fig. 1. Comparison of Tauc-Lorentz to Cody Lorentz optical model for high K films. Figure first published in Ref. [1] and used with permission.

used to extract the dielectric function has proven well suited for this purpose. Takeuchi, Ha, and King first used spectroscopic ellipsometry to observe below band gap defect states in HfO_2 [13]. A 9.4 nm pure Hf layer was annealed in oxygen resulting in HfO_2 films that were $>$ 10 nm thick. The C-V determined shift in flat band voltage indicated that the films had decreasing electron trap densities as anneal temperature increased [$\sim 4x10^{11}$ cm^{-2} at 600 C and $\sim 3x10^{-9}$ cm^{-2} at 700 C] [13]. The defect states were observed between 4.5 and 5.0 eV, well below the band gap of 5.7 eV of their films [13]. The intensity of the defect absorption increased with electron trap density. This study also observed defect states and a change in change in the dielectric function in the energy range at the 5.7 eV band gap due to crystallization [13]. In another study, Nguyen et al., attributed an absorption feature 0.2–0.3 eV below the band gap in polycrystalline HfO_2 [14] to defect states and observed that addition of 20% silicon to the HfO_2 prevented crystallization. Although the Cody-Lorentz model is very useful for film thickness determination, it does not account for discrete below band-gap defect states. The dielectric function of HfO_2 films was determined from ellipsometry data assuming that the dielectric function of silicon is unperturbed. When ϵ_2 is determined in this manner, several absorption peaks are observed below the HfO_2 band gap of ~6eV [15]. This is shown in Fig. 3. The origin of these peaks is the subject of numerous publications. Some publications identify all these peaks (2.9, 3.4, 4.2, 4.75, and 5.3 eV) as transitions between defect states or between the valence band and defect states as listed in Table 1 [16]. However, the peaks at 3.4, 4.2, and 5.3 eV have also been assigned to changes in the dielectric function of silicon at critical points induced by the electric field from electrons trapped at defect sites. Changes in the dielectric function of Si would appear in the HfO_2 data because the data inversion uses reference data for silicon dielectric

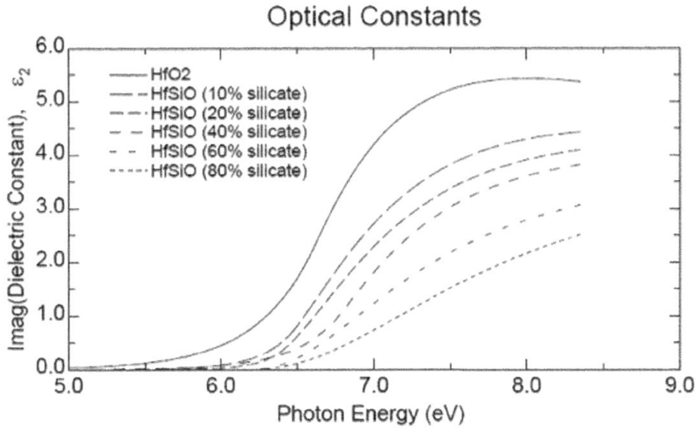

Fig. 2. Optical response, ϵ_2, of halfnium silicate. Figure first published in Ref. [1] and used with permission.

function. It has not been corrected for the effects of the electric field from charges trapped in the dielectric film. This is further discussed below in the Section E. The peaks at 2.9 and 4.75 eV have been the subject of further study. All of the peaks (2.9, 3.4, 4.2, 4.75, and 5.3 eV) are absent when the same process is used to deposit HfO2 on a fused silica substrate. Only the band edge defects states 0.5 eV below the conduction band edge are observed. That is an indication that the 2.9, 3.4, 4.2, 4.75, and 5.3 eV peaks are not due to defects inside the high k. This suggests that the defect states are at or in the SiO_2 layer or the Si–SiO_2 or SiO_2–HfO_2 interface and may be induced by the high k. The peak at 2.9 eV has been attributed to oxygen vacancies and the peak at 4.75 eV has been attributed to an interstitial defect [15].

The assignment of the peaks occurring at critical point energies to changes in dielectric function of silicon and the peaks at 2.9 and 4.75 eV to interface states is controversial. Another means of observing a field induced change in the silicon dielectric function is photoreflectance spectroscopy. Observation of a photoreflectance signal from silicon requires band bending [17–19]. A photoreflectance signal will not be observed when a high quality oxide layer is grown on silicon [17–19]. When ion bombardment damages the oxide layer, band bending occurs and photoreflectance can be used to determine the amount of trapped charge. Thus the defects in the HfO_2 can induce band bending and photoreflectance can be observed [19]. However, quantitative determination is difficult and great care must be used when interpreting data [17–19]. Another aspect of photoreflectance spectroscopy is that the signal at the E_1 critical point is more intense than that at the E_2, and E'_1 critical points. The impact of the difference in magnitude of the electric field of the light between ellipsometry and the laser used for photoreflectance is not

Table 1. HfO_2 defect state transition energies.
Between the HfO_2 conduction band (CB) and valence band lie edge states at both bands and two conduction band defect states $D(CB)_1$ amd $D(CB)_2$ and two valence band defect levels $D(VB)_1$ and $D(VB)_2$. Since these transitions are excited by the light, the transition is from the low to high energy level. The relevant critical point energies occurring at similar energies are also listed [16].

Transition	Energy	CP an Energy
$D(VB)_1$ to $D(CB)_2$	2.7 eV ± 0.4 eV	
$D(VB)_1$ to $D(CB)_1$	3.4 eV ± 0.4 eV	E_1 3.4 eV
$D(VB)_2$ to $D(CB)_2$	3.4 eV ± 0.4 eV	E_1 3.4 eV
Edge VB to $D(CB)_2$	4.2 eV ± 0.4 eV	E_1 4.2 eV
$D(VB)_2$ to $D(CB)_1$	4.3 eV ± 0.4 eV	
Edge VB to $D(CB)_2$	5.1 eV ± 0.4 eV	E'_1 5.3 eV
Edge VB_1 to Edge CB_2	5.6 eV ± 0.3 eV	

presently understood. This could impact the relative intensities at the E_1, E_2, and E'_1 critical points. As the investigations into the correct identification of the defect states observed by ellipsometry continue, it important to note that regardless of the identification, the information can be used to evaluate the quality of high k films. It is important to remember that high k deposited on quartz does not show the 2.9, 3.4, 4.2, 4.75, and 5.3 eV peaks.

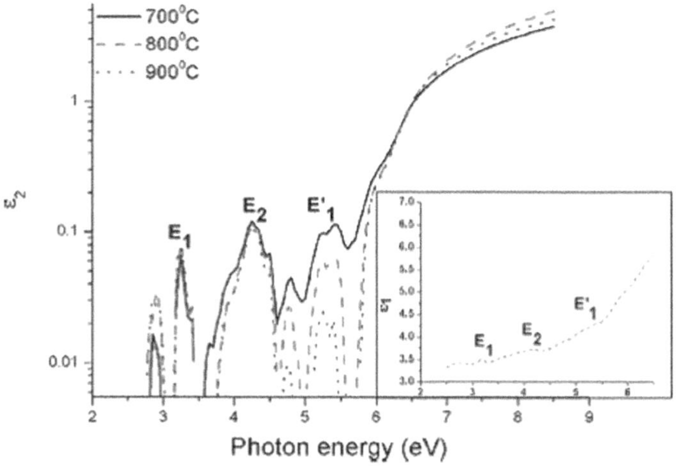

Fig. 3. The imaginary and real parts of the dielectric function of HfO_2 are plotted vs photon energy. The peaks associated with electric field induced changes in the dielectric function of the silicon are labeled by the associated critical point designation, E_1, E_2, and E'_1. Figure from Ref. [10] and used with permission.

3 Optical model for metal films

Many metal films less than ~30 nm can be measured optically despite the opaque nature of thicker films [20]. Although the optical properties of metal films are well understood, measurement of thin metal film thickness is considered to be difficult. The classical Drude model of an electron gas describes optical absorption from the Far IR to the Plasmon energy. The Drude model reflects the optical properties of metals that are highly reflective at lower energies and become transparent at the plasmon frequency. Typical plasmon frequencies for these metals are in the UV. For example, Cu **[Ar] 3d**10 **4s**1, Ag **[Kr] 4d**10 **5s**1, and Au **[Xe].4f**14.**5d**10.**6s**1 all have filled d shells and the same number of free electrons. They also have the same plasmon frequency of $\sim 9\,\text{eV}$. At energies above the plasmon frequency ω_p, these metals absorb light, and thus are less reflective. The Drude dielectric function for a metal is given by:

$$\epsilon(\omega) = 1 - \omega_p^2/(\omega^2 - i\omega/\tau_D) \tag{6}$$

Where ω_p is the plasmon frequency and τ_D is Drude electron relaxation time Because these transitions from the d-band to the empty states above the Fermi level can be occur over a fairly narrow band of energies, the optical response in that energy range can be modeled by a Lorentz oscillator. This approach needs to be modified for thin films where a thickness dependence has been observed.

Collins and coworkers have related the electron mean free path to grain size R_g and a reflection coefficient R [21].

$$\lambda \approx 2[(1 - R)/3R]R_g \tag{7}$$

Film thickness comes into the equation through the relationship between grain size and thickness. Light absorption in the free electron (Drude) model of metals is due to the relaxation time. The expression for the relationship between mean free path and relaxation time is:

$$\tau^{-1} = \tau_{bulk}^{-1} + (\nu_F/\lambda) \tag{8}$$

Very recently, Collins and co-workers found that if $R_g \sim$ firm thickness and $R = 0.5$ a reasonable fit to the dielectric function of Mo is obtained [21]. This approach works well for Ni and NiSix thin films [22].

4 Charge pumping based capacitance – voltage measurements

Electrical measurement of transistor gate dielectric films has long been a critical part of semiconductor metrology. Detection and quantification of charge in the dielectric layers is a critical part of dielectric metrology. As mentioned above, high k materials are prone to defects such as oxygen vacancies that trap charge this making electrical measurement of high k materials more difficult. Young described the measurement techniques used to study and quantify charge trapping: Capacitance-Voltage (C-V) hysteresis, alternating stress and sense Vfb/Vt instability, charge pumping, and fast transient Id-Vg measurement in his thesis [3]. All of these methods measure a different property (e.g., flat band voltage, the density of trapped charge, etc.). All of these methods are extensions of traditional C-V and I-V measurements.

C-V hysteresis measurements can be used to monitor flat band voltage shifts, ΔVfb. Constant voltage gate dielectric stress (CVS) with interspersed limited-voltage-range C-V measurements around flat band have proven to be more reliable for determination of ΔVfb [3]. CVS results in the de-trapping of some of the charge between the stress and sense sequence [3]. There are two types of charge pumping C-V measurement of defect states in high k films. The distinction between these is the location of the defect states: interface layer vs bulk high k. Fixed-amplitude charge pumping (FA CP) measures interface state densities, whereas variable-amplitude (VA CP) measures trap densities in the high-k bulk [3]. Figure 4 shows FA-CP and VA-CP. These measurements use a transistor test structure, and data interpretation is difficult due to the gate and source/drain leakage. Current (I) – Voltage (V) measurements often provide information not available from C-V measurements. The shift of the drain current (Id) vs gate voltage (Vg) is one example. Fast transient Id-Vg curves measured in the microsecond regime using the up and down swing of a trapezoid pulse (i.e., ΔVt) determine the amount of the trapped charge.

5 Optical measurement of charge trapped in high k and interface

The semiconductor industry has long relied on the ability to fabricate silicon dioxide – silicon interfaces with relatively low amounts of interfacial states ($\sim 10^9/\text{cm}^2$) [23]. The interface state density must be kept as low as possible to maintain transistor electrical properties such as threshold voltage. In addition, charge trapped at these defect sites can scatter electrons and reduce mobility. High k dielectric films such as hafnium oxide and hafnium silicates have considerably higher densities of interface and other defect states ($\sim 10^{12}$ to $10^{15}/\text{cm}^2$). The thin silicon dioxide or silicon oxynitride layer between the high k and silicon substrate reduces the density of the interface states. Optical measurements can be used to characterizing these defect states by measuring the effect of trapped charges on the dielectric function of the silicon below the high k film stack.

The effect of an electric field on optical measurements of semiconductor properties is well documented [4]. The increase in the tunneling probability of an electron between the valence and conduction bands due to an applied electric field is known as the Franz-Keldysh effect. These electric field induced

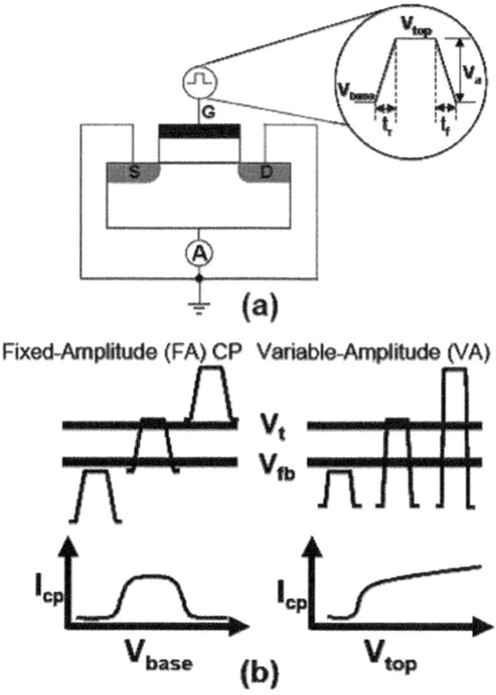

Fig. 4. Fixed – amplitude charge pumping and variable amplitude charge pumping figure from Ref. [3].

changes at light absorption are most noticeable at critical points such as the E_1 critical point of silicon. Aspnes derived a third derivative form for the electric field induced changes in the dielectric function observed by both photoreflectance and electro-reflectance.

$$\Delta\epsilon \propto |E|^2 \frac{\partial(\epsilon_0)}{\partial(\hbar\omega)^3} \tag{9}$$

E is the amplitude of the electric field, ϵ is the dielectric function in the absence of the electric field, and $\hbar\omega$ is the energy of the light field.

Photoreflectance intensity has been shown to be correlated to amount of trapped charge in HfO2 films deposited directly on n-type Si(100) substrates [24]. In this study, the intensity of photoreflectance spectra (PRS) at the E_1 critical point inversely correlated to the C-V determined charge in the oxide between 0.5×10^{12} cm^{-2} and 2×10^{12} cm^{-2} [24]. Simulations of the decrease in intensity with increase positive charge in the oxide layer showed that the most sensitivity was between 10^{11} and 10^{12} cm^{-2} [24].

The functional form of Eq. (9) should also apply to ellipsometric measurement of changes in the silicon dielectric function due to the electric field from trapped charge in the high k [4]. As shown in Fig. 5, annealing of the high k film stack changes the amount of trapped charge as observed by the electric field induced change in the dielectric function of silicon around the E_1 critical point. The substrate silicon is p-type Si(100) in these studies.

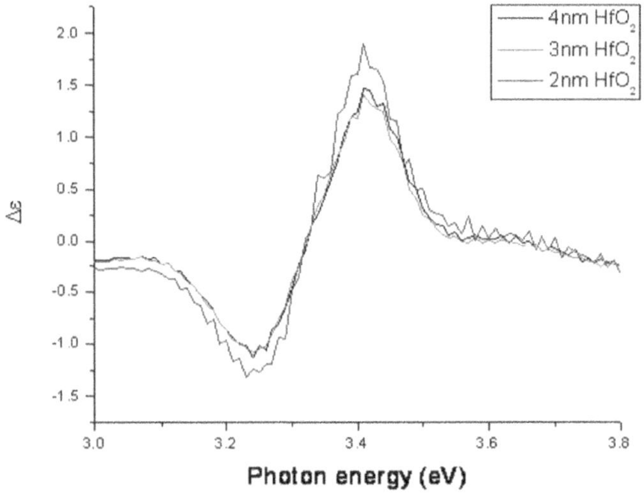

Fig. 5. Change in ϵ_2 of silicon around the E_1 critical point due to trapped charge buildup: $\Delta\epsilon$ vs photon energy figure courtesy Jimmy Price [4].

The electric field dependence of second harmonic generation (SHG) is well documented [22]. This effect has been used to demonstrate differences in the amount of charge trapped in high k films as anneal conditions are varied [23]. The electric field breaks the translation symmetry of centrosymmetric silicon. SHG serves as a means of verifying ellipsometric observations.

6 Measurement of ultra-thin SOI and observation of quantum confinement

The measurement of film thickness for ultra-thin semiconductor layers is complicated by the change in optical properties due to quantum confinement. Recently, we showed that quantum confinement shifts the energy of critical points which have a strong excitonic character such as the E1 critical point of silicon while other critical points remain at the same energy [6, 7]. Quantum confinement has two effects on the optical transition at E1. The energy of E1 is increased, and the shape of absorption edge follows from the changes in the density of states. Recent work has shown that the particle in a one-dimensional box model predicts the experimentally observed $1/L^2$ dependence (where L is the thickness of the silicon on insulator layer):

$$\Delta E = E_{\mathrm{Gap}}^{\mathrm{SOI}} - E_{\mathrm{Gap}}^{\mathrm{bulk}} = \frac{(\pi\hbar)^2}{2m^* L^2} \tag{10}$$

Here, $\hbar$ is Plank's constant, and m^* is the effective mass. Accounting for the effect of quantum confinement on silicon enables the measurement of thin SOI films less than 10 nm in thickness.

One interesting point is that shifts in E_1 less than KT (Boltzman's constant times the temperature ~ 25 meV) in energy can be observed at room temperature. Aspnes group observed ~ 10 meV difference between the critical point of hydrogen terminated (111) and (001) orientations at room temperature [26–28]. He has shown that the reason that these shifts can be observed has to do with the lifetime of the final state in the optical transition [25, 26]. Therefore, thermal energy does not obscure observation of changes in a transition energy less than KT.

7 Conclusions

Optical measurements have advanced to meet the challenges associated with high , metal gate, and thin SOI materials. In this review, we have covered the use of the Cody Lorentz model for high films, Drude-Lorentz models for thin metal films, charge pumping based C-V measurement of defects in high k, ellipsometric and second harmonic generation measurement of trapped charge, and the observation and impact of quantum confinement in ultra-thin SOI.

8 Acknowledgements

The author gratefully acknowledges critical discussion about photoreflectance with Ed Seebauer. The author also acknowledges the efforts of James Price and numerous discussion with David Aspnes. In addition, the author also acknowledges discussions about metal film measurement with Robert Collins and James Hilfiker.

References

1. J. Price, P.Y. Hung, T. Rhoad, B. Foran, and A.C. Diebold, Spectroscopic ellipsometry characterization of HfxSiyOz films using the Cody-Lorentz parameterized model, Appl. Phys. Lett. **85**, (2004), pp. 1701–1703.
2. G.E. Jellison, Jr. and F.A. Modine, Parameterization of the optical functions of amorphous materials in the interband region, Appl. Phys. Lett. **69**, (1996), pp. 371–373 and erratum in Appl. Phys. Lett. **69**, (1996), p. 2137.

3. C. Young, Charge Trapping Characterization Methodology for the Evaluation of Hafnium-based Gate Dielectric Film Systems, PhD diss. North Carolina State University, 2003.
4. J. Price, P.S. Lysaght, S.C. Song, A.C. Diebold, Y.Q. An, and M.C. Downer, Observation of interfacial electrostatic field-induced changes in the silicon dielectric function using spectroscopic ellipsometry, Phys. Stat. Sol. **a 205**, (2008), pp. 1–4
5. R. Carriles, J. Kwon, J.C. Miller, Y.Q. An, M.C. Downer, J. Price, and A.C. Diebold, Second-harmonic generation from Si/SiO2/Hf(1-x)SixO2 structures, Appl. Phys. Lett. **88**, (2006), pp. 161120–161120-3.
6. J. Price and A.C. Diebold, Spectroscopic ellipsometry characterization of ultrathin silicon-on-insulator films, J. Vac. Sci. Technol. **B24**, (2006), pp. 2156–2159.
7. A.C. Diebold and J. Price, Observation of quantum confinement and quantum size effects, Phys. Stat. Sol. A **205**, No. 4, (2008), pp. 896–900.
8. N.V. Nguyen, C.A. Richter, and Y.J. Cho, Effects of high-temperature annealing on the dielectric function of Ta2O5 films observed by spectroscopic ellipsometry, Appl. Phys. Lett. **77**, (2000), pp. 3012–3014.
9. A.C. Diebold, J. Canterbury, W. Chism, C.A. Richter, N.V. Nguyen, J. Ehrstein, and C. Weintraub, Characterization and production metrology of gate dielectric films, Mater. Sci. Semi. Process **4**, (2001), pp. 3–8.
10. Y.J. Cho, N.V. Nguyen, C.A. Richter, J.R. Ehrstein, B.H Lee, and J.C. Lee, Spectroscopic ellipsometry characterization of high-k dielelctric HfO2 thin films and the high temperature annealing effects on their optical properties, Appl. Phys. Lett. **80**, (2002), pp. 1249–1251.
11. G.D. Cody, in Semiconductors and Semimetals, edited by J.I. Pankove (Academic, Orlando, Fl, 1984), Vol. **21**B.
12. A.S. Ferlauto, G.M. Ferreira, J.M Pearce, C.R. Wronski, R.W. Collins, X. Deng, and G. Ganguly, Analytical model for the optical functions of amorphous semiconductors from the near-infrared to ultraviolet: Applications in thin film photovoltaics, J. Appl. Phys. **92**, (2002), pp. 2424–2436.
13. H. Takeuchi, D. Ha, and T.-J. King, Observation of bulk HfO2 defects by spectroscopic ellipsometry, J. Vac. Sci. Technol. A **22**, (2004), pp. 1337–1341.
14. N.V. Nguyen, A.V. Davydov, D. Chandler-Horowitz, and M.M. Frank, Subbandgap defect states in polycrystalline hafnium oxide and their suppression by admixture of silicon, Appl. Phys. Lett. **87**, (2005), pp. 192903-1–192093-3.
15. J. Price, H.-J. Li, P. Lysaght, and A.C. Diebold, Identification of sub-bandgap absorption features at the High-κ/silicon interface and their relation to charge trapping states, Appl. Phys. Lett. **91**, (2007), pp. 61925-1–61925-3.
16. Y.M. Strzhemechny, M. Batalev, S.P. Tumakha, H.P. Goss, C.L. Hinkle, C.C. Fulton, and G. Lucovsky, L.J. Brillson, Low energy electron-excited nanoscale luminescence spectroscopy studies of intrinsic defects in HfO2 and SiO2-HfO2-SiO2-Si stacks, J. Vac. Sci. Technol. B **26**, (2008), pp. 232–243.
17. K. Dev, M.Y.L. Jung, R. Gunawan, R.D. Braatz, and E.G. Seebauer, Mechanism for coupling between properties of interfaces and bulk semiconductors, Phys. Rev. B **68**, (2003), pp. 195311-1–195311-6.
18. K. Dev and E.G. Seebauer, Band bending at the Si(1 1 1)-SiO2 interface induced by low-energy ion bombardment, Surf. Sci. **550**, (2004), pp. 185–191.
19. E.G. Seebauer, private communication.

20. A.C. Diebold, Optical measurement of metal film thickness, Conference Proceedings Advanced Metallization Conference, Albany, October 9–11, 2007.
21. S. Marsillac, N. Barreau, H. Khatri, J. Li, D. Sainju, A. Parikh, N.J. Podraza, and R.W. Collins, Spectroscopic ellipsometry studies of top window and back contacts in chalcopyrite photovoltaics technology, Proceedings of the 4th International Conference on Spectroscopic Ellipsometry, Stockholm, June 2007, Physica Status Solidi (c) 5, (2008), pp. 1244–1248.
22. V. Kameni and A.C. Diebold, in preparation.
23. R.R. Razouk and B.E. Deal, Dependence of interface state density on silicon thermal oxidation process variables, J. Electrochem. Soc. **126**, (1979), pp. 1573–1581.
24. M. Sohgawa, M. Yoshida, T. Naoyama, T. Tada, K. Ikeda, T. Kanashima, A. Fujimoto, and M. Okuyama, Contactless characterization of fixed charge in HfO2 thin film from photoreflectance, Jpn. J. Appl. Phys. **44**, (2005), pp. 2409–2414.
25. G. Lupke, Characterization of semiconductor interfaces by second-harmonic generation, Surf. Sci. Rep. **35**, (1999), pp. 75–161.
26. D.E. Aspnes, L. Mantese, K.A. Bell, and U. Rossow, Coherence effects and time dependences of the optical response of surfaces and interfaces of optically absorbing materials, Phy. Stat. Sol. B **220**, (2000), pp. 709–715.
27. L. Mantese, K.A. Bell, U. Rossow, and D.E. Aspnes, Evidence of near-surface localization of excited electronic states in crystalline Si, J. Vac. Sci. Technol. B **15**, (1997), pp. 1196–1200.
28. L. Mantese, U. Rossow, and D.E. Aspnes, Surface-induced optical anisotropy of oxidized, clean, and hydrogenated vicinal Si(001)surfaces, Appl. Surf. Sci. **107**, (1996), pp. 35–41.

Index

GPSR Compliance

The European Union's (EU) General Product Safety Regulation (GPSR) is a set of rules that requires consumer products to be safe and our obligations to ensure this.

If you have any concerns about our products, you can contact us on ProductSafety@springernature.com

In case Publisher is established outside the EU, the EU authorized representative is:

Springer Nature Customer Service Center GmbH
Europaplatz 3
69115 Heidelberg, Germany

Batch number: 10761599

Printed by Printforce, the Netherlands